MICROBIOLOGY RESEARCH ADVANCES

BIOLOGY AND COMMUNICATION IN THE INFORMATION AGE

MICROBIOLOGY RESEARCH ADVANCES

Additional books in this series can be found on Nova's website
under the Series tab.

Additional e-books in this series can be found on Nova's website
under the e-book tab.

MICROBIOLOGY RESEARCH ADVANCES

BIOLOGY AND COMMUNICATION IN THE INFORMATION AGE

RAFIQUL ISLAM
EDITOR

nova publishers
New York

For permission to use material from this book please contact us:
Telephone 631-231-7269; Fax 631-231-8175
Web Site: http://www.novapublishers.com

Library of Congress Cataloging-in-Publication Data

ISBN: 978-1-62808-252-4

Published by Nova Science Publishers, Inc. † New York

CONTENTS

PREFACE

The information age came with the promise of transparency greater than anything witnessed heretofore by humanity. Of course, transparency is the essence of foresight and knowledge. The hope was for a greater accountability that would follow increased transparency, compelling policy makers to depend on knowledge and foresight rather than disinformation and hidden agendas. Recent events of the new millennium indicate that the increase has been in opacity and disinformation. The information age, often dubbed as the 'knowledge era', has become the antithesis of knowledge, however; even its most ardent proponents admit that. Despite globalization, the information age has failed to generate knowledge-based decision-making tools. The failures of the information age to generate knowledge and remove or attenuate much of the basis of international conflicts are symptomatic of a deeper illness. Science has become the captive of the 'New Science' that focuses on short-term, tangible, and status-quo, and doesn't even allow the real science to face in a direction that would allow the knowledge-model to launch. This science is incompatible with nature, even in competition with nature, and is not even facing the same direction as the science that is needed to increase knowledge — the only process on which rational hopes of discovering the truth may be placed. Consequently, all the promises of the information age have been failing spectacularly. Today, we have medicines that don't cure, an economy that doesn't economize, education systems that don't educate, purification techniques that do not purify, and justice systems that award plaintiffs or deliver discriminatory treatment of an accused depending on ethnic origin, religious belief, or skin colour. Today, we do collect data at an unprecedented rate, and, yet, we do not process a single set of data with any technique that is truly natural. Today, in the name of simulation, airplanes do not emulate how birds fly, computers do not emulate how the brain computes, cameras do not emulate how the human visual cortex captures and fixes an image, electricity disdains to emulate lightning (insufficiently continuous) or the electric eel (too dependent on the material properties of the receiving medium), commercial ships forego natural sources of locomotion like wind, and submarines do not emulate how fish navigate beneath the water's surface. If nature is the truth, our New Science must be falsehood.

Chapter 1 – Electromagnetic radiation of mobile devices causes the changes in catalytic activity of membrane-bound acetylcholinesterase of human erythrocytes (Yt-antigen) during the adaptation to chronic exposure of electromagnetic radiation. Increased activity of AChE of plasma irradiation in vitro corresponds to the activity of AChE quiet plasma in the body and decrease the activity of erythrocyte AChE in vitro irradiation corresponds to the high activity of erythrocyte AChE adaptation in the body, which is close to a state of decompensation, and is likely to shift to stress.

Chapter 2 – Methods of synthesis and biological activities of unsubstituted and substituted epimer aminoalcohols of 5α-androstans are described.

Chapter 3 – On the basis of Berkeley-Mach-Poincaré principle of relativity and within the assumption of absence of any specific (inertial) reference systems in empty space the new mechanics of bodies motion has been constructed. This motion is determined only by the relative distances and velocities of bodies. It has been shown that the mutual rotation of a pair of particles related to almost immobile "distant stars" (galaxies) leads to centrifugal force origin.

Chapter 4 – The status of the total content of water soluble antioxidants, ceruloplasmin and uric acid were determined in 45 patients with alcohol abuse and 37 health donors before and after UV affect of blood serum. All of them were a male, aged 22 to 35 years old. The mean value of water soluble antioxidant after 5 min of UV was significantly higher ($p<0,05$) in health donors, compared with the health donors before UV and patients with alcoholic abuse respectively. The water soluble antioxidant level before UV was significantly lower ($p<0,05$) in the alcoholic abuse group compared to the health donors group. Ceruloplasmin and uric acid level significantly higher ($p<0,05$) in the alcoholic abuse group compared to the health donors group. However, the level of ceruloplasmin in both group decrease under the UV affect ($p<0,05$). But uric acid level in both groups before and after UV affect was within a very close range and the differences were not significant ($p>0,05$). These results show that the antioxidant system of healthy donors provided a response to UV affect as a stressor, and the antioxidant system of patients with alcoholic abuse is too weak for this. These dates may be used in diagnosis, treatment and prevention of alcohol abuse and in rehabilitation after the disease as well.

Chapter 5 – Dihydroquercetin (taxifolin, DHQ) is a natural flavonoid, which possesses antioxidant activity. Recently, in addition to the protective effect of antioxidants against lipid peroxidation, increasing attention is paid to the possibilities of antioxidants, including dihydroquercetin, to prevent oxidation of proteins. Fibrinogen is a plasma protein that serves an important role in clotting. In this work we study the ability of dihydroquercetin to inhibit oxidative modification of fibrinogen. We used ozone as the oxidant. Assessment of functional activity of fibrinogen, both before and after oxidation with ozone, was conducted by determining the time of formation of a fibrin clot after addition of thrombin in a solution of fibrinogen. We found that in the presence of dihydroquercetin ozone oxidation of fibrinogen is negligible, also in the presence of dihydroquercetin fibrinogen retains its functional properties after ozone oxidation, whereas in the case of ozone oxidation of fibrinogen without dihydroquercetin, the functional activity of fibrinogen is completely lost.

Chapter 6 – This paper proposes an Improved Particle Swarm Optimization (IPSO) for extracting knowledge from data for the fuzzy rule based classifier model. In the standard PSO, if the swarm uses a small inertia weight or constriction coefficient, then its convergence is premature. Further, if the global best is found early in the searching process, then it may get struck in poor local minima. To overcome this issue, the concept of selection from Genetic Algorithm (GA) is incorporated in addition to standard velocity and position updating of PSO. The performance of the proposed approach is evaluated using ten publicly available bench mark data sets. From the simulation study, it is found that the proposed approach improves the convergence and generates a compact and comprehensible fuzzy rule based system with high classification accuracy for all the data sets. Statistical analysis of the test result shows the superiority of the proposed method over other approaches reported in the literature.

Chapter 7 – The tissue specificity of a new bioregulators group, isolated from different mammalian tissues, is studied in this article. It was shown that protein-modulators which are serum albumin isoforms are responsible for tissue specificity of MHTBs membranotropic action. The tissue specificity of biological action of MHTBs-containing fractions isolated from liver and lung mammalian tissues are demonstrated on the model of roller organotypic culturing of newt liver tissue *in vitro*.

Chapter 8 – The present work concerns the general conceptions of marketing, marketing research, its goals, problems, categories, objectives. Plus in the given work are mentioned the ways of marketing program evaluation, the ways to cut the impact of reduced spending, different steps important to take in order to help the consumer feel comfortable. A special attention is pained to the basics of marketing, different points of views on them. The present work may also help to navigate away from the marketing and business mistakes that will cost money, time and future growth potential.

Chapter 9 – In different plants there are new group of bioregulators that can classified into the group of regulatory molecules that function as 'tuners' of organ and tissue homeostasis. These bioregulators with physical, chemical properties and biological activity similar to those of MHTBs which had been found earlier in different animal tissues. It was also shown that these bioregulators exhibited the biological effect characteristic of these plants.

Chapter 10 – The software product intended for the integrated monitoring of the human health status by means of express-evaluation techniques of emotional state, central and autonomic nervous system, and system of hemodynamics. It can be used by a wide range of users without age restrictions is presented and described here. The designed software complements our earlier information system "Universal monitoring of ecological health."

Chapter 11 – In this paper a loop antenna is designed at 2 GHz and its performances is compared with fractal loop antennas. It has been seen that as the order of iteration of the loop antenna increases so does the values of gain, standing wave ratio and reflection coefficient. Standing wave ratio reaches the coveted value of 1 and reflection ratio reaches -10dB which are required for proper matching. It is also found that the fractal loop antenna exhibits improved impedance and Standing Wave Ratio (SWR) performances in a reduced physical area as compared to non-fractal Euclidean geometries. Therefore it is very suitable for direction finding in Radars applications.

Chapter 12 – Reliable data storage has been recognized as one of the main dimensions of information technology. Traditionally, the storage systems achieve higher fault tolerance and reliability by introducing redundancy. Mirroring technology is a simple and reliable method but would result in low storage efficiency. In recent years, researchers have revealed that many XOR (exclusive-OR) erasure codes can be applied to storage systems. This paper presents a survey on state-of-the-art development in XOR based erasure codes for the construction of data storage technologies, and discusses their characteristics. In this paper, we have divided XOR erasure codes into two categories: namely Specialized Tolerance Codes and more powerful General Tolerance Codes. The characteristics and encoding/decoding algorithms are described in detail. We hope this survey will inspire more active research in this area.

Chapter 13 – The parameters of the physicochemical regulatory system of lipid peroxidation (LPO) in liver of female outbreed mice at the autumn and spring-summer seasons after action of X-radiation at doses less than 1.5 mGy under changing dose rate. The initial values result in significant effect on the change in the relative weight of the liver, the relative content of phosphatidylcholine and lysoforms of phospholipids in the liver of mice after X-rays exposure. Significant decrease and substantial influence of the dose rate dynamics during irradiation are revealed for the molar ratio of [sterols]/[phospholipids], the ratio of phosphatidylcholine / phosphatidylethanolamine, and the ratio of the sums of the more easily oxidizable to the more poorly oxidizable fraction of phospholipids. Experimental data allow us to conclude that the biological effects of X-rays at low doses under changing dose rate have a complex, nonlinear character.

Chapter 14 – The effect of synthetic antioxidants and bioregulators — 5-hydroxybenzimidazole derivatives on the erythrocyte membrane structure and the erythrocyte morphology was studied by the spin probe method and scanning electron microscopy. Analysis of the EPR spectra and electron micrographs showed that derivatives containing side substituents differing in hydrophobic properties can have a concentration-dependent varying-intensity modifying effect on the structural organization of the plasma membrane of erythrocytes and on their shape. Within the concentration range 10-7-10-3 M, all the substances studied have an echinocytogenic effect, which is most pronounced for the derivative containing benzyl and ethoxy substituents at positions 2 and 5, respectively. The data obtained suggest that these nonelectrolytes are distributed in the outer monolayer of the erythrocyte membrane.

Chapter 15 – Membrane protective properties of a number of isobornylphenols have been studied on model of the erythrocyte hemolysis induced by hydrogen peroxide. It has been shown that their ability to protect the erythrocyte membrane against oxidative hemolysis substantially depends both on the structure and the concentration of the compound. The highest membrane protective activity has been found for substance having isobornyl substitutes at both *ortho* positions. It has been found that all the isobornylphenols studied inhibit the accumulation of oxidation products, which indicates their ability to inhibit lipid peroxidation in cells. The cytotoxicity of a number of isobornylphenols at high concentrations and the absence of the linear dependence of "biological effect - dose" make it possible to suggest that the biological activity of isobornylphenols may be associated not only with their antiradical or antioxidant activity.

Chapter 16 – Toxicity of 27 phenols with different 2-, 6-positions of alkyl substitutes and functional groups NH_2 , CN, OH in 4-posicion was studied. The least toxicity has phenols with two tert-butyl substitutes of benzene nuclei, which increases with mowing p-off benzene nuclei. Obtained results can help to synthesize new nontoxic bioactive phenols.

Chapter 17 – Semisynthetic isobornylphenols (IBP) was shown to exhibit the antioxidant properties in biological model systems reducing the intensity of lipid peroxidation. IBP can possess both membrane protective and membrane active properties in depend on their structure. Surface-active properties defined by their influence on the aggregation of lecithin in hexane are correlate with their effect on the architectonics of erythrocytes. The data obtained allow us to conclude that the biological activity of IBP is due to their antioxidant and surfactant properties, that is necessary to keep in mind under the analysis of their biological activity in systems *in vivo*.

Chapter 18 – With the help of spatial-energy concept it is demonstrated that the formation and change of high-energy bonds in ATP take place at the functional transitions of valence-active orbitals of the system “phosphorus-oxygen”. These values of energy bonds are in accord with experimental and quantum-mechanical data.

Chapter 19 – It has been done the quantum−chemical modeling of the methyl acetate hydrolysis reaction in the processes of enzymatic and homogeneous catalysis by Chemotrypsin and Serine respectively It was shown, that the structure of an active center of the Chemotrypsin taken in references is proved by our calculations. It were estimated the kinetic parameters for the reactions of acyl compounds formation and also of their hydrolysis for both types of the catalysis. It was shown, that the formation of acylchemotrypsin proceeds via a series of the intermediate complexes, caused by the conformation transition, which, in a great measure, caused by the hydrogen bonds of the Chemotrypsin and depend on the journey of Serine active group hydrogen atom. It was established, that the stabilization of the activated complexes of the acidulating and deacidulating reactions in enzymatic and homogeneous catalysis in a great measure depends on a value of the charges of the reactive center, that is indicated on the values of the kinetic parameters of such reactions. It has been calculated the thermodynamics of the elementary stages for the acidulating and deacidulating reactions.

Chapter 20 – pH-dependent detection of the EPR spectra of atomic hydrogen $H^{\bullet}$ (g = 2.0023, A = 50.6 mT) in frozen aqueous solutions of oxyacids AOH initiated by photoinjected low energy electrons upon energy impact below the excitation thresholds of water or oxyacids (photon energy of not more than 4.5 eV) was crucial for identification of the dissociative electron capture by bound acidic proton. The data on direct electron attachment to an oxyacid that can create an acidic bound [H+] network under certain conditions allows one to consider direct participation of the acidic bound proton in the intramolecular ET as one electron relay being highly effective in mediating electron transport in acid catalysis of redox reactions.

Chapter 21 – The AFM method has been used for research of possibility of the stable supramolecular nanostructures formation on the basis of nickel $Ni_xL^1_y(L^1_{ox})_z(L^2)_n(H_2O)_m$ and $Ni^{II}(acac)_2{\cdot}NaSt{\cdot}PhOH$ complexes with the assistance of intermolecular H-bonds.

Chapter 22 – In this article we attempt to consider mechanism of origin of rearrangement of chromosomes in different phases of mitotic cycle. We are showing the appearance of rearrangements of chromosome under the influence of antineoplastic medicament «phosphemid» — di- (ethylene imid)-pyrimidyl-2-amidophosphoric acid, named mutagen. Seeds *Crepis capillaris* and fibroblasts of human and mouse were treated by this mutagen. Rearrangements of chromosomes were analyzing in seedlings on metaphase plates. The rearrangements were analyzed at the scheduled periods of time after appearance of seedlings named «arisings». Also: chromosomal rearrangements were examined at anaphases and telophases of fibroblasts.

Chapter 23 – The article considers regularities of meniscus height changing in the capillary and ultraviolet spectra of ascorbic acid and paracetamol high-diluted solutions, and also solutions, containing both compounds. There have been detected concentration ranges of these compounds action in supersmall doses.

Chapter 24 – Para-aminobenzoic acid (PABA) by way of the well-known activator of growth and development was applied for treatment of ears of winter wheat and winter barley. The varieties of these cultures are destined for North-Caucasian region. Also panicles of selection samples of millet – the varieties are introduced in North Ossetia - were subject to treatment by PABA. PABA was dissolved in ac. a., but not in water, as usual. Advantages of this mode are described. This method allows accelerating a plant-breeding process in case of a long period of harvesting.

Versions of these chapters were also published in *Journal of Information, Intelligence and Knowledge,* Volume 4, edited by Rafiqul Islam, published by Nova Science Publishers, Inc. They were submitted for appropriate modifications in an effort to encourage wider dissemination of research.

In: Biology and Communication in the Information Age
Editor: Rafiqul Islam
ISBN: 978-1-62808-252-4

Chapter 1

THE CHANGES IN CATALYTIC ACTIVITY OF MEMBRANE-BOUND ACETYLCHOLINESTERASE OF HUMAN ERYTHROCYTES DUE TO CHRONIC EXPOSURE TO ELECTROMAGNETIC RADIATION OF MOBILE DEVICES

Mikhail Shlapak * ***and Sergey Y. Rybalko***
Department of Medical Physics of the S. I. Georgievsky Crimean State Medical University, Simferopol, Ukraine

ABSTRACT

Electromagnetic radiation of mobile devices causes the changes in catalytic activity of membrane-bound acetylcholinesterase of human erythrocytes (Yt-antigen) during the adaptation to chronic exposure of electromagnetic radiation. Increased activity of AChE of plasma irradiation in vitro corresponds to the activity of AChE quiet plasma in the body and decrease the activity of erythrocyte AChE in vitro irradiation corresponds to the high activity of erythrocyte AChE adaptation in the body, which is close to a state of decompensation, and is likely to shift to stress.

Keywords: Adaptation, acetylcholinesterase, mobile devices and health, Yt-antigen and radiation

INTRODUCTION

Influence of electromagnetic radiation of mobile phones (EMR MP) on the human body is actively studied in many research groups around the world, but currently available data are insufficient to create a satisfactory theory explaining the biological effects of EMR MP.

* E-mail: mihail.shlapak@gmail.com.

These effects include increased permeability of the blood-brain barrier, changes in the expression of heat shock proteins, impaired calcium homeostasis, reduction of melatonin synthesis, etc. These effects can alter the adaptive capacity of an organism, in this connection arises the problem of studying the biochemical and physiological mechanisms of adaptation, which occurs only with a change in regulatory systems, such as enzymes. [1] Without studying of these effects we can't penetrate into the mechanisms of adaptation of whole organisms. This is especially true of studying neurotransmitter mechanisms of stress and adaptation, in particular, cholinergic, which belong to neuroendocrine stresslimitiruyuschim systems.

One of the major components of the cholinergic system and cholinergic marker enzyme elements is acetylcholinesterase (AChE). It is known that AChE is unique among the set of enzymes. This is explained by the fact that it is important not only in the transmission of nerve impulses, but also in the regulation of metabolism.[2, 3]. The foregoing relates to the nervous tissue AChE. As regards erythrocyte AChE, its physiologic role is not fully established. Until recently, erythrocyte AChE, practically not been studied in terms of participation in metabolism during adaptation. We chose the erythrocyte as an object of study because it is accessible to experimental studies, on the other hand, the erythrocyte plasma membrane, as an essential element of the cell contains a large number of biomolecules with different functions and have a variety of classification, such as membrane-bound erythrocyte AChE plays the role of Antigen Groups blood Cartwright (Yt).

Experimental Part

The purpose of our experiment was to study the activity of membrane-bound AChE of human erythrocytes by the action of electromagnetic radiation of mobile phones. For this purpose, was monitored individual dose of electromagnetic loads (IDEL), a group of student volunteers, 80 men, they have also been determined AChE activity in plasma and erythrocyte membranes.

Results and Discussion

As a result of statistical analysis was established a significant correlation with the activity of AChE IDEL as in plasma and in erythrocytes (Table 1).

Table 1. Correlation between the activity of AChE and IDEL in plasma and erythrocyte membranes (N = 80)

	IDEL	AChE activity of the plasma	The activity of erythrocyte AChE
IDEL	1	-0,632**	0,184
AChE activity of the plasma	-0,632*	1	0,520*
The activity of erythrocyte AChE	0,184	0,520*	1

Note **- Pmist. <0.01 *- Pmist. <0.02.

As seen from the results shown in Table 1, reliable and a negative correlation between AChE activity and IDEL blood plasma (r =- 0,632) shows a decline in AChE activity of the plasma with increasing electromagnetic load, and significant positive correlation between activity and IDEL erythrocyte AChE (r =- 0,520) a tendency to increase with an increase in erythrocyte AChE IDEL. Based on the apparent inconsistency of the data we have carried out a further study of the same blood sample under irradiation with EMR MP in vitro. Blood samples were subjected to electromagnetic radiation in the MT for 30 min, then measured the activity of AChE in plasma and erythrocytes. The results are shown in Table 2.

Table 2. Changes in the activity of AChE plasma and red blood cells irradiated in vitro (n = 80)

	control (M ± m)	Experience (exposure 30 minutes) (M ± m)
AChE activity of the plasma (mcmol / s • L)	83,58 ± 4,438	45,98 ± 3,816 **
The activity of erythrocyte AChE (mol / s • L)	85,51 ± 6,968	113,2 ± 5,363 *

Note **- Pmist. <0.0001 *- Pmist. <0.01.

Based on the results presented in Table 2 shows that irradiation of electromagnetic radiation of mobile devices activity of AChE significantly reduced plasma and erythrocyte AChE activity conversely increased, consistent with previous observations (Table 1).

Performed our studies on the effect of electromagnetic radiation on the activity of AChE MT human erythrocytes may help to trace the changes in stressors studied the enzyme at different stages of its operation.

According to the theory of adaptive reactions, using the results obtained during the experiments and calculations, we can assume the system's transition to the next level, associated with an increase in IDEL. Increased activity of AChE of plasma irradiation in vitro corresponds to the activity of AChE quiet plasma in the body and decrease the activity of erythrocyte AChE in vitro irradiation corresponds to the high activity of erythrocyte AChE adaptation in the body, which is close to a state of decompensation, and is likely to shift to stress.

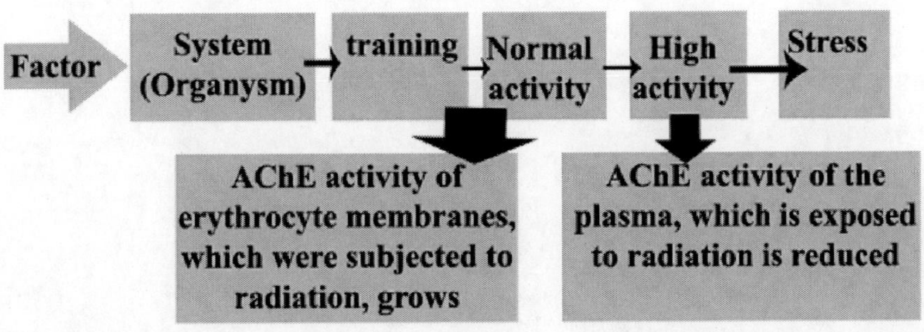

Figure1. Interpretation of the results according to the theory of adaptive responses (LH Harkavy and EB Kvakin).

Conclusion

1. A significant inverse correlation (-0.632) between the activity of AChE and plasma IDEL students involved in monitoring the biological effects of EMR MP.
2. A direct significant correlation between the activity of AChE plasma and erythrocytes (0.520).
3. Show significant reduction in the activity of AChE of plasma irradiation EMR MP 30 minutes from 83.58 to 45.98 mmol / sxl.
4. Show a significant increase in the activity of AChE of human erythrocytes irradiated in vitro with 85.51 to 113.2 mmol / sxl.

References

[1] The Role of Erythrocytic Acetylcholinesterase in Hormonal Regulation of Adaptation to Physical Exercise. I.V.Men'shikov. *Udmurt State University.*

[2] Acetylcholinesterase Adaptation to Voluntary Wheel Running is Proportional to the Volume of Activity in Fast, but not Slow, Rat Hindlimb Muscles. Victor Gisiger, Manon Bélisle, Phillip F. Gardiner. *European Journal of Neuroscience,* Volume 6, Issue 5, pages 673–680, May 1994.

[3] Blood Groups and Red Cell Antigens. Laura Dean.National Center for Biotechnology Information (NCBI), National Library of Medicine, *National Institutes of Health, Bethesda*, MD 20892-6510.

In: Biology and Communication in the Information Age
Editor: Rafiqul Islam
ISBN: 978-1-62808-252-4

Chapter 2

SYNTHESIS AND BIOLOGICAL ACTIVITY OF 17-AMINO-5α-ANDROSTAN-3-OLS†

N. Sh. Nadaraia and M. I. Sikharulidze*
Tbilisi State Medical University Iovel Kutateladze Institute of Pharmacochemistry, Tbilisi, Georgia

ABSTRACT

Methods of synthesis and biological activities of unsubstituted and substituted epimer aminoalcohols of 5α-androstans are described.

Keywords: Aminosteroids, synthesis, biological activities

INTRODUCTION

Chemical modification of natural compounds of plant origin by introduction of pharmacophore groups and modeling of therapeutic preparations of needed physiological activity is an important task of chemistry and pharmacology.

Steroids widely distributed in plants and animals. They are potential carriers of biological activity and by this are extensively used for development of new medicinal remedies with specific action on functions of different organs of humans and animals.

Aminosteroids from this point of view are of great interest. There is information in scientific literature that this class of steroids exhibit antiarrhythmic, hypotensive, cardiotonic and antitumor activities as well as ability to stimulate the central nervous system having at it no hormonal effect [1-13]. Recommended for use in medicinal practice at cardiac arrhythmia are 3α-amino-2β-hydroxi-5α-androstan-17-on(1) and methyl ester of 3α-hydroxy-11α-(3-methylbutylamino)-2β-ethoxy-5α-androstan-17β-carboxylic acid [14,15]. The high fungicide activity was revealed by 17β-amino-5α-androstan (3) [16].

* E-mail: docmadona@yahoo.com

† Dedicates to Prof. N. N. Suvorov and PhD. V. I. Sladkov.

It needs to be noted that the degree and direction of physiological activity of drugs are largely determined by the nature and steric configuration of the substitute.

In the present paper, the results of study on synthesis and pharmaceutical activities of isomeric 17-amino-5α-androstan-3-ols and their derivatives are generalized in order to elucidate the influence of confguration at C-3 and C-17 as well as that of substitutes at nitrogen atom on biological properties.

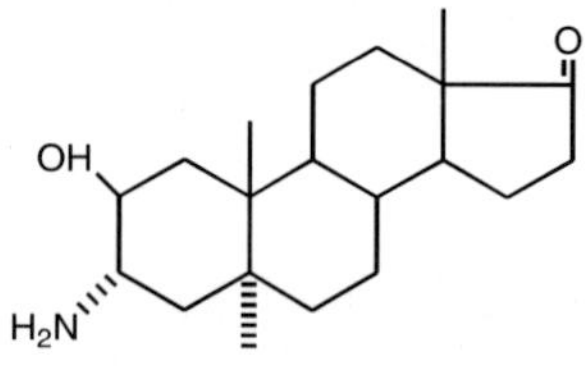

1

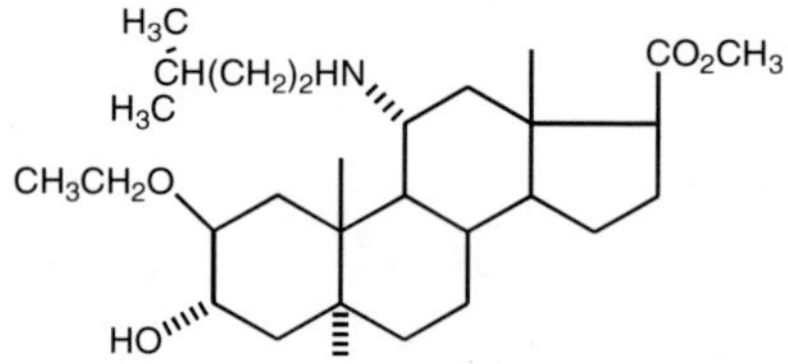

2

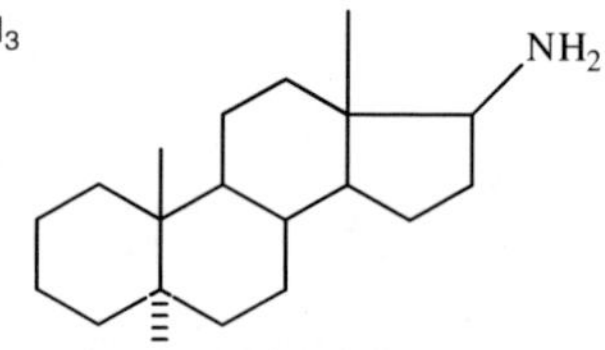

3

I. SYNTHESIS OF 17B-AMINO-5A-ANDROSTAN-3B-OL

The first statement about synthesis of 17β-aminon-5α-androstan-3β-ol(4) in the process of synthesis of androstandion from oxime of 3β-acetoxypregnenolon (5) was made in USA patent [17]. It was shown that in the process of boiling of oxime(5) with thionyl chloride in benzole, Beckman regrouping takes place that leads to formation of 3β-acetoxy-17-acetylamino-5α-androstan(6) (scheme 1.1).

It was pointed out [18-20] that desacetylation by amidic acid(6) was a complex task. Yield of aminosteroid (4) from compound (6) did not exceed 30%. The authors attributed such low yield of amine(4) to instability of steroid(4) in strong acidic environment at temperatures >120^{0}C. At alkaline hydrolysis of amide(6) in different alcohols at temperatures 160-180^{0}C, increase of yield of amine(4) up to 90% was managed. Beckman regrouping of O-oxime tosylate (5) that takes place during the processing of steroid(5) by TsCl in pyridine at heating (100°C) in ethanolamine is accompanied by hydrolysis of intermediate acetamide(6) and leads to formation of amine(4), which was described in [21] as benzilidene derivative.

Firm “Faberwerk Hoechst AG” has developed techniques of obtaining of amine(4) via Beckman regrouping of oxime(5) under influence of phosphorus oxychloride in pyridine. Considering conservation of configuration at the migrating group carbon atom during Beckman, Curtius and Schmidt regroupings [20-26], one may assume that obtained by Beckman regrouping oxime(5) and amine(4) exist in 17β configuration. However, low yield of steroid(4) at hydrolysis of amide(6) by acid and the strict conditions of alkaline hydrolysis are the limiting factors for use of this method of 17β-amino-5α-androstan -3β-ol synthesis.

Method of obtaining of 17β-(1$'$-alkoxyethylidenamino)steroids(7,8) developed by [25-27], implying interaction of pregnanolon (9) with waterless nitrous acid in the presence of hydrochloric acid in the corresponding alcohol, enables transition to the primary amine(4). Realisation of Schmidt reaction is complicated by the safety reasons.

The authors of [28-30] synthesized amine(4) by reduction of oxime(10) with sodium in absolute ethanol, or propanol. It was shown by Shoppe in [31] that reduction of 17-ketoximes with both sodium in ethanole and hydrogen with platinum in acetic acid leads to formation of 17β-aminosteroids. The authors in [32] obtained amine of 5α-androstan line by hydration of

17-hydroximinoandrost-5-en-3β-ols with platinum oxide in acetic acid. At this, they were not able to determine configuration at C-17. It is a widely recognized [33] fact that catalytic hydration of oximes to amines is a fast reaction and the nature of the obtained products is determined by kinetic control that applies to less stabile axial amines, i.e. 17α-amines in the given case. Formation of axial amines after reduction of steroid oximes with palladium, platinum or nickel catalysts is demonstrated in [34-37]. According to Bertch [40], after reduction by lithium aluminium hydride [38-39] or sodium in liquid ammonia, the steroid oximes transfer into mixture of axial and equatorial amines. Reductive amination according Leuckart-Wallach [41-43] leads preferably to equatorial amines.

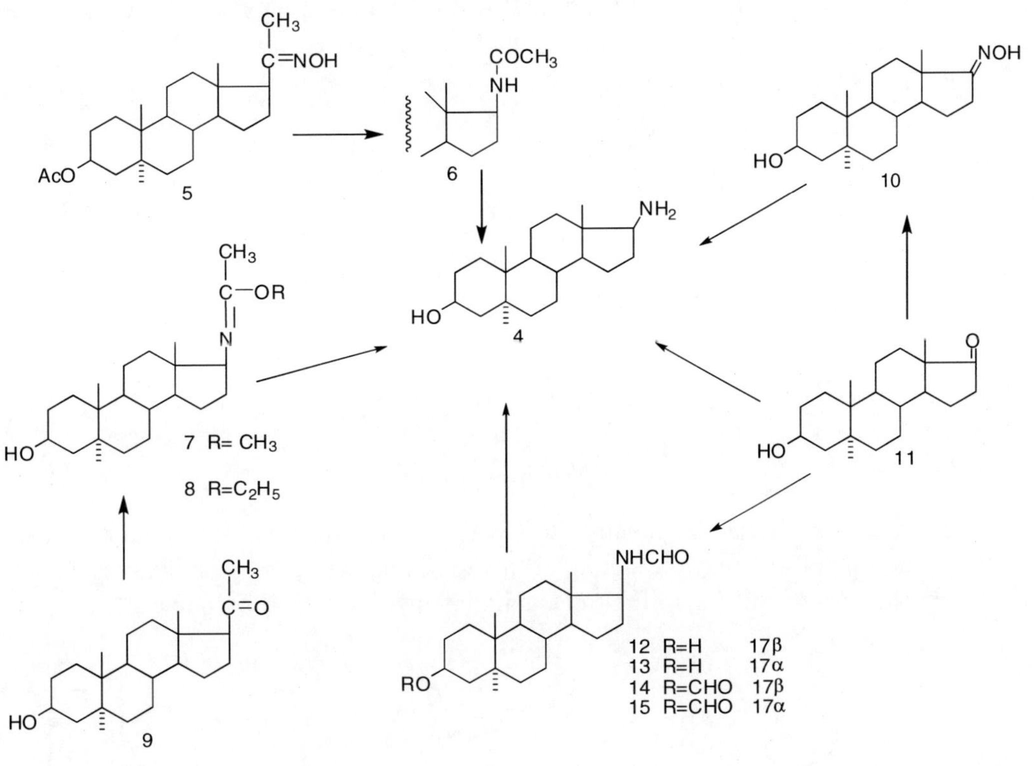

Scheme 1.1.

In [44], the authors, expanding the method of reductive amination elaborated by French scientists [45] on 5α-androstan line on 3β-hydroxyandrost-5-en-17-on, synthesized amine(4). This method [45] implies interaction of ketone(11) with amalgamed aluminium in ethanol in the presence of hydrazine hydrate. In the course of elaboration 17β-amine (4) synthesis, the authors of [46] compared two ways of inclusion of nitrogen containing substitute in C-17 position in order to study their stereoselectivity. The first of them was a known method consisting in oximation of epiandrosterone(11) ketone group and reduction of oxime(10) to amine(4) by sodium in alcohol and the second one, applied for epiandrosteron, consisted in reductive amination by Leuckart-Wallach reaction. Interaction of ketone(11) with combination of ammonium, formamide and formic acid at 180^0C leads to formation of mixture of 17α- and 17β- formyl derivatives(12-15). Acidic hydrolysis of the reaction product gave ~10:1 ratio of 17β- and 17α-amines (based on data of NMR^{1}H spectra). After recrystallization, pure 17β-epimer (4) was obtained. The first method gives 17α- and 17β epimers (4,16) in the same ratio, but the second one is more preparative and simple in use.

II. SYNTHESIS OF 17A-AMINO-5A-ANDROSTAN-3B-OL

In 1962 Henbest and Jackson [47] showed that traditional activation of α-carbon atom of 17β-hydroxil group in testosterone (17) to nucleophilic attack by azide-ion through its transformation in sulphonic acid ester (18) allows to provide the process of S_N2-substitution and obtain 17α-azidoandrost-4-en-3-on (19) (scheme 2.1). Hydric reduction is a more simple transition from azido- to aminogroup. This well-known method proposed by Streitwieser and Schaeffer [48] makes a basis for synthesis of 17α-aminosteroids from 17β-hydroxyderivatives.

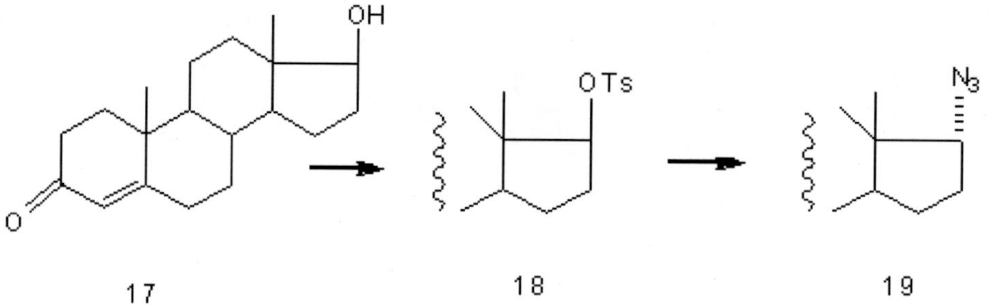

Scheme 2.1.

Using Streitwieser-Schaeffer method, Robinson [49] synthesized 17 α-aminoandrost-5-en-3β-ol by heating of 17β-tozyloxysteroid (20) with sodium azide in N-methylpyrrolidone to 130^0C and consequent reduction of lithium aluminium hydride in THF (scheme 2.2).

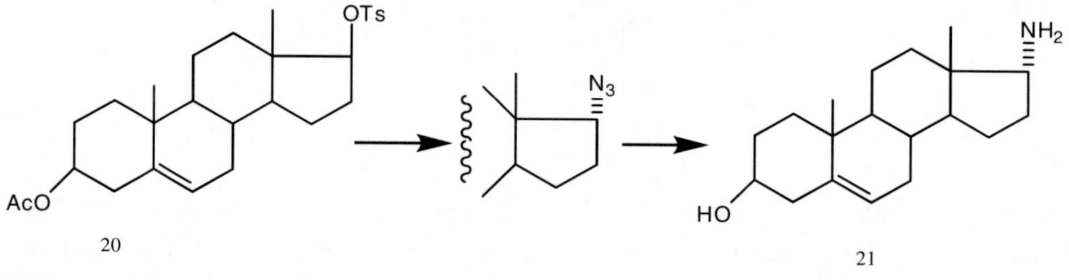

Scheme 2.2.

Davis and his collaborators [50] expanded research in the direction of biologically active derivatives of 17-aminosteroids and synthesized 17α-isomer (16) transferring 17β-tozyloxysteroid (22) in 17α-azide(23) and reducing the latest (scheme 2.3). The authors also managed to synthesize some of 17α-derivatives from 17β-steroid(22) by nucleophilic substitution of 17β-tozyloxygroup with diethylamino- , pyrrolidino- and piperidinogroups. Synthesis of 17α-aminosteroids via azides was exercised also by other authors [51].

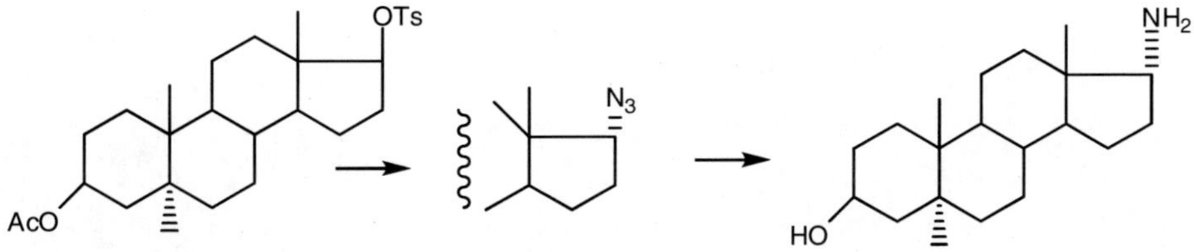

22 23 16

Scheme 2.3.

A way of obtaining of androstan line 17α-aminosteroid using inversion of 17β-alcohol sulphonic esters by their reaction with amoníaco or lower alkylamines was described in French patent [52].

In spite of the fact that sulphonic acid esters are more inclined to substitution rather than elimination [53], in the conditions applied by the authors in [31, 47, 50-52], the process of substitution by S_N1 mechanism and elimination by E1 and E2 mechanisms cannot be excluded. However, neither side effects nor data about stereochemical yield in the process of reconfiguration at C-17 are declared in these studies.

In [46], the authors studied dependence of stereochemical yield of reaction on experimental conditions and found out that at identical conditions [49,50], 17α-azide(23), ketone (24) and 17α-alcohol(25) are formed in ratio 3:2:1 from 17β- tosyloxysteroid (22) with 93% yield (scheme 2.4).

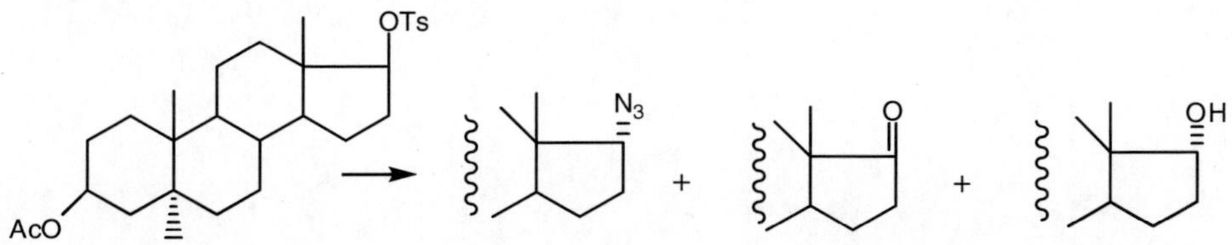

22 23 24 25

Scheme 2.4.

The yield of the target aside (23) was 42%. During execution of DMFA reaction at containment of 15% water and reduction of temperature down to 100-105^{0}C, the authors of [46] managed to reduce the yield of solvolysis (25) product, exclude formation of oxidation(24) product and increase the yield of the target 17α- azide (23) up to 72%. Reduction of azide (23) to amine(16) by excessive lithium aluminium hydride in waterless diethylalcohol leads to formation of 17α-amine(16).

III. SYNTHESIS OF 17B-AMINO-5A-ANDROSTAN-3A-OL

The first description of 17β,3α-epimer (26) synthesis from oxime of androsterone(27) by reduction with sodium in ethanol was given in 1935 [54]. It is interesting that during reduction of oxime(27) synthesized from androsterone(28) the same amine(26) is obtained by both hydrogen with platinic oxide and sodium in ethanol.

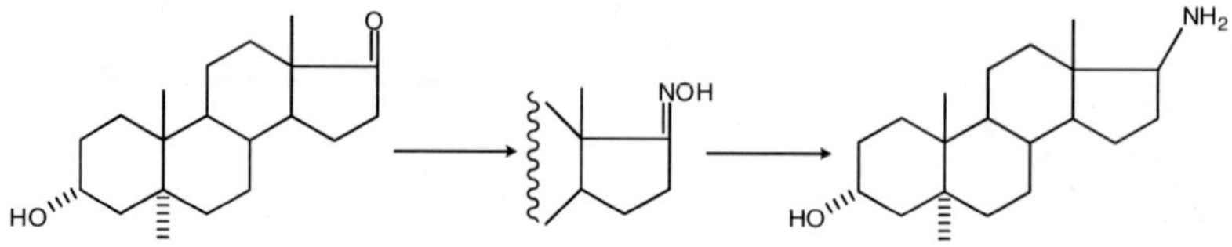

28 27 26

Scheme 3.1.

Along the works [55-58], synthesis of 3α, 17β-epimer(26) from 3-epihydroxybisnorallocholanic acid was described in USA patent [59]. This method is quite labour consuming for epimerization of 3β-hydroxil as well as for inclusion of aminogroup in C-17 position. Later on [60], amine(26) was synthesized from androsterone(28) by reductive amination [45].

It needs to be mentioned that the methods of amine(26) synthesis on the basis of animal, or plant materials are connected with the necessity of epimerization at C-3. In this regard, the authors [61] used a simple method of Leuckart-Wallach [62] for epimerization of 3β-hydroxyl group and synthesized androsterone(28) from epiandrosterone(11) with 50% yield. The yield was cut down by formation of 5α-androst-2-en-17-on, the products of β-elimination, and 5α-androstandion-3,17, the product of oxidation. Androsterone(28) similarly to [46] epiandrosterone(11) transformed in 3α, 17β-epimer(26) by reaction of reductive amination [61].

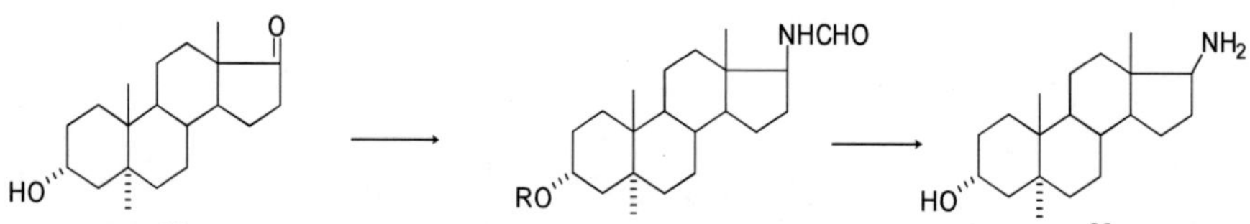

28 26

29 R=CHO 17β
30 R=H 17β
31 R=CHO 17α
32 R=H 17α

Scheme 3.2.

Application of method by Mitsunobu[63], allowed[64] to attain the complete transformation of configuration of the secondary hydroxyl group by C-3 of epiandrosterone(11). Extending this method on 17- acetylaminoderivative (12), the authors [61] developed more convenient scheme of amine(26) synthesis.

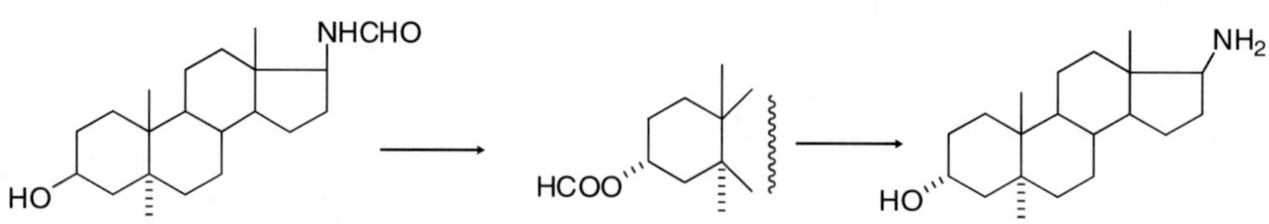

12 29 26

Scheme 3.3.

IV. SYNTHESIS OF 17A-AMINO-5A-ANDROSTAN-3A-OL

To obtain never described before in literature epimer(33), the authors [65] started from 3β, 17α-epimer(13). Using method by Mitsunobu[64] for transformation of configuration of 3β-hydroxil group applied earlier [61], they synthesized 3α, 17α-epimer(31) for 17β-amide (12). Hydrolysis of diformyl derivative (31) by 20-% hydrochloric acid led to 3α, 17α-epimer(33).

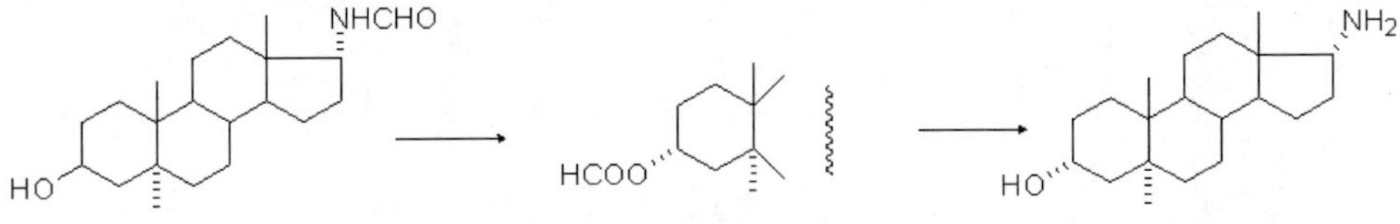

13 31 33

Scheme 4.1.

Data of ^{1}H NMR for compound (4,16,26,33)
(δ, $CDCl_3$, J/Hz)

Compound	3-H		17-H		18-CH_3	19-CH_3	J/Hz
	α	β	α	β			
4	3.59m	-	2.64t	-	0.63s	0.82s	$J_{17}a'/_{16}e'=8.5$ $J_{17}a'/_{16}a'=8.8$
16	3.59m	-	-	2.90d	0.68s	0.81s	$J_{17}e'/_{16}e'\sim0$ $J_{17}e'/_{16}a'=7.3$
26	-	4.04m	2.65t	-	0.63s	0.79s	$J_{17}e'/_{16}e'=8.9$ $J_{17}a'/_{16}a'=8.9$
33	-	4.04m	-	2.89d	0.68s	0.79s	$J_{17}e'/_{16}e'\sim0$ $J_{17}e'/_{16}a'=6.8$

V. SOME DERIVATIVES OF 17-AMINO-5A-ANDROSTAN-3-OLS

There are many derivatives of 17β-aminoalcohol(4) for pharmacological studies described in literature. 17β-aminoacetamid derivatives (34,35) were synthesized by the authors [1] to investigate anticholingeric, antiarithmic, anti-inflammatory and other types of activities.

Synthesis of steroid(36) is described and its antibacterial activity is elucidated in [28].

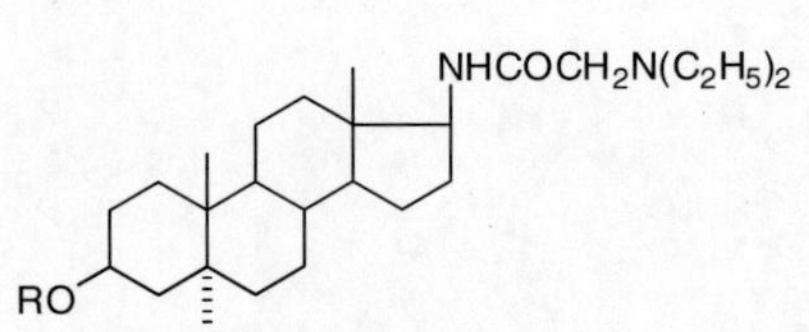

34 R=H
35 R=Ac

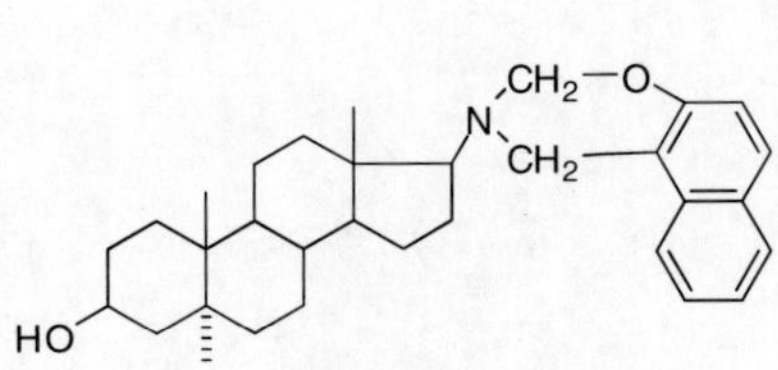

36

The authors of works [5,6,8,9] synthesized different 17β–[N-(aminoalkil)amino]-5α-androstan-3β-ols (37-49) from corresponding androst-5-ens, or 17β–[N-(aminoalkil)amino]-derivatives for investigation of their hypocholesterinemic and antiarithmic activities.

Compounds (50-54), the syntheses of which from corresponding Δ^5-analogues is described in patents [10, 11], act on central nervous system. Aminosteroids (55-57), obtained by Leuckart-Wallach reaction, were studied in [66,67] as miorelaxants.

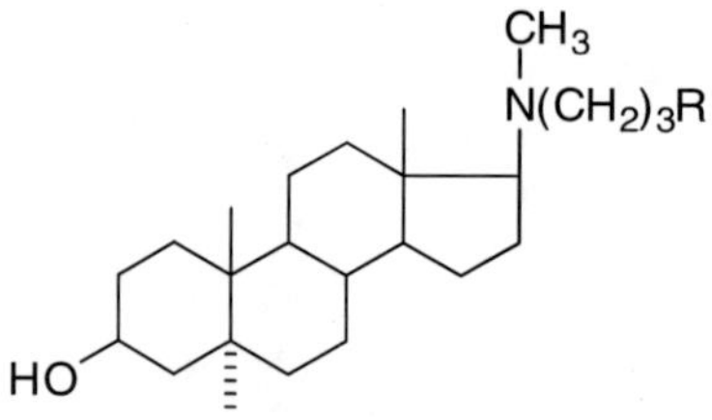

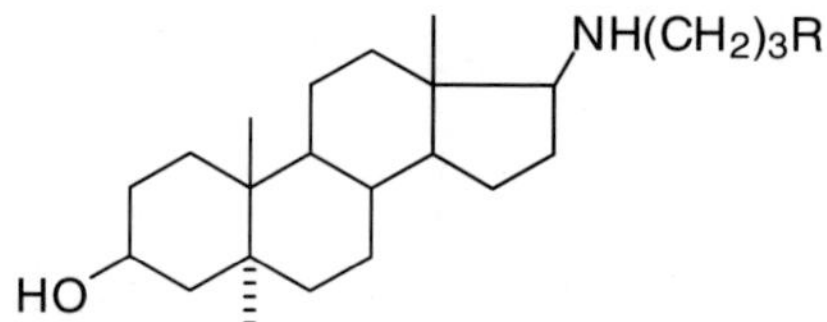

37 R=$N(CH_3)_2$ 45

38 R=$N(C_2H_5)_2$ 46

39 R= 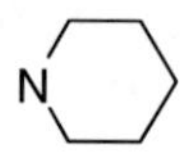

47

40 R= 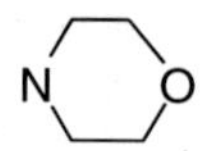

48

41 R= 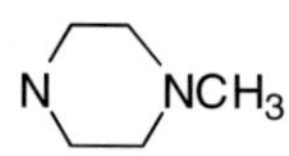

49

42 R= 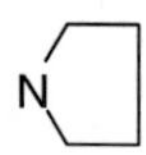

43 R=CN

44 R=NH_2

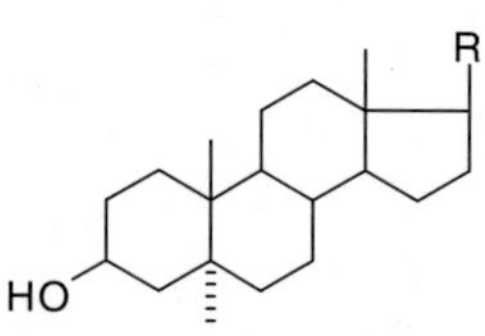

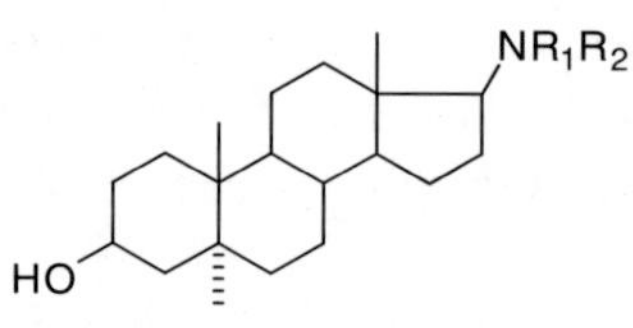

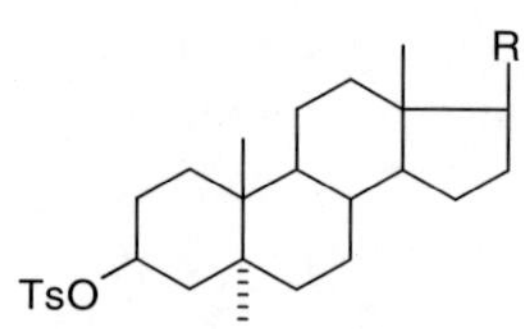

50 R= 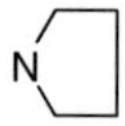

51 R= 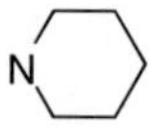

52 R_1=R_2=C_2H_5

53 R_1=CH_3, R_2=$CH_2C_6H_5$

54 R_1=H, R_2=C_6H_5

55 R_1=R_2=CH_3

56 R= 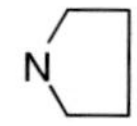

57 R=$N(CH_3)_2$

The authors of [68] have synthesized compounds (58,59) using N-alkylation of amine(4) to study its antitumor activity. Furthermore, cardiotonic activity of steroid maleinimides(60,61) was studied in [3].

Some 3β-(β-D-glucosil)-, 3β-(β-D-arabinosil)- and 3β(L-arabinosil)-17β-amino-5α-androstans are synthesized to elucidate their antihypertension activity. Pettit and collaborators [4, 69-71] realized the synthesis of some steroid peptides such as 17β-(L-prolil)- and 17β-(L-arginil-L-prolil)amino-5α-androstan-3β-ols in order to study their corticotropic activiry. Later, the authors of [72, 73] synthesized also some steroid peptides in order to reveal new antiarrhythmics. Data exist in scientific literature [3, 6-10, 50, 74] on the synthesis of only several 3β, 17α- and 3α, 17β-epymer derivatives (62-70) for pharmacological studies.

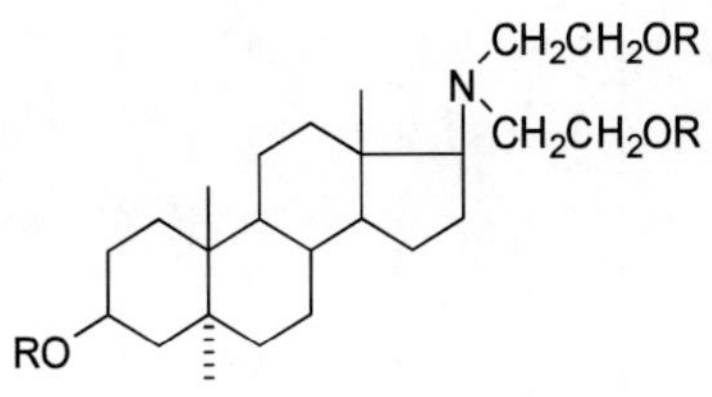

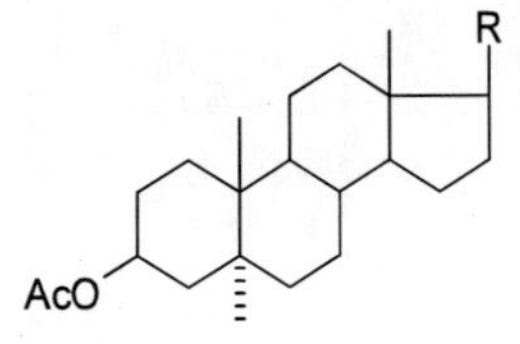

58 R=H

59 R=Cl

60 R=

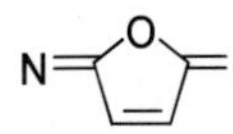

61 R=

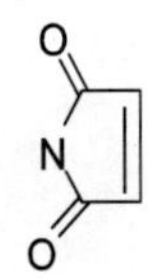

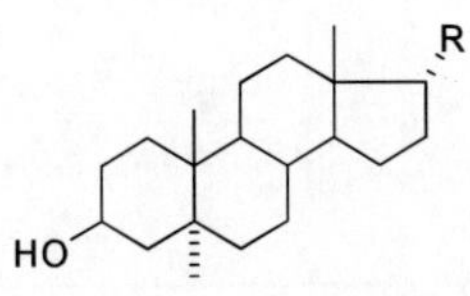

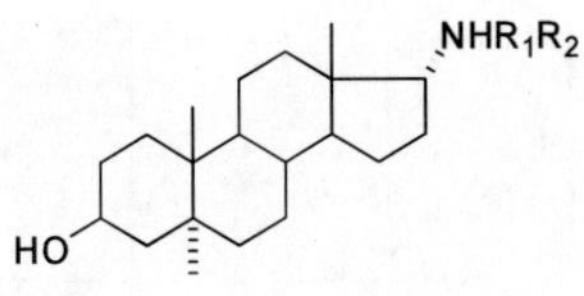

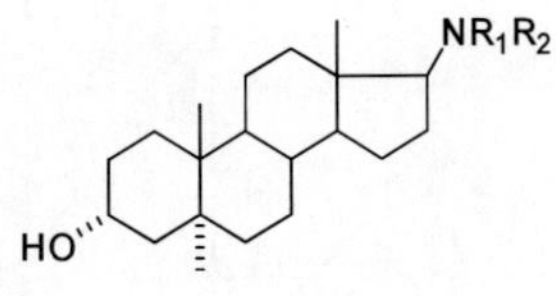

62 R=

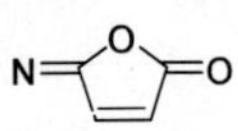

67 R_1=H, R_2=$(CH_2)_3N(CH_3)_2$ 69

68 R_1=CH_3, R_2=$(CH_2)_3N(CH_3)_2$ 70

63 R=

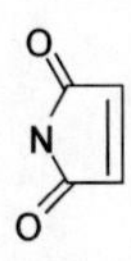

64 R=

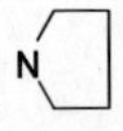

65 R=

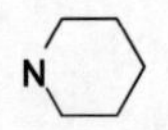

66 R=

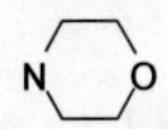

Table 5.1. ^{1}H NMR Data of compound (34,71-75) ($CDCl_3$, ppm, J/Hz)

Compound	3-αH	17-H		$18\text{-}CH_3$	$19\text{-}CH_3$	R	J/Hz
		α	β				
34	3.59m	3,85q	-	0,69s	0.81s	$NHCOCH_2N(CH_2\,CH_3)_2$ 7.39d; 3.00s; 2.53q; 1.02t	$J_{17}a'/_{16}e'$=8.8 $J_{17}a'/_{16}a'$=9.0 $J_{17}a'/_{NH}$=9.3 J_{CH2CH3}=7.0
71	3.59m	3,87q	-	0,69s	0.82s	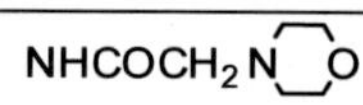 7.10d; 2.91d; 3,08d; 2,52m; 3.72m	$J_{17}a'/_{16}e'$=8.6 $J_{17}a'/_{16}a'$=9.0 $J_{17}a'/_{NH}$=9.5 J_{CH2CH2}=4.4 $J_{CH2(hem)}$=15.9
72	3.59m	3,86q	-	0,69s	0.81s	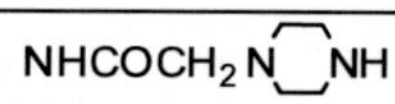 7.19d; 2.88d; 3.06d; 2.89t; 2.50m	$J_{17}a'/_{16}e'$=8.8 $J_{17}a'/_{16}a'$=9.5 $J_{17}a'/_{NH}$=9.3 J_{CH2CH2}=4.7 J_{CH2NH}=5.1 $J_{CH2(hem)}$=17.1
73	3.59m	3.86q	-	0.69s	0.81s	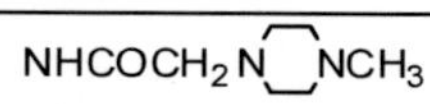 7.16d; 2.95d; 3.05d; 2.57t; 2.45t; 2.30s	$J_{17}a'/_{16}e'$=9.1 $J_{17}a'/_{16}a'$=9.2 $J_{17}a'/_{NH}$=9.5 J_{CH2CH2}=4.5 $J_{CH2(hem)}$=16.6
74	3.59m	-	3.9 5t	0.80s	0.81s	$NHCOCH_2N(CH_2\,CH_3)_2$ 7.47d; 2.91d; 3.08d; 2.56q; 1.04t	$J_{17}e'/_{16}e'$=0 $J_{17}e'/_{16}a'$=7.8 $J_{17}e'/_{NH}$=8.7 J_{CH2CH3}=7.0 $J_{CH2(hem)}$=17.5
75	3.59m	-	3.9 5t	0.82s	0.82s	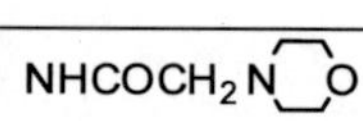 7.06d; 2.91d; 3.08d; 2.55m; 3.74m	$J_{17}e'/_{16}e'$~0 $J_{17}e'/_{16}a'$=8.1 $J_{17}e'/_{NH}$=8.5 J_{CH2CH2}=4.5 $J_{CH2(hem)}$=14.7

Table 5.2. ^{1}H NMR Data of compound (76-88) ($CDCl_3$, ppm, J/Hz)

Compound	3-H		17-H		$18\text{-}CH_3$	$19\text{-}CH_3$	R	J/Hz [a)]
	α	β	α	β				
76	3.59m	-	2.53t	-	0.69s	0.81s	$NHCH_2CH_3$ 2.65q; 1.08t	$J_{17}a'/_{16}e'=8.5$ $J_{17}a'/_{16}a'=8.6$ $J_{CH2CH3}=7.1$ $J_{CH2NH}=2.2$
77	3.59m	-	2.51t	-	0.69s	0.81s	$NHCH_2CH_2OH$ 2.79t; 3.60t	$J_{17}a'/_{16}e'=8.5$ $J_{17}a'/_{16}a'=8.5$ $J_{CH2CH2}=5.7$
78	3.59m	-	2.48t	-	0.69s	0.81s	$NHCH_2CONH_2$ 3.24d; 3.33d; 5.44s	$J_{17}a'/_{16}e'=9.0$ $J_{17}a'/_{16}a'=9.4$ $J^{CH2}=17.6$
79	3.58m	-	2.48t	-	0.69s	0.81s	$NHCH_2CH_2\ NH_2$ 2.72m	$J_{17}a'/_{16}e'=8.6$ $J_{17}a'/_{16}a'=8.9$ $J_{CH2CH2}=5.0$ $J_{CH2NH}=2.2$
80	3.59m	-	2.40t	-	0.69s	0.81s	$NHCH_3$ 2.34s	$J_{17}a'/_{16}e'=7.8$ $J_{17}a'/_{16}a'=9.4$
81	3.59m	-	-	2.47d	0.71s	0.81s	$NHCH_3$ 2.34s	$J_{17}e'/_{16}e'\sim0$ $J_{17}e'/_{16}a'=6.8$
82	-	4.04m	2.41t	-	0.69s	0.78s	$NHCH_3$ 2.43s	$J_{17}a'/_{16}e'\sim9.0$ $J_{17}a'/_{16}a'\sim9.0$
83	-	4.02m	-	2.48d	0.72s	0.78s	$NHCH_3$ 2.35s	$J_{17}e'/_{16}e'=1.5$ $J_{17}e'/_{16}a'=6.8$
84	3.59m	-	2.71t	-	0.69s	0.81s	$NHCH_2CN$ 3.53d; 3.64d	$J_{17}a'/_{16}e'=8.3$ $J_{17}a'/_{16}a'=8.5$ $J_{CH2(hem)}=17.3$
85	3.59m	-	-	2.84d	0.75s	0.81s	$NHCH_2CN$ 3.51d; 3.62d	$J_{17}e'/_{16}e'\sim0$ $J_{17}e'/_{16}a'=6.8$ $J_{CH2(hem)}=17.5$
86	-	4.04m	2.71t	-	0.69s	0.78s	$NHCH_2CN$ 3.53d; 3.64d	$J_{17}a'/_{16}e'=8.3$ $J_{17}a'/_{16}a'=8.6$ $J_{CH2(hem)}=17.6$
87	-	4.04m	-	2.84d	0.75s	0.78s	$NHCH_2CN$ 3.50d; 3.61d	$J_{17}e'/_{16}e'\sim0$ $J_{17}e'/_{16}a'=6.6$ $J_{CH2(hem)}=17.6$
88 [b)]	3.33m	-	3.46t	-	0.76s	0.98s	$N^+(CH_3)_3I^-$ 3.08s	$J_{17}a'/_{16}e'=9.5$ $J_{17}a'/_{16}a'=9.5$ $J_{3H/OH}=4.9$

Note:

a) $J_{17/NH}$ absent of rapid exchange.

b) The spectra were surved in d_6-DMSO; Chemical shift of OH groups δ 4.44 m.p.

Taking into account the fact, that the search for new antiarrhythmics are carried out primarily amoung substituted amides, the space shielded by functional groups [75], the authors [12,13] sinthesized some epimeric alkilamino- and alkilaminoacetylaminoderivatives (34,55,71-89) for study the structure-activity relationships.

Data exist [76] on synthesis of some 3β- esters of 17β-acetamido-5α-androstan, which are considered as antitumor agents.

It needs to be noticed that derivatives of 3β, 17β-epimer (4) are studied quite widely when the derivatives of the rest three epimers(16,26,33) remain less investigated.

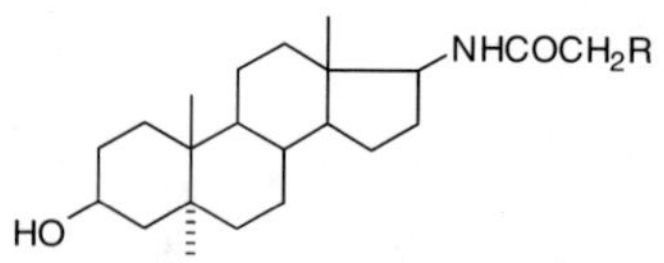

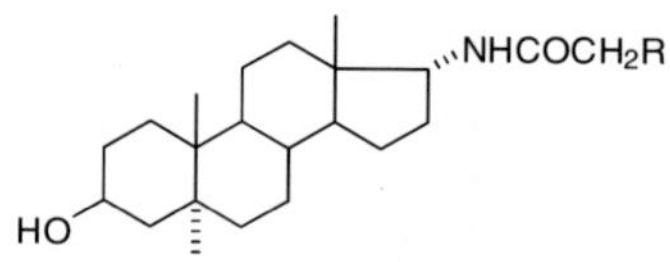

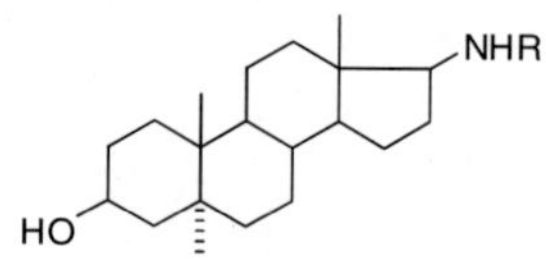

34 $R=N(C_2H_5)_2$ 74

71 R= N O 75

72 R= N NH

73 R= N NCH_3

76 $R=CH_2CH_3$

77 $R=CH_2CH_2OH$

78 $R=CH_2CONH_2$

79 $R=CH_2CH_2NH_2$

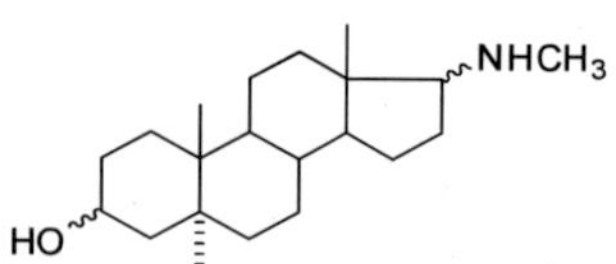

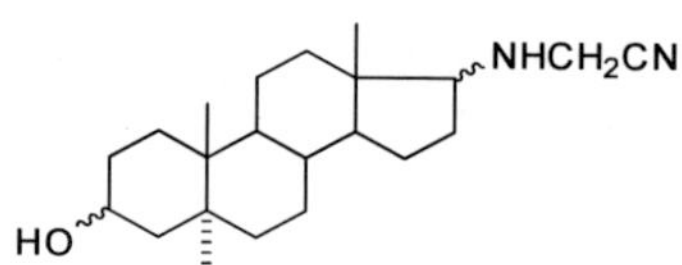

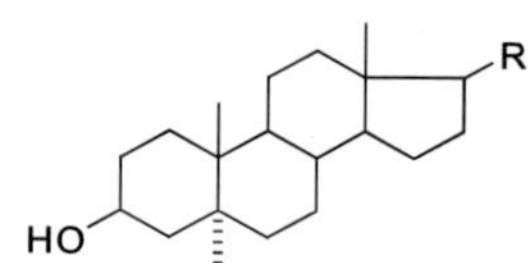

80 3β, 17β 84

81 3β, 17α 85

82 3α, 17β 86

83 3α, 17α 87

88 $R=N^+(CH_3)_3I^-$

89 $R=N^+(CH_3)_2C_2H_5Br^-$

VI. BIOLOGICAL STUDY OF EPIMERS AT C-3 AND C-17 17-AMINO-5A-ANDROSTAN-3-OLS

Information on phatmacological study of synthesized earlier compounds is of fragmental character and often limited by just indication of activities anticipated by the authors. Until recently, the data on biological activities of unsubstituted aminoalcohols (4, 16, 26, 33) were absent. In [12], the authors showed that at aconitive arrhythmia, 3β, 17β- (4) and 3β, 17α-epimers(16) out of four theoretically plausible unsubstituted aminoalcohols (4,16,26,33) exhibit higher activity in prevention of cardiac rhythm disorder. Alteration of configuration at C-3(26) leads to decrease of antiarrhythmic activity and 3α, 17α-epimers (33) have no such activity at all.

The influence of monomethylation of 17β-aminogroup(80 and 82) on activity in not significant. Transformation of primary 17β-aminogroup in the tertiary (55) or quadruple(88) one is accompanied by insignificant decrease of antiarrhythmic activity. Introduction of hydroxyl-β-ethyl group (77) leaves the activity at the same rate (88) but introduction of amino- β-ethyl group (79) leads to its elimination. Aminoacetyl derivatives (34,72,73) lose their activity.

Pharmacological study [13] of four epimers of aminoderivatives (84-87) showed that the mentioned compounds have hypnosedative, antifever and antihypoxic activities. The highest activities were shown by 3β, 17β- (84) and 3α, 17α-epimers (87).

Synthesis of other four isomers at C-3 and C-17 3-amino-5α-androstan-17-ols for elucidation of their antiarrhythmic activities is described in [77]. The study has shown that neither of them is active at aconitive arrhythmia.

Conclusion

We can conclude that not only configuration at C-3 and C-17 and the substitute at these atoms, but the order of substitution of hydroxyl- and aminogroups also affect the biological activity. In order to reveal the antiarrhythmic and general narcotic activities in aminoalcohols of 5α-androstan line, diequatorial orientation of hydroxyl- and aminogroups at C-3 and C-17, i.e. 3β, 17β in this case, are the most convenient.

References

[1] Patent 3001988 USA;

[2] R. A. Lukas, D.F. Dickel, R.L. Dzeimian, M.J. Ceglowski, B.L. Hensle, H.B. Macphillamy. *J. Am. Chem. Soc*. 82, 21, 5688 (1960)

[3] T. Nambara, T. Shibata, M. Mumura, H. Hosoda. *Chem. Pharm. Bull*. 19, 5, 954 (1971)

[4] G.R. Pettit, A. K. Gupta, R.L. Smith. *Can. J. Chem*. 44, 17, 2023 (1966)

[5] R.E. Counsell, P.D. Klimstra, R.E. Ranney. *J. Med. Pharm. Chem*. 5, 6, 1224 (1962)

[6] P.D. Klimstra, R.E. Ranney, R.E. Counsell. *J. Med. Chem*. 9, 3 323 (1966)

[7] C.M. Zeng, B.D. Manion, A. Benza. *J. Med. Chem*. 48, 8, 3051 (2005)

[8] Patent 3133915 US.

[9] Patent 615634 Belgium.

[10] Patent 642020 Belgium.

[11] Patent 3097200 USA.

[12] N.Sh. Nadaraia, V.I. Sladkov, E.V. Dorodnikova, M.D. Mashkovskiy, E.P. Kemertelidze, N. N. Suvorov. *Chem.Pharm.J*. 22, 3, 288 (1988) (Russian).

[13] Author Certificate №1519190, USSR (1989).

[14] M.M. Campbell, R.C. Craig, A.C. Boyd, I.M. Gilbert, R. T. Logan, J. Redpath, R.G. Roy, D.S. Savage, T. Sleigh. *J. Chem. Soc. Perk. Trans*. I. 9, 2235 (1979).

[15] T.J. Campbell, E.M.V. Williams, Brit. I. Pharmacol. 76, 337 (1982).

[16] M.I. Merlani, M.G. Davitishvili, N.Sh. Nadaraia, M.I. Sikharulidze, K. Papadopulus, *Chem. of Nat. Compounds*, 40, 2, 144 (2004).

[17] Patent 2212363 US.

[18] Patent 2655519 US.

[19] Patent 2655520 US.

[20] J. Schmidt-Thomé. *Chem. Ber.* 88, 6, 895 (1955).
[21] Patent 2531441 US.
[22] J. Schmidt-Thome. *Angew. Chem.* 67, 22 ,715 (1955)
[23] J. Kenyon,D.P. Young. *J. Chem. Soc.* 5, 263 (1941)
[24] Campbell, J. Kenyon. *J. Chem. Soc.* 1, 25, (1946)
[25] P. Ruggieri, C. Ferrari, C. Gandolfi. *Gazz. Chim. Ital.* 91, 6-7, 655 (1961).
[26] Patent 3029262 US.
[27] Patent 3137710 US.
[28] Patent 3063990 US.
[29] M.E. Kuehne, E.A. Konopka, B.F. Lambert. *J. Med. Pharm. Chem.* 5, 2, 281 (1962).
[30] G.R. Petit, A.K. Gupta, R.L. Smith. *Can. J. Chem*. 44, 17, 2023 (1966)
[31] C.W. Shoppe, J.C.P. Sly, *J. Chem. Soc.* 1, 345 (1959)
[32] Patent 3189597 US.
[33] Determiantion of the Structure of Organic Compounds by Physical and Chemical Methods. Moscow, V. 2, 650 (1967)
[34] A.F. Hofmann, *Acta Chem. Scand.* 17, 1, 173 (1963)
[35] D.P. Dodgson, R.D. Haworth. *J. Chem. Soc.* 1, 67, (1952)
[36] C.W. Shoppe, D.E. Evans, H.C. Richards, G.H.R. Summers. *J. Chem. Soc.* 6, 1649 (1956)
[37] M.M. Janot, Q. Khuong-Huu, R. Goutarel. *Bull. Soc. Chim.* Fr. 8, 1640 (1960)
[38] R.A.B. Bannard, A.F. Mckay. *Can. J. Chem*. 33, 6, 1166 (1955)
[39] D.E. Evans, C.W. Shoppe, G.H.R. Summers. *Chem. Ind.* London, 50, 1535 (1954)
[40] Patent 3009925 US.
[41] S. Tanabe, M. Onda. *J. Pharm. Soc*. Japan. 72, 7, 944 (1952)
[42] R.R. Sauers. *J. Am. Chem. Soc.* 80, 18, 4721 (1958)
[43] J.J. Panouse, J. Schmitt, P.J. Cornu, A. Hallot, H. Pluchet, P. Comoy. *Bull. Soc. Chim.* Fr. 8, 1753 (1963)
[44] H. E. Smith, C.A. Taylor, A.F. Mcdonagh, F.M. Chen. *J. Org. Chem.* 47, 13, 2525 (1982)
[45] J. Schmitt, J.J. Panouse, A. Hallot, H. Pluchet, P. Comuy, P.J. Cornu. *Bull. Soc. Chim.* Fr. 4, 771 (1964)
[46] N.N. Nadaraia, V.I. Slodkov, L.N. Kuleshova, N.N. Suvorov. *J. Org.Chem.* 23, 3, 533 (1987) (Ruissian)
[47] H.B. Henbest, W.R. Jackson. *J. Chem. Soc*. 3, 954 (1962)
[48] Streitwieser, W.D. Schaeffer. *J. Am. Chem. Soc.* 78, 21, 5597 (1956)
[49] C.H. Robinson, C. Ermann. *Steroids,* 6, 509 (1965)
[50] M. Davis, E.W. Parnell, D. Warburton. *J. Chem. Soc*. (C). 19, 1698 (1966)
[51] M. Alauddin, B. Caddy, J.J. Lewis, M. Martin-Smith, M.F. Sugrue. *J. Pharm. Pharmacol* 17, 1, 55 (1965)
[52] Patent 1365225 France.
[53] P. Veerravagdu, R.T. Arnold, E.W. Eigenmann. *J.Am. Chem. Soc*. 86, 15, 3072 (1964);
[54] Patent 182205 Switzerland.
[55] R.E. Marker. *J. Am. Chem. Soc*. 58, 3, 480 (1936)
[56] Patent 2292080 US.

[57] L. Ruzicka, M.W. Goldberg. *Helv. Chim. Acta*. 19, 107, 1936.
[58] L. Ruzicka, M.W. Goldberg, J. Meyer, H. Brungger. *Helv. Chim. Acta,* 17, 1395 (1934)
[59] Patent 2289235 US.
[60] H.E. Smith, B.G. Padilla, J.R. Neergaard, F.M. Clen. *J. Org. Chem.* 44, 10, 1690 (1979)
[61] N.Sh. Nadaraia, V.I. Sladkov, N. N. Suvorov. *J. Org. Chem J.* 24, 4, 760 (1988) (Russian)
[62] B. Raduchell. *Synthesis,* 4, 292 (1980)
[63] O. Mitsunobu, M. Yamada. *Bull. Chem. Soc*. Japan. 40, 10, 2380 (1967)
[64] A.K. Bose, B. Lal, W.A. Hoffman, M.S. Manhas. *Tetrahedron Lett. Perg*. Press. 18, 1619 (1973)
[65] N.Sh. Nadaraia, V.I. Sladkov, N. N. Suvorov. *J. Org. Chem* J. 24, 7, 1472 (1988) (Russian)
[66] Patent 6400710 Holland.
[67] Patent 629268 Belgium.
[68] J.B. Jones, D.J. Adam, J. D. Leman. *J. Med. Chem*. 14, 9, 827 (1971)
[69] G.R. Pettit, R.L. Smith, A.K. D. Gupta. *Can. J. Chem*. 45, 5, 501 (1967)
[70] G.R. Pettit, R.L. Smith, H. Klinger. *J. Med. Chem*. 10, 2, 145 (1967)
[71] G.R. Pettit, A.K.D. Gupta. *Can. J. Chem*. 45, 5, 567 (1967)
[72] Vincze, L. Hackler, Z. Szendi, G. Schneider. *Steroids*, 61, 697 (1996)
[73] M. Mokotoff, M. Zhao, R.J. Marshall, E. Winslow, K.L. Wong, Q.J. Liao. *Steroids*, 55, 399 (1990)
[74] V.V. Ranade, F. Kohen, R.E. Counsell. *J. Med. Chem*. 14, 1, 38 (1971)
[75] L.N. Yankhotov, S.S. Liberman. *Chem.Pharm.J*. 19, 11, 1306 (1985) (Russian)
[76] C. Camoutsis, C. Sambani, D.T. Trafalis, P. Peristeris. *Eur. J. Med. Chem.* 34, 7-8, 645 (1999)
[77] L.SH. Amiranashvili, V.I. Sladkov, N.I. Menshova, E.P. Kemertelidze, N.N. Suvorov. *Bull.. Georgian Acad. Sci*., 138, 1, 85 (1990)

In: Biology and Communication in the Information Age
Editor: Rafiqul Islam
ISBN: 978-1-62808-252-4

Chapter 3

THE UNIVERSAL MECHANICS

Alexandre Yu. Kyrie *

Physics Department, M.V. Lomonosov Moscow State University
Moscow, Russian Federation

ABSTRACT

On the basis of Berkeley-Mach-Poincaré principle of relativity and within the assumption of absence of any specific (inertial) reference systems in empty space the new mechanics of bodies motion has been constructed. This motion is determined only by the relative distances and velocities of bodies. It has been shown that the mutual rotation of a pair of particles related to almost immobile "distant stars" (galaxies) leads to centrifugal force origin.

Keywords: Gravitation, Relativity, Causality

1. INTRODUCTION

The concept of absolute space (aether) was introduced by Newton at build-up of a mechanics [1]. This concept was undergone to criticism from outside Newton's many contemporaries, in particular Berkeley [2] who came to a conclusion about impossibility of existence of absolute space (aether) with its inertial reference systems, and also to understanding of that inertia should be caused by motion of bodies related to "distant stars". To a similar deduction later Mach [3] and Poincaré [4] have come who have specified in necessity of creation of a *relativistic mechanics*, i.e. a mechanics of the relative motion of bodies.

The given problem has been solved neither in special, nor in common theories of relativity which still operated with concepts of absolute space. Numerous attempts to construct the theory, in which inertia of rotation would be caused by influence of remote

* E-mail: bk.novosadov@mail.ru.

galaxies ("far stars"), have appeared unsatisfactory [5]. Thus, the problem of Berkeley-Mach-Poincaré is not solved in existing physical theories.

The theory named as a *universal mechanics* which initial principles essentially differ from usually used is stated below. It in particular concerns space-time properties of bodies motion.

Firstly, only relative motions of bodies are considered. In this connection it is supposed that there does not exist in the space any preferential (inertial) reference systems. Thereby the space is not considered as something absolute one (aether) relative to which bodies move.

Secondly, the causality principle is used for description of interactions of bodies, i.e. it is supposed that action of one body onto another, located on distance r, happens in time r/c, where c is the velocity of propagation of interactions (light velocity). Thus the velocity of propagation of interaction is constant and does not depend on character of motion of bodies that also means a space relativity. Causal character of interaction leads to irreversibility of motion of bodies in time. There is not also any equality of actions of bodies against each other as bodies act on each other in different moments of time and, accordingly, at different distances.

Use of a principle of least action has allowed to discover a Lagrangian for a pair of particles and, further, for system of interacting particles.

In the limit of small velocities of bodies in comparison with c it has been shown that a mutual rotation of two particles related to massive, remote, slowly moving bodies (galaxies) leads to origin of a centrifugal force. Also the plane motion of a pair of particles with conservations of the mechanical moment of momentum and the velocity of the inertia centre is realized, - i.e. we obtain the equations of the Newtonian mechanics. These results are a consequence of assumption of isotropy of space allocation of large remote bodies, provided three various conditions of isotropy being required.

2. Main Principles

Space Relativity: A Causality Principle

The space relativity means, in particular, that one particle in empty space cannot be characterized by a motion, and the only pair of particles cannot rotate relatively each other because of the impossibility of relative rotation in the absence of absolute space and other bodies.

Further it is supposed that the space does not have got any specific systems of reference and directions, and also that action of any particle 2 onto a particle 1 depends only on the relative distance r_{21} and the relative velocity v_{21} of those particles.

The causality principle means that if the reference system is associated with particle 1 and the time t_1, the corresponding time t_2 when the particle 2 acts ("sends a signal") onto the particle 1, is equal to

$$t_2 = t_1 - \frac{r_{21}(t_1)}{c}. \tag{2.1}$$

With the space relativity also means that the velocity of interaction c does not depend on motion of particles.

In the reference system associated with the particle 1 it is possible to define the velocity of the particle 2 related to the particle 1:

$$v_{21}(t_1) = -\frac{dr_{21}(t_1)}{dt_1}. \tag{2.2}$$

Let's consider that acting of the particle 1 is characterized by a definite direction so that, for example, at acting of two particles 2 and 3 on the particle 1 it is possible to determine an angle between directions of acting for simultaneously coming signals to the particle 1 at the moment (Figure 2.1).

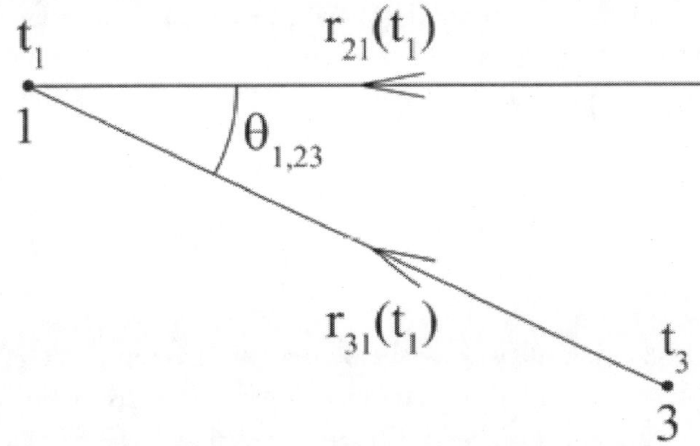

Figure 2.1. Scheme of action of particles 2 and 3 onto a particle 1.

Arrows show the direction of the action.

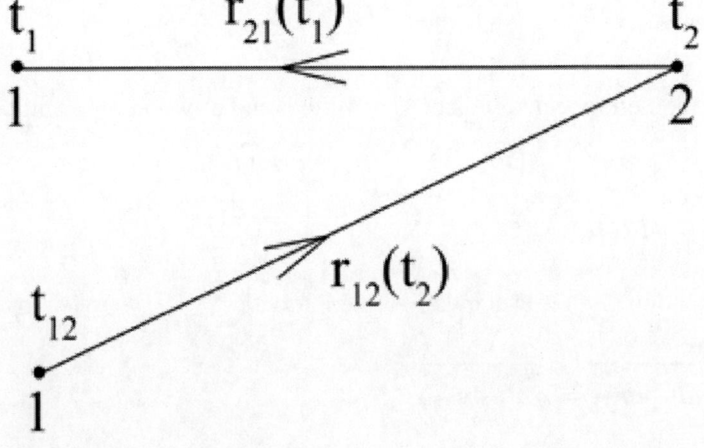

Figure 2.2. An action of particle 2 onto particle 1.

The causality principle leads to difference between the times of action of the particle 2 onto the particle 1 and contrary the particle 2 onto the particle 1 (Figure 2.2), so that their actions against each other will appear, generally speaking, various that leads to an inequality of the action and counteraction, and also to irreversibility of motion in time. Note also that further in drawings it is necessary to view only the angles having physical sense. For example, on Fig 2.2 "angle between directions" and has no physical sense, unlike Fig 2.1 where the angle has physical sense.

Constancy of Propagation Velocity of a Gravitational Interaction (Light Velocity)

This principle used in a special theory of relativity, in this case means that in the reference system of a particle 1′ coinciding with the particle 1 and moving with velocity V in a direction from the particle 2 to the particle 1, light propagation is effected with the same velocity c, as in the system of the particle 1. Coordinates transformation from system 1′ to system 1 can be written down in the form of two-dimensional Lorenz-Poincaré transformation

$$dr_{21}' = \kappa\left(dr_{21} + Vdt_1\right), \quad dt_1' = \kappa\left(dt_1 + \frac{V}{c^2}dr_{21}\right), \quad \kappa = \left(1 - \frac{V^2}{c^2}\right)^{-\frac{1}{2}}, \tag{2.3}$$

provided usage of the restriction of particles velocities by the magnitude c, the symmetry of transitions $1' \leftrightarrow 1$ and $V \to -V$, and also group properties of transformations at transition in a particle reference system 1′, and further in a reference system of a particle 1" moving with velocity V′ related to the particle 1′ [6]. It is possible to present the transition (2.3) in the form of unitary transformation:

$$\left(dr_{21}^i\right)' = P^{ij}\, dr_{21}^j, \quad P^{ij} = \kappa\begin{bmatrix} 1 & -i\frac{V}{c} \\ i\frac{V}{c} & 1 \end{bmatrix}, \tag{2.4}$$

where the summation on repeated indices is admitted, and a two-component vector is defined by

$$dr_{21}^i = \left(dr_{21};\, icdt_1\right). \tag{2.5}$$

The transformation (2.4) leaves invariant the interval

$$ds_{21} = \sqrt{-dr_{21}^i\, dr_{21}^i} = \sqrt{c^2 - v_{21}^2}\, dt_1. \tag{2.6}$$

The two-component velocity vector can be written down as follows

$$u_{21}^i = -\frac{dr_{21}^i}{ds_{21}} = \left(1 - \frac{v_{21}^2}{c^2}\right)^{-1/2} \left(\frac{v_{21}}{c}; -i\right). \tag{2.7}$$

The two-component vector of coordinate differentials (2.5) suggests about existence of an integral vector

$$r_{21}^i = \left(r_{21}; ic\left(t_1 - t_2\right)\right),$$

where is defined by the relationship (2.1) with the help of which one obtains

$$r_{21}^i = \left(r_{21}; i r_{21}\right). \tag{2.8}$$

It is possible to form from the vectors (2.7) and (2.8) the following invariants concerning transformation (2.4):

$$r_{21}^i r_{21}^i = 0, \quad u_{21}^i u_{21}^i = -1, \quad r_{21}^i u_{21}^i = r_{21}\sqrt{\left(1 + \frac{v_{21}}{c}\right) \Big/ \left(1 - \frac{v_{21}}{c}\right)}. \tag{2.9}$$

The formulae (2.3) and (2.2) lead to the following transformation of the relative velocity:

$$v_{21}' = \frac{v_{21} - V}{1 - \frac{v_{21} V}{c^2}}, \tag{2.10}$$

coinciding by the form with a velocity transformation rule of velocities addition in special relativity theory. However the transformation of the distance (2.8) differs essentially from transformations of "rule lengths" in a special relativity theory:

$$r_{21}' = r_{21}\sqrt{\left(1 + \frac{V}{c}\right) \Big/ \left(1 - \frac{V}{c}\right)}. \tag{2.11}$$

This expression is symmetrical relative to replacement $r_{21} \to r_{21}'$, $V \to -V$.

Lagrangian and Action Function

The Lagrangian characterizes action of the particle 2 onto the particle 1, provided with the definition of the action function

$$S_{21} = \int L_{21}\left(v_{21}, r_{21}\right) dt_1, \tag{2.11}$$

which should have minimum on real trajectories and be invariant relative to the transformation (2.4). We shall be interested only in a gravitational interaction, provided there being absent any other fields and the action function being a functional of but invariants (2.6) and (2.9). Then the invariance requirement of leads to the following Lagrangian

$$L_{21}(v_{21},\, r_{21}) = \psi\left(r_{21}^{i}\, u_{21}^{i}\right)\sqrt{1-\frac{v_{21}^{2}}{c^{2}}}, \tag{2.12}$$

where is a function of the unique nontrivial invariant (2.9), which is restricted at $r_{21} \to \infty$. Further for entry simplification the index 21 will be omitted and we shall put $t_1 \equiv t$.

The minimum condition for the action function S leads to the requirement of positivity of the second variation of the action function

$$\delta^2 S = \int dt \left\{ \tfrac{1}{2}(\delta r)^2 \frac{\partial^2 L}{\partial r^2} + \tfrac{1}{2}(\delta v)^2 \frac{\partial^2 L}{\partial v^2} + \delta r\, \delta v \frac{\partial^2 L}{\partial r\, \partial v} \right\}, \tag{2.13}$$

$$\delta v = -\frac{d}{dt}\delta r. \tag{2.14}$$

Taking into account vanishing the variations δr and on the boundaries of integration interval provided the minimum condition, one obtains

$$\int dt\, \delta r\, \delta v \frac{\partial^2 L}{\partial r\, \partial v} = \frac{1}{2}\int dt\, (\delta r)^2 \left(-v \frac{\partial^3 L}{\partial v\, \partial r^2} + \dot{v} \frac{\partial^3 L}{\partial r\, \partial v^2} \right). \tag{2.15}$$

The presence of the term proportional to in the expression for (2.13) and (2.15) makes impossible minimizing the function S. For the absence of such term it is necessary fulfillment of condition

$$\frac{\partial^3 L}{\partial r\, \partial v^2} = 0. \tag{2.16}$$

Thus with the condition we arrive at inequalities

$$\frac{\partial^2 L}{\partial v^2} > 0, \tag{2.17}$$

$$\frac{\partial^2 L}{\partial r^2} - v \frac{\partial^3 L}{\partial v\, \partial r^2} > 0. \tag{2.18}$$

From the condition (2.16) we obtain a general expression for L

$$L(v,r)=A(v)+B(r)+v\,D(r). \tag{2.19}$$

Comparing this expression with the formula (2.12) and taking into consideration (2.9) we arrive at the Lagrangian of the form

$$L(v,r)=\psi\left(r\sqrt{\frac{1+\frac{v}{c}}{1-\frac{v}{c}}}\right)\sqrt{1-\frac{v^2}{c^2}}, \tag{2.20}$$

from which it is easy to see that the function satisfying to (2.19) is as follows

$$\psi(x)=E+Rx+T\frac{1}{x}, \tag{2.21}$$

where E, R, T are some constants. One has to put $R = 0$ in this function because of limitation L when $r\to\infty$. Then the expression for may be represented in the form

$$L(v,r)=-m_2\,c^2\left\{\sqrt{1-\frac{v^2}{c^2}}+\left(1-\frac{v}{c}\right)\frac{r_0}{r}\right\}. \tag{2.22}$$

Here m_2 and r_o are constants. As the Lagrangian transformation

$$L\to L+\frac{d}{dt}f(r,t)$$

leaves invariant the first variation of the S one can while obtaining motion equations use instead of the formula (2.22) the following expression

$$L_{21}(v_{21},r_{21})=-m_2\,c^2\sqrt{1-\frac{v_{21}^2}{c^2}}+m_2\,c^2\frac{r_0}{r_{21}}. \tag{2.23}$$

Note however that the expression given for L may be used but for specified purpose.

It will be shown below that both the first (kinematic), and the second (power) terms in the right hand side of the expressions (2.22), (2.23) describe action of the particle 2 onto the particle 1.

The inequalities (2.17) and (2.18) for the Lagrangian (2.22) and (2.23) lead to the requirement of constants to be positive

$$m_2 > 0, \quad r_0 > 0. \tag{2.24}$$

Let us assume that here for the considered case of gravitational interaction the action function of the particle 2 onto the particle 1 is defined by the mass of the particle 2 only, and that is an universal world length.

The constant of world length could define a space-time curvature. However, as it can be easy seen, any dependence leads to impossibility of satisfying the action minimum condition (2.16). This means that an introduction of a curved space-time is impossible in the theory given.

Minimum Action Principle and Equations of Motion

Equations of motion for a system of particles are determined by the minimum condition of corresponding action function. The special feature of the given theory is that it is impossible to define unified action function for all the system of particles. Thus special action function should be defined for each distance.

In this case we consider a motion of the particle 2 related to the particle 1 (i.e. distance $r_{21}(t)$) at the presence of other bodies k (Figure 2.3).

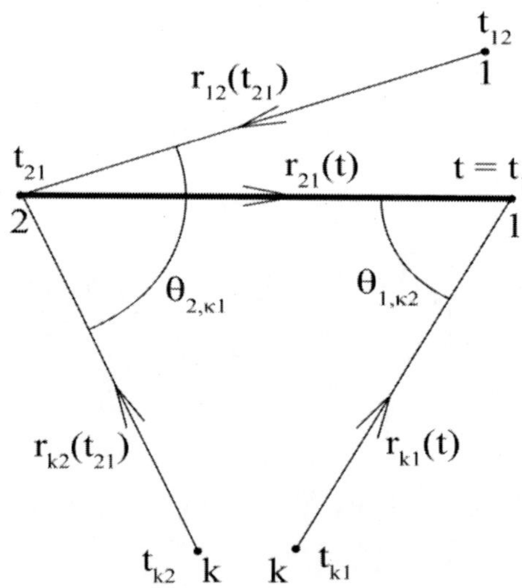

Figure 2.3. Scheme of action onto particles 1 and 2, defining the variation of the distance r21.

The action function which has to be varied by the distance obviously involves all pair action functions of influence on the particle 2 at the moment $t_2 \equiv t_{21}$, and the same functions of influence on the particle 1 at the moment $t = t_1$:

$$S[r_{21}] = S_{21} + S_{12} + \sum_{k \neq 1,2} (S_{k1} + S_{k2}), \quad S_{\alpha\beta} = \int L_{\alpha\beta}\, dt_\beta, \qquad t \equiv t_1, \quad t_2 \equiv t_{21}. \tag{2.25}$$

Here the Lagrangians are similar to (2.23). Corresponding moments of time are defined by causal relationships

$$t_{21} = t - \frac{r_{21}(t)}{c}, \quad t_{12} = t_{21} - \frac{r_{12}(t_{21})}{c}, \quad t_{k1} = t - \frac{r_{k1}(t)}{c}, \quad t_{k2} = t_{21} - \frac{r_{k2}(t_{21})}{c}. \tag{2.26}$$

We derive the equations for putting the first variation of (2.25) by the parameter to zero. Thus it is necessary to bring all integrals over time (2.25) to a uniform time t. Then the first variation of may be written down in the form

$$\delta S_{\alpha\beta} = -\int \delta r_{21} \cdot G_{\alpha\beta}\, dt, \quad G_{\alpha\beta} = m_\alpha \frac{\delta r_{\alpha\beta}}{\delta r_{21}} \left\{ \frac{d}{dt} \left[\dot{r}_{\alpha\beta} \left(\dot{t}_\beta^2 - \frac{\dot{r}_{\alpha\beta}^2}{c^2} \right)^{-1/2} \right] + \dot{t}_\beta \frac{c^2 r_0}{r_{\alpha\beta}^2} \right\}, \tag{2.27}$$

with

$$v_{\alpha\beta} = -\frac{d}{dt_\beta} r_{\alpha\beta},$$

and taking into consideration that it is similar to the expression (2.22). The points above symbols denote derivatives by the time t:

$$\dot{r}_{\alpha\beta} \equiv \frac{d}{dt} r_{\alpha\beta}, \quad \dot{t}_\beta \equiv \frac{d}{dt} t_\beta. \tag{2.28}$$

Thus the vanishing of the first variation of leads to the equation of motion for r_{21}:

$$G_{21} + G_{12} + \sum_{k \neq 1,2} (G_{k1} + G_{k2}) = 0. \tag{2.29}$$

In this equation functions (2.27) are defined by variations of distances by the quantity δr_{21}.

Dependent Variations of Distances

First calculate the variation of by r_{21}. Let the particle 1 move along a direction $2 \to 1$ (Figure 2.4), having passed distance at the time $t' - t$. As a consequence of motion of the particle 2 at the time $t' - t$, one obtains the corresponding increase of the distance r_{21}

$$\Delta r_{21}^{v} \approx -v_{21}\left[(t'-t'_{21})-(t-t_{21})\right] = -\delta r_{21}\frac{v_{21}(t)}{c}. \tag{2.30}$$

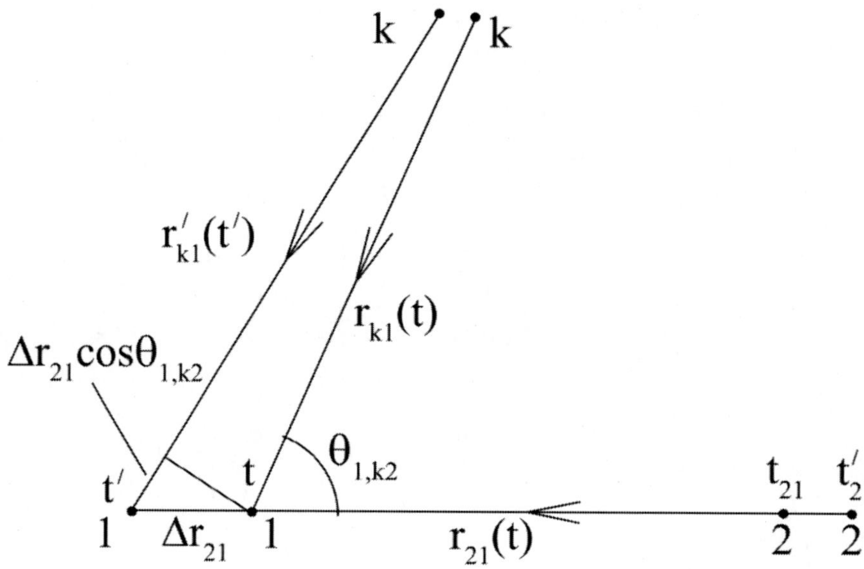

Figure 2.4. Scheme of influence of "priming" displacement of particle 1 on a modification of the distances and $r'_{k1} - r_{k1} = \delta r_{k1}$.

So the total change of is equal to

$$\delta r_{21} = \Delta r_{21} + \Delta r_{21}^{v} = \Delta r_{21} + \delta r_{21}\frac{v_{21}(t)}{c}. \tag{2.31}$$

Therefore we have got

$$\delta r_{21} = \frac{\Delta r_{21}}{1+\frac{v_{21}(t)}{c}}. \tag{2.32}$$

When the particle 1 moves along the direction $2 \to 1$ the distance will shift by $\Delta r_{21}\cos\theta_{1,k2}$, so the total change δr_{k1}, with taking into account a motion of a particle k with a velocity $v_{k1}(t)$, similar to (2.31), is equal to

$$\delta r_{k1} \approx \Delta r_{21}\cos\theta_{1,k2} - \delta r_{k1}\frac{v_{k1}(t)}{c},$$

from which we obtain

$$\delta r_{k1} \approx \frac{\Delta r_{21} \cos\theta_{1,k2}}{1+\frac{v_{k1}(t)}{c}}. \tag{2.33}$$

Comparing the expressions (2.32) and (2.33) we arrive at

$$\frac{\delta r_{k1}}{\delta r_{21}} = \frac{1+\frac{v_{21}(t)}{c}}{1+\frac{v_{k1}(t)}{c}} \cos\theta_{1,k2}. \tag{2.34}$$

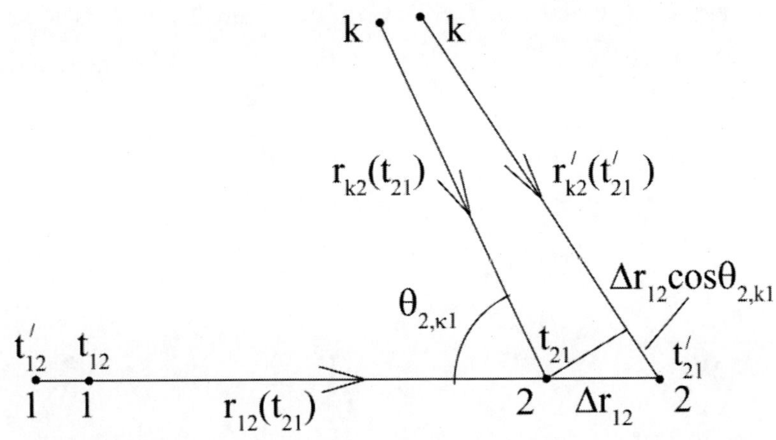

Figure 2.5. Scheme of influence of "priming" displacement of particle 2 on a modification of the distances and $r'_{k2} - r_{k2} = \delta r_{k2}$.

In an analogous manner we get a variation of by (not r_{21}!) (Figure 2.5):

$$\frac{\delta r_{k2}}{\delta r_{12}} = \frac{1+\frac{v_{12}(t_{21})}{c}}{1+\frac{v_{k2}(t_{21})}{c}} \cos\theta_{2,k1}. \tag{2.35}$$

Let us calculate a variation of by r_{21}. For this purpose it is necessary to calculate a variation of by and then to present a variation by as follows

$$\frac{\delta r_{k2}}{\delta r_{21}} = \frac{\delta r_{12}}{\delta r_{21}} \cdot \frac{\delta r_{k2}}{\delta r_{12}}. \tag{2.36}$$

Calculate the variational derivative. Let the particle 2 move along the direction $1 \rightarrow 2$, having passed distance (Figure 2.6).

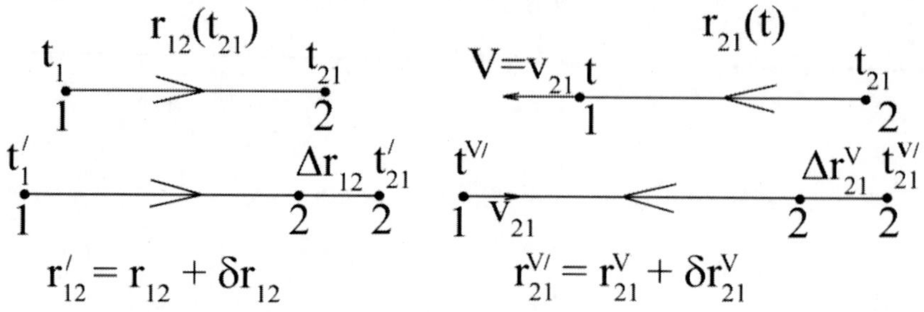

Figure 2.6. Scheme of influence of "priming" displacement of particle 2 in the reference system of this particle, while variating the distances and r_{21}.

Then the corresponding change of the distance r_{12}, equal to and analogous to (2.32), may be represented in the form

$$\delta r_{12} \approx \Delta r_{12} - v_{12}(t_{21}) \cdot \Delta t_{12}, \quad \Delta t_{12} = (t'_2 - t'_1) - (t_2 - t_1) = \frac{\delta r_{12}}{c},$$

whence we obtain

$$\delta r_{12} \approx \frac{\Delta r_{12}}{1 + \frac{v_{12}(t_{21})}{c}}. \tag{2.37}$$

While considering the variation δr_{21}, transfer to the reference system where the particle 2 is at rest, also. This reference system moves relative to the particle 1 with the velocity (Figure 2.6). We shall denote all the parameters in such reference system by the upper index V. As far as the particle 2 in this reference system is at rest, its displacement equals to the displacement Δr_{12}, with the distance variation being connected with by a formula analogous to (2.37):

$$\delta r^V_{21} = \frac{\Delta r^V_{21}}{1 + \frac{v_{21}(t)}{c}} = \frac{\Delta r_{12}}{1 + \frac{v_{21}(t)}{c}}. \tag{2.38}$$

The quantity is connected with by the distance transformation rule (2.11) at $V = v_{21}$:

$$\delta r^V_{21} = \delta r_{21} \sqrt{\left(1 + \frac{v_{21}(t)}{c}\right) \Big/ \left(1 - \frac{v_{21}(t)}{c}\right)}.$$

Thus, from this relation and expressions (2.37) and (2.38) we arrive at

$$\frac{\delta r_{12}}{\delta r_{21}} = \frac{1+\frac{v_{21}(t)}{c}}{1+\frac{v_{12}(t_{21})}{c}}\sqrt{\frac{1+\frac{v_{21}(t)}{c}}{1-\frac{v_{21}(t)}{c}}}, \qquad \frac{\delta r_{k2}}{\delta r_{21}} = \frac{1+\frac{v_{21}(t)}{c}}{1+\frac{v_{k2}(t_{21})}{c}}\sqrt{\frac{1+\frac{v_{21}(t)}{c}}{1-\frac{v_{21}(t)}{c}}}\cos\theta_{2,k1}. \tag{2.39}$$

As it is clear from the formulae (2.25) and (2.26), we have to use the following expressions for the quantities $\dot{t}_\beta$

$$\frac{dt_1}{dt} \equiv 1, \qquad \frac{dt_2}{dt} \equiv \frac{dt_{21}}{dt} = 1-\frac{\dot{r}_{21}}{c}, \tag{2.40}$$

together with the corresponding expressions for velocities

$$v_{21} = \dot{r}_{21}, \quad v_{k1} = -\dot{r}_{k1}, \quad v_{12} = -\frac{dr_{12}}{dt_{21}} = -\frac{\dot{r}_{12}}{1-\frac{\dot{r}_{21}}{c}}, \quad v_{k2} = -\frac{dr_{k2}}{dt_{21}} = -\frac{\dot{r}_{k2}}{1-\frac{\dot{r}_{21}}{c}}.$$

So the dependent variations of distances take forms

$$\frac{\delta r_{k1}}{\delta r_{21}} = \frac{1-\frac{\dot{r}_{21}}{c}}{1-\frac{\dot{r}_{k1}}{c}}\cos\theta_{1,k2}, \quad \frac{\delta r_{12}}{\delta r_{21}} = \frac{\left(1-\frac{\dot{r}_{21}}{c}\right)^2}{1-\frac{\dot{r}_{21}+\dot{r}_{12}}{c}}\sqrt{\frac{1-\frac{\dot{r}_{21}}{c}}{1+\frac{\dot{r}_{21}}{c}}},$$

$$\frac{\delta r_{k2}}{\delta r_{21}} = \frac{\left(1-\frac{\dot{r}_{21}}{c}\right)^2}{1-\frac{\dot{r}_{21}+\dot{r}_{k2}}{c}}\sqrt{\frac{1-\frac{\dot{r}_{21}}{c}}{1+\frac{\dot{r}_{21}}{c}}}\cos\theta_{2,k1}. \tag{2.41}$$

Substituting expressions (2.40) and (2.41) into the formulae for (2.27), we obtain

$$G_{21} = m_2\left\{\ddot{r}_{21}\left(1-\frac{\dot{r}_{21}^2}{c^2}\right)^{-3/2} + \frac{c^2 r_0}{r_{21}^2}\right\}, \tag{2.42}$$

$$G_{12} = m_1\frac{\left(1-\frac{\dot{r}_{21}}{c}\right)^3}{1-\frac{\dot{r}_{21}+\dot{r}_{12}}{c}}\sqrt{\frac{1-\frac{\dot{r}_{21}}{c}}{1+\frac{\dot{r}_{21}}{c}}}\left\{\left[\ddot{r}_{12}\left(1-\frac{\dot{r}_{21}}{c}\right)+\ddot{r}_{21}\frac{\dot{r}_{12}\,\dot{r}_{21}}{c^2}\right]\left[\left(1-\frac{\dot{r}_{21}}{c}\right)^2-\frac{\dot{r}_{12}^2}{c^2}\right]^{-3/2}+\frac{c^2 r_0}{r_{12}^2}\right\}, \tag{2.43}$$

$$G_{k1} = m_k \cos\theta_{1,k2} \frac{1-\frac{\dot{r}_{21}}{c}}{1-\frac{\dot{r}_{k1}}{c}} \left\{ \ddot{r}_{k1} \left(1-\frac{\dot{r}_{k1}^2}{c^2}\right)^{-3/2} + \frac{c^2 r_0}{r_{k1}^2} \right\}, \tag{2.44}$$

$$G_{k2} = m_k \cos\theta_{2,k1} \frac{\left(1-\frac{\dot{r}_{21}}{c}\right)^3}{1-\frac{\dot{r}_{k2}+\dot{r}_{21}}{c}} \sqrt{\frac{1-\frac{\dot{r}_{21}}{c}}{1+\frac{\dot{r}_{21}}{c}}} \left\{ \left[\ddot{r}_{k2}\left(1-\frac{\dot{r}_{21}}{c}\right)+\ddot{r}_{21}\frac{\dot{r}_{k2}\dot{r}_{21}}{c^2}\right]\left[\left(1-\frac{\dot{r}_{21}}{c}\right)^2-\frac{\dot{r}_{k2}^2}{c^2}\right]^{-3/2} + \frac{c^2 r_0}{r_{k2}^2} \right\}. \tag{2.45}$$

Thus the equations for (2.29), (2.42)–(2.45) contain derivatives not only of but of another distances which determine the influence onto the particle 1 (at the time moment *t*) and the particle 2 (at the time moment t_{21}). In such a way one can derive equations for distances r_{12}, r_{k1}, r_{k2}, etc.

3. Slow Motion of a Pair of Particles Interacting with Distant, Massive, Almost Immovable Bodies

Consider a motion of a pair of particles 1 and 2 with masses and between which the distance is much less then distances to massive bodies with numbers *k* $(m_k >> m_1, m_2)$

$$r_{21} << r_{k1}, r_{k2}. \tag{3.1}$$

A condition of slow motion of particles results in that the characteristic time of modification of the distance exceeds considerably time during which the light passes this distance

$$\tau_{21} >> r_{21}/c. \tag{3.2}$$

So, further we shall assume that the characteristic time when distances among distant bodies change considerably exceeds τ_{21}.

The Equation for r_{21}

In a limit of small relative velocities of all particles one may put $r_{12} \approx r_{21}$, then from expressions (2.29), (2.41) – (2.45) we obtain the following equation for r_{21}

$$(m_1 + m_2)\ddot{r}_{21} + \sum_k m_k \left(\ddot{r}_{k1} \cos\theta_{1,k2} + \ddot{r}_{k2} \cos\theta 2, k1\right) =$$

$$= (m_1 + m_2) F(r_{12}) + \sum_k m_k \left[F(r_{k1}) cos\,\theta_{1,k2} + F(r_{k2}) cos\,\theta_{2,k1}\right]. \qquad (3.3)$$

Here the force function is determined like

$$F(r) = -\frac{c^2 r_0}{r^2}. \qquad (3.4)$$

The second term in the left hand side of the equation (3.3) is of particular interest. This term corresponds to the action of far *k* bodies onto a pair of particles 1 and 2, with this effect bearing not “force”, but kinematical character caused by correlation of the distances r_{21}, and r_{k2}.

As shown below, this action of remote *k* bodies gives rise to centrifugal force origin when a pair of particles mutually rotates relative remote bodies.

Equations for r_{k2}

The equations for (2.29), (2.27) look similar to

$$G_{k1} + G_{21} + \sum_{n \neq k} G_{n1} + G_{1k}^{(1)} + G_{2k}^{(1)} + \sum_{n \neq k} G_{nk}^{(1)} = 0, \qquad (3.5)$$

$$G_{k2} + G_{12} + \sum_{n \neq k} G_{n2} + G_{2k}^{(2)} + G_{1k}^{(2)} + \sum_{n \neq k} G_{nk}^{(2)} = 0. \qquad (3.6)$$

Schemes of actions are given on Figure 3.1 and 3.2, correspondingly. From these schemes one can see that terms in the left hand side of the equation (3.5) and terms in the left hand side of the equation (3.6) depend on positions of the particles 1 and 2 at time moments and correspondingly.

These times precede to the moment *t* by quantities of orders $2r_{k1}/c$, $2r_{k2}/c$. Since we are interested in rather remote *k* bodies (for example, galaxies to which distances are much more than millions of light years), it is clear that phases of motions of the particles 1 and 2 at such early moments of time does not correlate in any way with the phase at the moment *t*, and furthermore they will essentially differ for different *k* bodies.

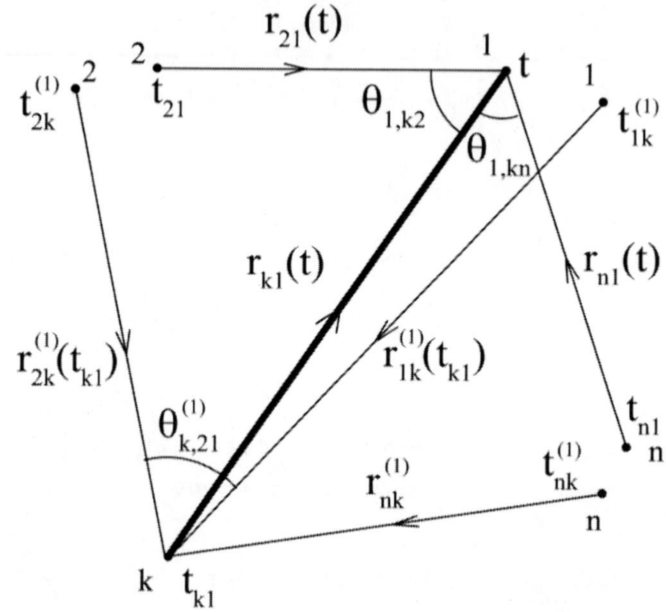

Figure 3.1. Scheme of acting onto particles k and 1, determining the modification of the distance r_{k1}.

Therefore yields and in the equations (3.5), (3.6) will provide an almost accidental influence onto motion of the particles 1 and 2 at the moment t. Further we shall neglect of such quasiaccidental influence, putting

$$G_{1k}^{(1)} = G_{2k}^{(1)} = G_{1k}^{(2)} = G_{2k}^{(2)} = 0 .$$

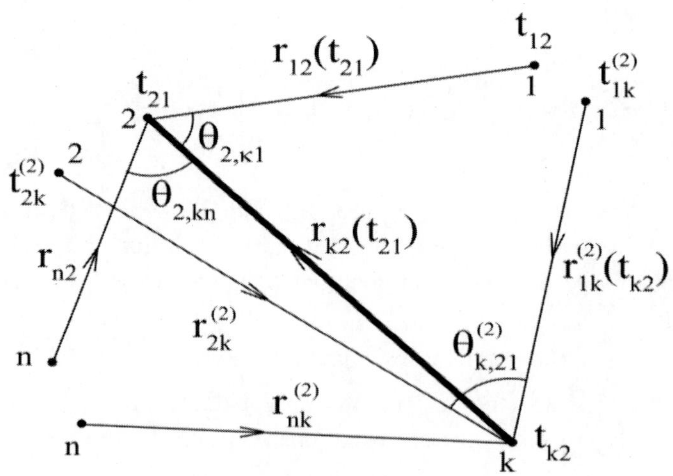

Figure 3.2. Scheme of acting onto particles k and 2, determining the modification of the distance r_{k2}.

Let us assume also that the characteristic times of the relative motion of remote k bodies considerably exceed time for the particles 1 and 2. Accordingly we shall neglect of the interaction among remote bodies putting in the equations (3.5), (3.6).

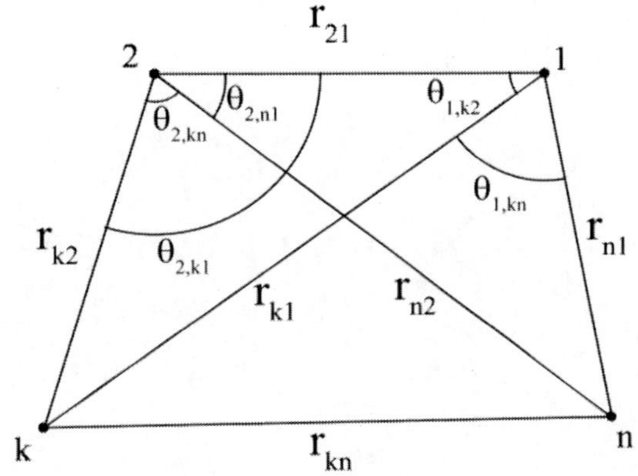

Figure 3.3. Scheme of acting onto particles 1 and 2 in case of slow motion when the relationships of angles and distances are defined by Euclidean geometry.

Consider as in the equation (3.3) above the case of small relative velocities when $\dot{r}_{\alpha\beta} << c$. In this limit a relation among angles and distances is defined by an Euclidean geometry. Then the schemes of influences of particles on Figure (3.1) and (3.2) must be replaced by the scheme of Fig 3.3, and the equations (3.5) and (3.6) become

$$m_k \ddot{r}_{k1} + m_2 \ddot{r}_{21} \cos\theta_{1,2k} + \sum_{n \neq k} m_n \ddot{r}_{n1} \cos\theta_{1,kn} =$$
$$= m_k F(r_{k1}) + m_2 F(r_{21}) \cos\theta_{1,2k} + \sum_{n \neq k} m_n F(r_{n1}) \cos\theta_{1,kn}, \tag{3.7}$$

$$m_k \ddot{r}_{k2} + m_1 \ddot{r}_{21} \cos\theta_{2,k1} + \sum_{n \neq k} m_n \ddot{r}_{n2} \cos\theta_{2,k1} =$$
$$= m_k F(r_{k2}) + m_1 F(r_{21}) \cos\theta_{2,k1} + \sum_{n \neq k} m_n F(r_{n2}) \cos\theta_{1,kn}. \tag{3.8}$$

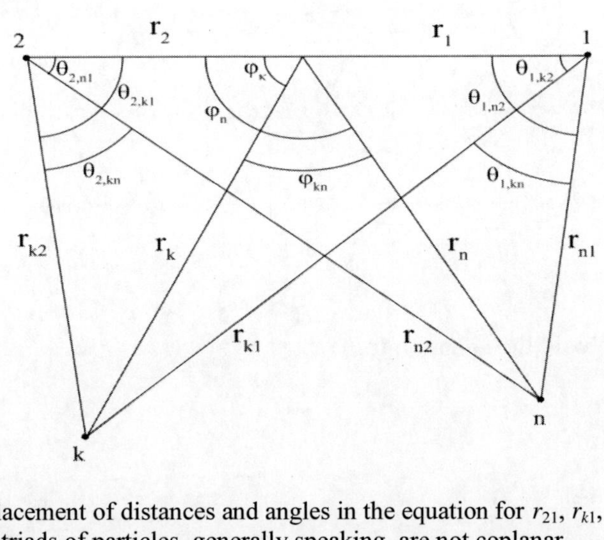

Figure 3.4. To the replacement of distances and angles in the equation for r_{21}, r_{k1}, r_{k2}. The planes formed by triads of particles, generally speaking, are not coplanar.

Now, introduce new variables (Figure 3.4) and make use of the following expansions for parameters at small ratios $r_{21}/r_k \ll 1$

$$r_{21} = r_1 + r_2, \quad r_{k1} \approx r_k + r_1 \cos\varphi_k, \quad r_{k2} \approx r_k - r_2 \cos\varphi_k, \quad r_{n1} \approx r_n + r_1 \cos\varphi_n,$$

$$\cos\theta_{2,k1} \approx -\cos\varphi_k + \frac{r_2}{r_k}\sin^2\varphi_k, \quad \cos\theta_{1,k2} \approx \cos\varphi_k + \frac{r_1}{r_k}\sin^2\varphi_k,$$

$$\cos\theta_{1,kn} \approx \cos\varphi_{kn}\left(1 - \frac{r_1}{r_k}\cos\varphi_k - \frac{r_1}{r_n}\cos\varphi_n\right) + \frac{r_1}{r_n}\cos\varphi_k + \frac{r_1}{r_k}\cos\varphi_n,$$

$$\cos\theta_{2,kn} \approx \cos\varphi_{kn}\left(1 + \frac{r_2}{r_k}\cos\varphi_k + \frac{r_2}{r_n}\cos\varphi_n\right) - \frac{r_2}{r_n}\cos\varphi_k - \frac{r_2}{r_k}\cos\varphi_n. \tag{3.9}$$

Substituting quantities (3.9) into the equation (3/3) and neglecting the terms quadratic by r_{21}, we arrive at an equation for r_{21}:

$$(m_1 + m_2)\ddot{r}_{21} + \sum_k m_k \left[r_{21}\frac{\ddot{r}_k}{r_k}\sin^2\varphi_k + \cos\varphi_k \overline{r_{21}\cos\varphi_k}^{\,\cdot\cdot} \right] =$$

$$= (m_1 + m_2)F(r_{21}) + r_{21}\sum_k m_k \frac{1}{r_k} F(r_k)\left(1 - 3\cos^2\varphi_k\right). \tag{3.10}$$

In the same limit the equation for takes the form

$$m_k \ddot{r}_k + m_k \overline{r_1\cos\varphi_k}^{\,\cdot\cdot} + m_2 \ddot{r}_{21}\cos\varphi_k +$$

$$+ \sum_{n\neq k} m_n \left\{ \ddot{r}_n \left[\cos\varphi_{kn}\left(1 - \frac{r_1}{r_k}\cos\varphi_k - \frac{r_1}{r_n}\cos\varphi_n\right) + \frac{r_1}{r_k}\cos\varphi_n + \frac{r_1}{r_n}\cos\varphi_k \right] + \cos\varphi_{kn}\overline{r_1\cos\varphi_n}^{\,\cdot\cdot} \right\} =$$

$$= m_k F(r_k) - 2m_k F(r_k)\frac{r_1}{r_k}\cos\varphi_k + m_2 F(r_{21})\left[\cos\varphi_k + \frac{r_1}{r_k}\sin^2\varphi_k\right] +$$

$$+ \sum_{n\neq k} m_n F(r_n)\left\{ \cos\varphi_{kn}\left(1 - \frac{r_1}{r_k}\cos\varphi_k - \frac{r_1}{r_n}\cos\varphi_n\right) + \frac{r_1}{r_n}\cos\varphi_k + \frac{r_1}{r_k}\cos\varphi_n - 2\frac{r_1}{r_n}\cos\varphi_{kn}\cos\varphi_n \right\}. \tag{3.11}$$

Accordingly we obtain the equation for r_{k2}

$$m_k \ddot{r}_k - m_k \overline{r_2 \cos\varphi_k}^{\cdot\cdot} - m_1 \ddot{r}_{21} \cos\varphi_k +$$

$$+ \sum_{n\neq k} m_n \left\{ \ddot{r}_n \left[\cos\varphi_{kn} \left(1 + \frac{r_2}{r_k}\cos\varphi_k + \frac{r_2}{r_n}\cos\varphi_n \right) - \frac{r_2}{r_n}\cos\varphi_k - \frac{r_2}{r_k}\cos\varphi_n \right] - \cos\varphi_{kn} \overline{r_2 \cos\varphi_n}^{\cdot\cdot} \right\} =$$

$$= m_k F(r_k) + 2 m_k F(r_k) \frac{r_2}{r_k}\cos\varphi_k + m_1 F(r_{21}) \left[-\cos\varphi_k + \frac{r_2}{r_k}\sin^2\varphi_k \right] +$$

$$+ \sum_{n\neq k} m_n F(r_n) \left\{ \cos\varphi_{kn} \left(1 + \frac{r_2}{r_k}\cos\varphi_k + \frac{r_2}{r_n}\cos\varphi_n \right) - \frac{r_2}{r_k}\cos\varphi_n - \frac{r_2}{r_n}\cos\varphi_k + 2\frac{r_2}{r_n}\cos\varphi_n \cos\varphi_{kn} \right\}. \tag{3.12}$$

Subtracting correspondingly left and right hand sides of the equations (3.11), (3.12), we arrive at

$$(m_1 + m_2)\ddot{r}_{21}\cos\varphi_k + \sum_n m_n \overline{r_{21}\cos\varphi_n}^{\cdot\cdot} \cos\varphi_{kn} +$$

$$+ r_{21} \sum_n m_n \ddot{r}_n \left[-\cos\varphi_{kn} \left(\frac{\cos\varphi_k}{r_k} + \frac{\cos\varphi_n}{r_n} \right) + \frac{\cos\varphi_n}{r_k} + \frac{\cos\varphi_k}{r_n} \right] -$$

$$= (m_1 + m_2) F(r_{21}) \cos\varphi_k +$$

$$+ r_{21} \sum_n m_n F(r_n) \left[-\cos\varphi_{kn} \left(\frac{\cos\varphi_k}{r_k} + \frac{\cos\varphi_n}{r_n} \right) + \frac{\cos\varphi_k}{r_n} + \frac{\cos\varphi_n}{r_k} - 2\frac{\cos\varphi_n \cos\varphi_{kn}}{r_n} \right]. \tag{3.13}$$

Adding correspondingly left and right hand sides of the equations (3.11) and (3.12), we obtain the following equation $(\cos\varphi_{kk} \equiv 1)$

$$2\sum_n m_n \ddot{r}_n \cos\varphi_{kn} - 2\sum_n m_n F(r_n) \cos\varphi_{kn} =$$

$$= -\ddot{r}_{21}(m_2 - m_1)\cos\varphi_k + (m_2 - m_1) F(r_{21}) \cos\varphi_k - \sum_n m_n \overline{(r_1 - r_2)\cos\varphi_k}^{\cdot\cdot} \cos\varphi_{kn} -$$

$$-(r_1 - r_2) \sum_n m_n \ddot{r}_n \left[-\cos\varphi_{kn} \left(\frac{\cos\varphi_k}{r_k} + \frac{\cos\varphi_n}{r_n} \right) + \frac{\cos\varphi_k}{r_n} + \frac{\cos\varphi_n}{r_k} \right] -$$

$$-(r_1 - r_2) \sum_n m_n F(r_n) \left[\cos\varphi_{kn} \left(\frac{\cos\varphi_k}{r_k} + \frac{\cos\varphi_n}{r_n} \right) - \frac{\cos\varphi_k}{r_n} - \frac{\cos\varphi_n}{r_k} + 2\frac{\cos\varphi_n \cos\varphi_{kn}}{r_n} \right], \tag{3.14}$$

where summation over n includes $n = k$.

It is easy to see that at choosing of the lengths ratio for intercepts and like

$$m_1 r_1 = m_2 r_2 , \tag{3.15}$$

which corresponds to the centre of masses in the Newtonian mechanics, the right hand side of the equation (3.14) becomes equal to zero in view of the equation (3.13). Then the equation (3.14) acquires the following form

$$\sum_n m_n \ddot{r}_n \cos\varphi_{kn} = \sum_n m_n F(r_n) \cos\varphi_{kn} . \tag{3.16}$$

Equations of Motion of a Pair of Particles 1 and 2 at the Isotropic Space Distribution of Massive Distant Bodies

Introduce a unit vector **e** in the three-dimensional space directed from a particle 1 to a particle 2, and a unit vector which directed from the center of mass to a body *k*. Then the sum in the right hand side of the equation (3.16) may be written down in the form

$$\sum_n m_n F(r_n) \cos\varphi_{kn} = \mathbf{e}_k \cdot \sum_n m_n F(r_n) \mathbf{e}_n .$$

If the isotropy condition is fulfilled then one has got

$$\sum_n m_n F(r_n) \mathbf{e}_n = 0 . \tag{3.17}$$

In this case the equation (3.16) becomes

$$\mathbf{e}_k \cdot \sum_n m_n \ddot{r}_n \mathbf{e}_n = 0 . \tag{3.18}$$

Consider a displacement of the centre of mass

$$\mathbf{r}_n(t) = \mathbf{r}_n^0 + \mathbf{R}(t) ,$$

where $R << r_n$. Therefore the vectors are little affected (of order r/r_n), and one may write down for r_n

$$r_n(t) \approx r_n^0 + \mathbf{e}_n \mathbf{R}(t) .$$

Substitution of this expression into the equation (3.18) gives

$$\sum_n m_n \left(\ddot{\mathbf{R}} \mathbf{e}_n\right) \mathbf{e}_n = 0 . \tag{3.19}$$

It is reasonable to assume that the isotropy of masses distribution of distant bodies satisfies to a relationship

$$\sum_n m_n e_{n,i} e_{n,j} = \delta_{ij} \tfrac{1}{3} \sum_n m_n . \tag{3.20}$$

Then from the relation (3.19) we obtain an equation of motion for the center of mass of the particles 1 and 2.

$$\ddot{\mathbf{R}} = 0, \quad \ddot{\mathbf{r}}_n = 0 . \tag{3.21}$$

Consider further the difference equation (3.13). Make use of the formula (3.21) for $\ddot{r}_n$, the isotropy conditions (3.17) and (3.19), and also one more isotropy condition

$$\sum_n m_n \frac{1}{r_n} F(r_n) e_{n,i} e_{n,j} = \delta_{ij} \tfrac{1}{3} \sum_n m_n \frac{1}{r_n} F(r_n) . \tag{3.22}$$

Then taking into consideration that the first adduct is small as compared by the second one in the left hand side of the equation (3.13) $(m_1, m_2 << m_k)$, one arrives at an equation

$$\ddot{\mathbf{r}}_{21} = \mathbf{e}(m_1 + m_2) f(r_{21}), \tag{3.23}$$

where

$$\mathbf{e} = \frac{\mathbf{r}_{21}}{r_{21}}, \quad f(r_{21}) = -\frac{\gamma}{r_{21}^2}, \quad \gamma = \frac{3c^2 r_0}{\sum_n m_n} . \tag{3.24}$$

Note that the equation for (3.10), provided the isotropy conditions (3.20) and (3.22), coincides with the equation (3.23) after having been scalar multiplied the latter from the left and from the right by the vector **e**. It is easy to see that relations (3.21) and (3.23) coincide with the Newton's equations of motion if one admits that (3.24) is the gravitation constant of the Newtonian mechanics. Thus, we obtain from the equation (3.23) the usual equation for the relative distance r (in cylindrical coordinates r, φ):

$$\ddot{r} - r(\dot{\varphi})^2 = (m_1 + m_2) f(r), \tag{3.25}$$

and the moment of momentum conservation law $r^2 \dot{\varphi} = \text{const}$. It must be emphasized that both the first term, and the second term (centrifugal force) in the left hand side of the equation (3.25) are caused by action of almost immobile, massive, distant bodies onto a pair of particles 1 and 2, with the gravitation constant γ (3.24) being inversely proportional to the total mass of those distant bodies.

Underline once again that the derivation of motion equations for a pair of particles (3.21) and (3.23) was based on three requirements (3.17), (3.20), (3.22) of isotropy of space allocation for distant bodies

It should be noted an interesting fact, also. As follows from the equation (3.25), both in the Newtonian, and in the theory presented here the relative motion is defined by the *sum* of masses of particles $m_1 + m_2$.

It means that it is in the Newton's equation for the relative distance r a "memory" remained about that the attraction of bodies is indeed the sum of mutual actions of particles each of which is proportional to a mass of an acting particle.

CONCLUSION

1. For the fist time the gravitation theory has been constructed on the relativity principle as it was understood by Berkeley, Mach, and Poincaré, i.e. without use of the concept of absolute space (aether), contrary to the theory of universal gravitation of Newton and the known theories of gravitation, including a general relativity theory of Hilbert-Einstein.
2. On the basis of principles of relativity, causality and limitation of velocities by the quantity c the attraction law has been obtained between particles which states that forces of mutual gravitational action are not equal and proportional to a mass of an acting particle. This law essentially differs from the Newtonian one, where the attractive force is proportional to the product of masses of particles.
3. The causality principle leads to time irreversible equations of motion for particles. For the considered problem of a pair of particles, moving in an environment of "far stars", time reversible equations have been obtained. However, it is the consequence of assuming small velocities of particles as compared the light velocity and small characteristic times of particle motion as compared the time of configuration change for "far stars", and also with neglecting of quasi-random action "from the past".
4. It has been discovered a new – kinematical – kind of action of bodies caused by not gravitational force, but the relative acceleration of particles (and bodies) when motions only depend on relative distances and velocities. It is this type of action with which the centrifugal force appears while accelerated motion and mutual rotation of a pair of particles relative to far massive bodies.
5. Appearance of the centrifugal force is brought about the isotropy of space distribution of far bodies provided fulfillment of the three different isotropy conditions. It is this case when the equations of the Newtonian kind for motion of a pair of particles can be obtained.
6. At a large removal from the centre of isotropy the centrifugal force may be proved to be much less than the calculated one with the help of the Newton's theory. It can

explain a possibility of rotation of nebular objects with an angular velocity much more than calculated in accordance with the Newton theory, without referring to ideas of "the latent masses".

7. The centrifugal force modification at large distances from the "Universe centre" can lead to essential modifications of structure and spectra of atoms and molecules.
8. With taking into consideration of anisotropy in allocation of distant bodies in kinematical and force terms of the equation for the relative distance between a pair of particles (eq. (3.10)) there appears a force proportional to interparticle distance and to anisotropy tensor. This force can influence on particle motion, with considerable increasing the distance between particles.
9. The gravitation constant in "the Newtonian" limit appears to be inversely proportional to the sum of masses of distant bodies of the Universe. Finiteness of points out to boundedness of the total mass of those bodies.

ACKNOWLEDGMENTS

I express my profound gratitude for fruitful controversies at all stages of performance of this work to Rukhadze A.A., Ignatov A.M., and to Novosadov B.K. for valuable notes.

REFERENCES

[1] Newton I. *Philosophiae Naturalis Principia Mathematica.* 1st Ed., London, 1686.

[2] Berkeley G. De Motu Sive de Motus Principio et Natura et de Causa Communicationis Motuum. 1721.

[3] Mach E. Die Mechanik in Ihrer Entwicklung. I Aufl. Leipzig, 1883.

[4] Poincaré H. *La science et l'hypothèse*. Paris: Flammarion, 1902.

[5] *Gravitation and Relativity*. Ed. H.-Y. Chiu, W.F. Hoffmann. W.A. Benjamin, N.Y.-Amsterdam, 1964.

[6] Poincaré H. *Sur la dynamique de l'electron*. Rendiconti del Circolo Matematico di Palermo. 1906. XXI. P. 129-176.

In: Biology and Communication in the Information Age
Editor: Rafiqul Islam
ISBN: 978-1-62808-252-4

Chapter 4

UV Influence on Total Content of Water Soluble Antioxidants and Other Biochemical Markers in Blood Serum of Patients with Alcohol Abuse

R. L. Poltavskaya**, G. N. Tchoupakhina and L. N. Skrypnik
Laboratory of Natural Antioxidants
Immanuel Kant Baltic Federal University, Kaliningrad, Russia

Abstract

The status of total content of water soluble antioxidant, ceruloplasmin and uric acid were determined in 45 patients with alcoholic abuse and 37 health donors before and after UV affect of blood serum. All of them were a man from 22 to 35 years old. The mean value of water soluble antioxidant after 5 min of UV was significantly higher ($p<0,05$) in health donors, compared with the health donors before UV and patients with alcoholic abuse respectively. Water soluble antioxidant level before UV was significantly lower ($p<0,05$) in the alcohol abuse group compared to the health donors group. Ceruloplasmin and uric acid level significantly higher ($p<0,05$) in the alcoholic abuse group compared to the health donors group. However, the level of ceruloplasmin in both groups decrease under the UV affect ($p<0,05$). But uric acid level in both groups before and after UV affect was within a very close range and the difference were not significant ($p>0,05$). These results show antioxidant system of healthy donors provided a response to UV affect as a stressor, and the antioxidant system of patients with alcoholic abuse is too weak for this. These dates may be used in diagnosis, treatment and prevention of alcoholic abuse and in rehabilitation after the disease too.

Keywords: UV influence, blood antioxidants, total water soluble antioxidants, ceruloplasmin, uric acid, alcoholic abuse

* E-mail: Poltavskaya.raya@gmail.com.

AIM AND BACKGROUND

To determine the status of the antioxidant system as an object of study was selected serum of blood, and as indicators of the antioxidant defence: the total content of water-soluble antioxidants (TCA); the concentration of ceruloplasmin (CP) and uric acid (UA); a group of healthy donors and investigated groups with social diseases (syndrome of alcohol abuse). But the main aim of the study was to determine the effect of UV light on these indices, like a stress-factor.

INTRODUCTION

In recent years, we've conducted a wide variety of research devoted to studying the role of free radicals, reactive oxygen species (ROS) and nitrogen, and antioxidants in experimental and clinical medicine, biology, chemistry, ecology and bioengineering [1]. Obviously, the formation of ROS depends mainly on the following environmental factors: exposure to UV radiation, X-ray radiation and atmosphere pollution [2].

In this regard, the recent interest in the determination of antioxidant status of people has increased. This purpose usually determines the antioxidant capacity (activity) of plasma, serum, saliva and other biological fluids [3-7].

Many studies also found that a low dietary intake of any antioxidants, vitamins and trace elements are associated to be a risk factor for the onset of the disease [8].

That's why the aim of this study was to investigate the possible changes in levels of endogenous antioxidants with indirect roles and estimate influence of environmental factors.

So the serum levels of the content of water soluble antioxidant were measured, ceruloplasmin and uric acid in a group of healthly volunteers and in the group of patients with alcoholic abuse before and after UV influence [9, 10].

EXPERIMENTAL PART

There were 337 people involved in this research. That was a group of men, classified by age: period from 22 to 35 years old. 37 men of them were healthy donors and 45 men were with the syndrome of alcohol abuse.

From each patient and healthy person, 6 ml of blood from the antecubital vein was taken. Blood was always collected while fasting during the morning hours, and centrifuged for 10-15 min at 2500 rpm. to obtain serum. Serum was not frozen or stored. All measurements was made in 1,5 hours after centrifuged.

The total Serum content of water-soluble antioxidants was measured by using the amperometric method described by Y. Yashin with TsvetYauza 01 – AA equipment [11].

Thus, in our case the total antioxidant activity was the integral characteristic which includes the potential antioxidant action of all the components in serum of blood in the aggregate of their interaction with each other in this complex system given the potential synergies, their cooperative antioxidant action and the contribution of the minor antioxidants [12].

It has been established that the amperometric method has many advantages: low detection limit, high selectivity (only those compounds whose molecules can be oxidized are determined; other compounds which are present even in large concentrations are not determined), low cell volume, and easy maintenance [13]. Serum antioxidants compounds were oxidized well under amperometric detection conditions. The antioxidant activity was measured by using the value of oxidation of such compounds on the working electrode of the amperometric detector. The signal was registered as differential dynamic curves. Special software calculated the peak areas (of the differential curves) of the analyzed and the standard substances [11]. In our case, we used quercetin (Quercetin hydrate 95%, Acros) as the standard substance. So this amperometric method was best suited for the evaluation of antioxidant activity as well as the total content of antioxidants (TAC) in this case.

Ceruloplasmin spectrophotometrically with p-phenylenediamine [14]. Uric acid was determined also spectrophotometry based on the formation of colored complex with urikazy and peroxidise (BioSystems S.A.).

All the samples of serum were irradiated with UV light at various doses from 4.5 to 22.5 kJ/m2 (exposure during various times from 5 to 25 minutes). Every 5 minutes was determined the total content of antioxidants, ceruloplasmin and uric acid.

All statistical analyses were conducted using "Statistica 6.0" (for Windows) statistical program. Student's t-test was used, and comparison of MS patients and controls. Correlations between parameters were studied using Person's test. Values are expressed as mean standard deviation (SD). A significance level (p-value) of 5% was used unless stated otherwise.

RESULTS AND DISCUSSION

Mean ages of the patients (n=45) and healthy control group (n=37) were 29,9 ± 2,9 and 28,8 ± 3,7 respectively. No significant differences appeared between the groups.

Characteristics of clinical and demographic parameters of the control and study groups before and after exposure to UV radiation in the table.

The table shows that after the fifth minute of irradiation with UV - light TAC increased significantly ($p < 0.05$), and then began to decline, although the period of exposure was 10-20 minutes TAC is higher than before exposure to UV radiation. The total content of antioxidants before exposure to UV light in the group with the syndrome of alcohol abuse was significantly lower than in the control group.

The level of concentration of ceruloplasmin in serum of patients with alcohol abuse was higher compared with healthy donors. Reducing the concentration of the protein in blood serum by UV-radiation was detected in both groups. Reducing the concentration of CP in both groups occurred to the same level of 0,20 ± 0,06 mg / l and 0,19 ± 0,04 mg / l in the control and study groups respectively.

The level of uric acid in the blood serum of patients with alcohol abuse higher than the uric acid level in the control group. Under the influence of UV-significant changes in uric acid concentration in the control and the study group were found ($p > 0.05$). The concentration of uric acid in blood serum of control group was positively correlated with the concentration of ceruloplasmin ($r = 0{,}98$; $p < 0{,}05$) (Figure 1).

Table 1. Effect of different exposure of UV light on total content of water-soluble antioxidants, the concentration of ceruloplasmin and uric acid in patients with alcoholic abuse

Parameters	Type of groupe	UV, min					
		0	5	10	15	20	25
Age	Control groupe	28,78 ± 3,7	28,78 ± 3,7	28,78 ± 3,7	28,78 ± 3,7	28,78 ± 3,7	28,78 ± 3,7
	groupe with alcoholic abuse	29,93 ± 2,9	29,93 ± 2,9	29,93 ± 2,9	29,93 ± 2,9	29,93 ± 2,9	29,93 ± 2,9
Gender	Control groupe	Male					
	groupe with alcoholic abuse	Male					
TAC, mg/ml	Control groupe	0,26 ± 0,03	0,30 ± 0,03	0,29 ± 0,03	0,28 ± 0,03	0,27 ± 0,03	0,25 ± 0,03
	groupe with alcoholic abuse	0,12 ± 0,03	0,07 ± 0,01	0,05 ± 0,01	0,05 ± 0,01	0,03± 0,01	0,03 ± 0,01
CP, mg/l	Control groupe	0,30 ± 0,05	0,29 ± 0,05	0,27 ± 0,05	0,26 ± 0,05	0,22 ± 0,05	0,20 ± 0,06
	groupe with alcoholic abuse	0,41 ± 0,07	0,34 ± 0,05	0,31 ± 0,05	0,27 ± 0,05	0,22 ± 0,04	0,19 ± 0,04
Fe(III), mkmol/l	Control groupe	17,66± 3,72	17,66± 3,72	17,66± 3,72	17,66± 3,72	17,66± 3,72	17,66± 3,72
	groupe with alcoholic abuse	19,73 ± 8,95	19,73 ± 8,95	19,73 ± 8,95	19,73 ± 8,95	19,73 ± 8,95	19,73 ± 8,95
UA, mkmol/l	Control groupe	283 ± 35	278 ± 38	272 ± 37	267 ± 38	257 ± 38	255 ± 38
	groupe with alcoholic abuse	411 ± 66	385 ± 56	373 ± 54	350 ± 52	330 ± 49	325 ± 48

In the group of patients found a positive correlation between the total content of antioxidants, and ceruloplasmin (0,94; p <0,05) (Figure 2), and uric acid (0,93; p <0,05)(Figure 3).

The reaction of the blood serum of the control group on the UV light as a stressor, expressed that the level of TAC in the initial period (5 min exposure) increased, only 25 minutes back to the original value. And there was no reaction on ultraviolet light in group of patients with alcoholic abuse. That can be explained by low levels of TAC in that group [15].

Inhibition of ceruloplasmin with UV light in both groups studied indicates its high sensitivity to UV, which is probably associated with the confirmation of the holo-ceruloplasmin in apo - CP by UV and / or changing its compactness protein globule.

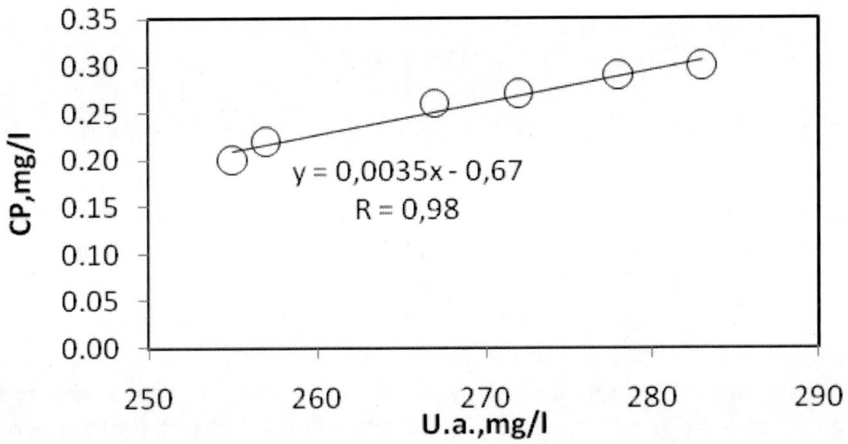

Figure 1. Correlation between concentration of uric acid and ceruloplasmin in blood serum of control group (r = 0,98 ; p <0,05).

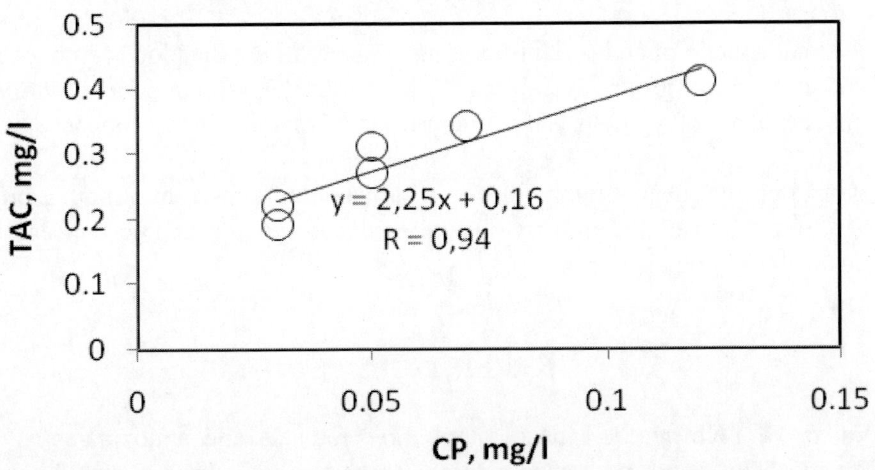

Figure 2. Correlation between concentration of ceruloplasmin and antioxidants in blood serum of patients with alcohol abuse (r = 0,94 ; p <0,05).

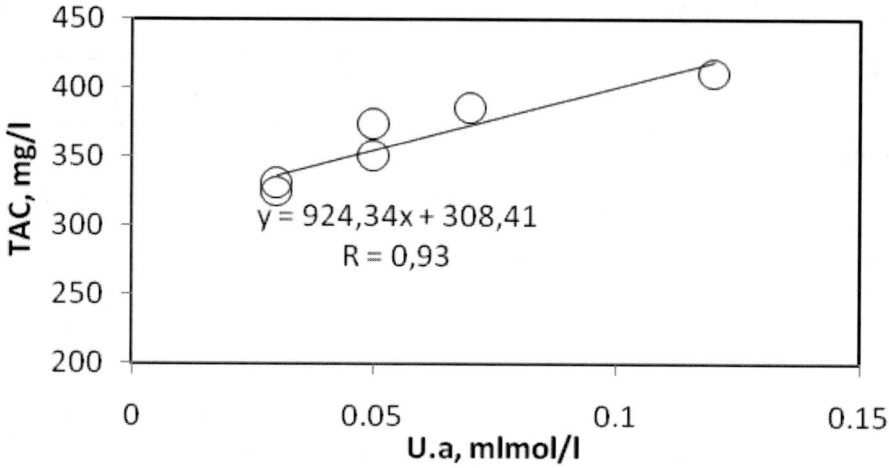

Figure 3. Correlation between concentration of uric acid and antioxidants in blood serum of patients with alcohol abuse (r = 0,94 ; p <0,05).

Studies on the effect of visible light (632.8 nm) confirm the reversible nature of the processes of unfolding and folding of molecules glycoprotein, as evidenced by photo induced changes in the chromatographic and spectral properties of the CP. Changes in the electrophoretic properties indicate the loss of copper and holo - ceruloplasmin confirmation of protein molecules in the form apo - CP that accompanied the unfolding of the globule [16].

Despite the absence of significant changes in the concentration of uric acid in the serum of both groups exposed to UV radiation, an excess of uric acid in the blood serum of the group with alcohol dependence compared to the control may be a marker of oxidative stress that develops in alcoholic abuse.

Conclusion

Thus, the total content of antioxidants in terms of quercetin in the blood serum of healthy donors conditioned control group several times higher than the TAC in patients with alcohol abuse. While the level of ceruloplasmin and uric acid is higher in the opposite treatment group.

UV light had no significant effect on levels of uric acid in the serum of both groups. UV light (at doses from 4.5 to 22.5 kJ/m2) significantly reduces the TAC and the concentration of CP in patients with alcoholic abuse.

References

[1] M. Valko, D. Leibfritz, J. Moncol at al. Free radicals and antioxidants in normal physiological functions and human disease. *The International Journal of Biochemistry and Cell Biology*. 2007. Vol. 39, P. 44–84.

[2] P. M Clarkson, H. S Thompson. Antioxidants: what role do they play in physical activity and health? *American Journal of Clinical Nutrition*. 2000. Vol. 72(suppl). P. 637–646.

[3] R. P. Singh, S. Sharad, S. Kapur Free radicals and oxidative stress in neurodegenerative diseases: relevance of dietary antioxidants. *Journal, Indian Academy of Clinical Medicine*. 2004 Vol. 5, No. 3

[4] C. M. da Costa, R. C.C dos Santos, E. S. Lima A simple automated procedure for thiol measurement in human serum samples. *J.Bras. Med. Lab*. 2006. Vol. 42, N.5, P. 345-350.

[5] G. K. Ziyatdinova, H. C. Budnikova, V. I. Pogoreltzev Determination of Total Antioxidant Capacity of Human Plasma from Patients with Lung Diseases Using Constant-Current culonometry. *Eurasian Journal of Analytical Chemistry,* 2006. Vol. 1, N. 1, P. 19-30.

[6] D. Koracevic, G. Koracevic, V. Djordjevic, A. Andrejcevic, V. Cosic Method for the measurement of antioxidant activity in human fluids. *J. Clin. Pathol*. 2001.Vol. 54. P. 356-361.

[7] Ch. P. Day Treatment of Alcoholic Liver Disease. *Liver Transplantation*. 2007. N. 13. P. 69-75.

[8] Kwun, K. Park, H. Jang at al. Lower antioxidant vitamins (A, C and E) and trace minerals (Zn, Cu, Mn, Fe and Se) status in patients with cerebrovascular disease. *Nutritional Neuroscience*. 2005. Vol. 8, N. 4. P. 251-257.

[9] N. Devipriya, M. Srinivasan et. al. Effect of ellagic acid, a natural polyphenol, on alcohol-induced prooxidant and antioxidant imbalance: a drug dose dependent study. *Singapore Med. J*. 2007. Vol. 48 N4. P. 311.

[10] Arun Kumar Mahto, Sailal M. Study of the Efficacy and safety of Liv. 52 Dstablets in alcoholic hepatitis: clinincal, biochemical and ultrasonographic evaluation, *Indian Medical Journal*. 2009. Vol. 103, N 5.

[11] V. N. Titov, V. V. Krylin, V. A. Dmitriyev, I. Yashin Plasma antioxidant activity is a test for impaired biological functions of endoecology, exotrophy, and inflammation reactions. *Clinical Laboratory Diagnosis* 2010. №7.

[12] Sies H. Total antioxidant capacity: appraisal of a concept. *The Journal of Nutrition*. 2007. Vol. 137. P. 1493 – 1495.

[13] L. Varvari, I.C. Popescu New method for antioxidant activity evaluation using a H_2O_2 amperometric sensor. *Rev. Roum. Chim*. 2010. Vol. 55. P. 851-857.

[14] Noyer M., Dwulet F.E., Hao Y.L., Putnam F.W. Purification and characterization of undegraded human ceruloplasmin // *Anal. Biochem*. 1980.Vol. 102. P. 450-458.

[15] Buico, C. Cassino, M. Ravera, P.-Giacomo, B. D. Osella Oxidative stress and total antioxidant capacity in human plasma. *Redox Report*. 2009. Vol. 14. N. 3. P. 125 – 131.

[16] P. Vachette, M. H.J. Koch, D. I. Svergun. Looking behind the Beamstop: X-Ray Solution Scattering Studies of Structure and Conformational Changes of Biological Macromolecules. *Methods in Enzymology*. 2003. Vol. 374, P. 584–615.

In: Biology and Communication in the Information Age
Editor: Rafiqul Islam
ISBN: 978-1-62808-252-4

Chapter 5

ABILITY OF DIHYDROQUERCETIN TO INHIBIT OZONE OXIDATION OF FIBRINOGEN

V. S. Rogovsky*[*1], M. A. Rosenfeld[2], V. B. Leonova[2], M. L. Konstantinova[2], S. D. Razumovsky[2], A. I. Matyoushin[1] and N. L. Shimanovsky[1]

[1]The Russian State Medical University (RSMU).
Moscow, Russia
[2]Emmanuel Institute of Biochemical Physics,
Russian Academy of Sciences.
Moscow, Russia

ABSTRACT

Dihydroquercetin (taxifolin, DHQ) is a natural flavonoid, which possesses antioxidant activity. Recently, in addition to the protective effect of antioxidants against lipid peroxidation, increasing attention is paid to the possibilities of antioxidants, including dihydroquercetin, to prevent oxidation of proteins. Fibrinogen is a plasma protein that serves an important role in clotting. In this work we study the ability of dihydroquercetin to inhibit oxidative modification of fibrinogen. We used ozone as the oxidant. Assessment of functional activity of fibrinogen, both before and after oxidation with ozone, was conducted by determining the time of formation of a fibrin clot after addition of thrombin in a solution of fibrinogen. We found that in the presence of dihydroquercetin ozone oxidation of fibrinogen is negligible, also in the presence of dihydroquercetin fibrinogen retains its functional properties after ozone oxidation, whereas in the case of ozone oxidation of fibrinogen without dihydroquercetin, the functional activity of fibrinogen is completely lost.

Keywords: Dihydroquercetin, taxifolin, oxidative modification of fibrinogen, ozon oxidation, reactive oxygen species

---

[*] E-mail: qwer@yasenevo.ru.

AIM AND BACKGROUND

Oxidative modification of fibrinogen inhibits thrombin-catalyzed clot formation [1], inhibits platelet adhesion and aggregation [2], enhances the activity of tissue plasminogen activator, cause negative effect on hemorheological parameters [3], and is actively involved in the process of inflammation of the vascular wall. Also oxidized fibrinogen is able to induce increased production of interleukin-8 in primary culture of human endothelial cells, which causes chemotaxis of immune cells [4-6]. Oxidation of fibrinogen causes an imbalance in the blood coagulation system.

On the one hand it increases thrombin clotting time and inhibites platelet adhesion and aggregation, on the other hand, it cause formation of neutrophil-platelet aggregates [7]. Thus, the learning of ability of various compounds to prevent oxidative damage to blood plasma proteins, particularly to fibrinogen, is relevant. The objective of this work was to investigate the ability of dihydroquercetin to inhibit the oxidative modification of fibrinogen.

INTRODUCTION

Flavonoids, or bioflavonoids, are a class of plant polyphenols with a wide spectrum of biological activities. These substances give color to plants. Compounds with antioxidant activity are increasingly used in the treatment of diseases in which the induction of lipid peroxidation is one of the key elements of pathogenesis. Such diseases include ischemic attacks, ischemic heart disease, vasculitis of various etiologies and many others [8]. Recently, in addition to the protective effect of antioxidants against damage to cell membranes, increasing attention paid to the possibilities of antioxidants, including dihydroquercetin, to prevent oxidation of proteins. It is generally recognized today that many amino acid residues of proteins are susceptible to oxidation by reactive oxygen species (ROS). Free-radical oxidation of proteins may be accompanied by the splitting of the polypeptide chains, a modification of amino acids and transformation of proteins to compounds which are highly sensitive to proteolytic degradation [9].

Oxide-modified proteins accumulate in the body with age, under oxidative stress and a number of diseases, particularly diabetes mellitus [10-12]. It is well known that fibrinogen is a plasma protein that serves an important role in clotting. Many diseases caused increased oxidation of fibrinogen, which significantly reduces its functional activity. It is shown that fibrinogen, at least 20 times more sensitive to oxidative modification in comparison with other major plasma proteins: albumin, immunoglobulins, transferrin, ceruloplasmin [13, 14]. That leads to the formation of oxidized forms of the protein, different from the native form in the chemical composition and the structural organization. Therefore fibrinogen changes its functional properties.

Formation of non-covalent bonds enable making of macromolecular fibrinogen clusters [14]. Oxidative modification of fibrinogen inhibits thrombin-catalyzed clot formation [1], inhibits platelet adhesion and aggregation [2], enhances the activity of tissue plasminogen activator, cause negative effect on hemorheological parameters [3], and is actively involved in the process of inflammation of the vascular wall. Also oxidized fibrinogen is able to induce increased production of interleukin-8 in primary culture of human endothelial cells, which causes chemotaxis of immune cells [4-6]. Oxidation of fibrinogen causes an imbalance in the blood coagulation system.

On the one hand it increases thrombin clotting time and inhibites platelet adhesion and aggregation, on the other hand, it cause formation of neutrophil-platelet aggregates [7]. Thus, the learning of ability of various compounds to prevent oxidative damage to blood plasma proteins, particularly to fibrinogen, is relevant. The objective of this work was to investigate the ability of dihydroquercetin to inhibit the oxidative modification of fibrinogen.

EXPERIMENTAL PART

In this work we carried out the oxidation of fibrinogen solution under standard conditions [4], and in the presence of dihydroquercetin. In the study we used bovine fibrinogen obtained from purified citrated blood plasma [13]. Fibrinogen was transferred to 0.06 M phosphate buffer (pH 7.4) containing 0.15 M NaCl, using Sephadex G-25 gel filtration.

To determine the effect of dihydroquercetin on the oxidation of fibrinogen we compared the UV spectra of test solutions before and after oxidation. Functional activity of fibrinogen was assessed by controlling its ability to form a fibrin clot after addition of thrombin.

In the studying of the effect of dihydroquercetin on the oxidation of fibrinogen we used ozone as the oxidant. Ozone was chosen for the following reasons. This natural agent belongs to the reactive oxygen species, and it is convenient to carry out research in model systems with ozone. Its half-life in aqueous solution in the presence of oxidation substrates does not exceed 1-2 min.

The oxidation state of the object is strictly regulated, because the amount of ozone that reacts with a reducing agent, is precisely measured spectrophotometrically at 254 nm. Ozonation of fibrinogen solution was carried out in a reactor (volume 3.3×10^{-3} L), equipped with quartz windows, ozone-oxygen mixture was blown through a free volume of the reactor. The amount of ozone in the reactor was 2.0×10^{-7} M. Ozone was synthesized in a discharge device by passing a stream of oxygen between the electrodes. The UV spectra of proteins before and after ozonation were recorded on SF-2000 spectrophotometer (Russia) in a 1 cm quartz cells.

To study kinetics of absorption of ozone by dihydroquercetin we used bubbling method in which ozone is passed through a layer of a solution containing the test compound, fixing the change in ozone concentration before and after passing through the solution. Assessment of functional activity of fibrinogen, both before and after oxidation with ozone, was conducted by determining the time of formation of a fibrin clot after addition of thrombin («Roche», France) in a solution of fibrinogen. Thrombin solution (0.05 ml, 2.5 units NIH) was added to 0.3 ml test sample.

RESULTS AND DISCUSSION

In the first stage bubbling method of ozone oxidation was used to study the kinetics of the interaction of ozone with dihydroquercetin. Concentration of DHQ was 10^{-5} mol / L. Kinetics of ozone absorption by DHQ is shown in Figure 1.

According to the data presented in Figure 1, using the formula 1, we calculated the rate constant for the interaction of ozone with dihydroquercetin and stoichiometric coefficient of this reaction, which is determined by the ratio of the amount of absorbed ozone to the amount of substance reacted. Stoichiometric coefficient of the reaction of ozone with dihydroquercetin is 3. The rate constant of the interaction of ozone with DHQ is 1.0×10^3 l/mol.s.

Formula 1. The equation for determining the rate constant for the interaction of ozone with dihydroquercetin [15].

$$k = W \times ([O_3]_0 - [O_3]_{gas}) / \alpha [O_3]_{gas} [reagent]$$

W - flow rate of the gas mixture (50-200 L/min); α - solubility of ozone in water (0.32, experimental determination); $[O_3]_0$ - initial concentration of ozone; $[O_3]_{gas}$ - concentration of ozone, which leaves the reactor; [reagent] - initial concentration of reagents.

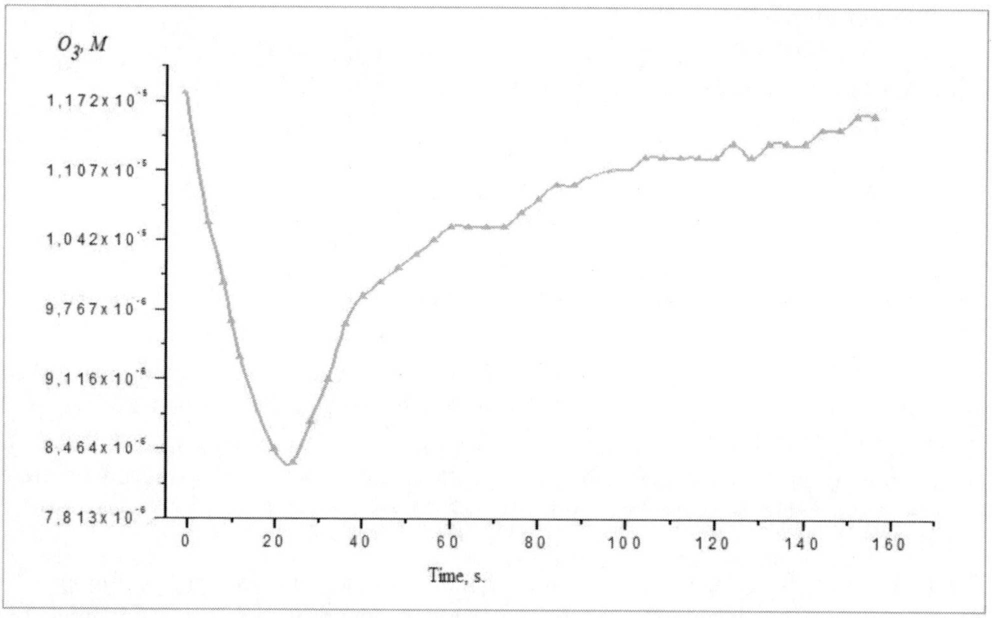

Figure 1. Kinetics of ozone absorption by dihydroquercetin.

Before studying of combined fibrinogen and dihydroquercetin oxidation by ozone we previously studied the interaction of native fibrinogen with ozone. For this purpose we carried out the oxidation of fibrinogen in a closed reactor for 10 minutes. The intensity and character of the oxidation process were determined using UV spectra of unoxidized (native) and oxidized fibrinogen.

Figure 2 shows the spectra of native and oxidized fibrinogen. Ozone oxidation of fibrinogen was assessed by the change in optical density at a wavelength of 280 nm (absorbance of the amino acid residues containing phenolic groups, such as tyrosine, phenylalanine, and unsaturated residues of tryptophan and histidine).

The spectrum of oxidized fibrinogen shows that ozone oxidation of the protein decreases the amplitude of the fibrinogen absorption peak, and the peak shifts to shorter wavelengths.

Also noted the appearance of a new minor absorption peak at 325 nm, which may be due to the formation of quinoid structures of the phenoxyl and imidazole core of amino acid residues under the influence of ozone. Next, we studied the ozone oxidation of fibrinogen in the presence of dihydroquercetin (Figure 3).

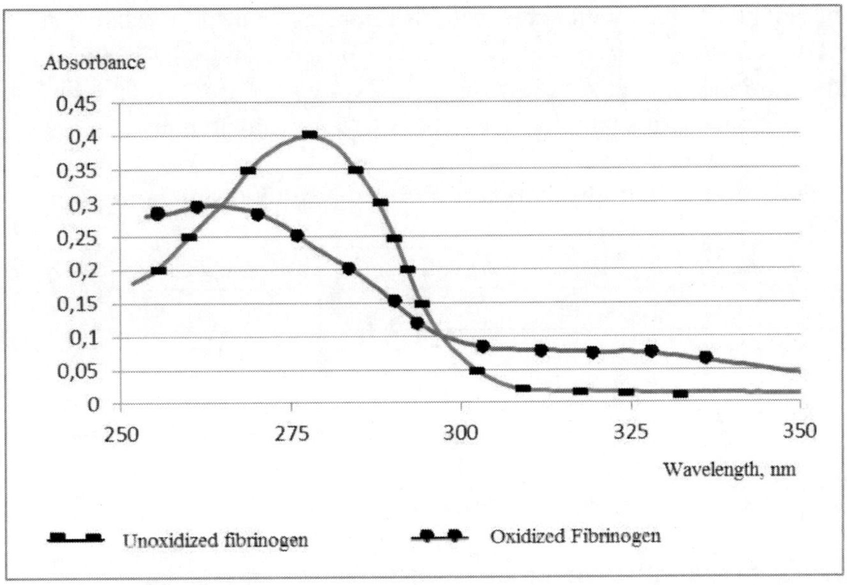

Figure 2. UV spectra of fibrinogen (1.87 mg/ml) before and after ozone oxidation.

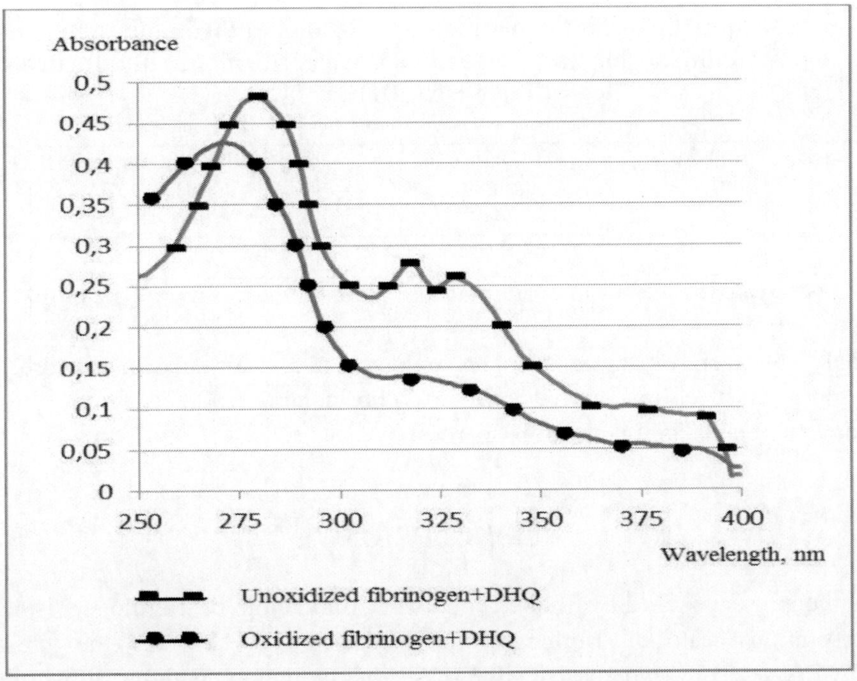

Figure 3. UV spectra of fibrinogen (1.87 mg/ml) in the presence of dihydroquercetin (0.8×10-4 M) before and after ozone oxidation.

By comparing the spectra presented in Figure 2 and Figure 3, we see that in the case of the oxidation of fibrinogen in the presence of dihydroquercetin absorption peak of fibrinogen is decreased by 12%, whereas in the absence of dihydroquercetin we can see a significant decrease in absorption peak of fibrinogen by 27%. Thus, we can conclude that dihydroquercetin in this case has a protective effect against ozone oxidation of fibrinogen, because it has ability to react with ozone and, therefore, prevent the oxidative degradation of fibrinogen. For convenience in comparison we composed a table that shows an absorption peak of native fibrinogen, oxidized fibrinogen, a mixture of fibrinogen and dihydroquercetin (fibrinogen+DHQ), and oxidized mixture of fibrinogen and dihydroquercetin.

Table 1. The absorption peaks of the studied substances

Studied compounds	Absorbance at absorption peak	Wavelength, nm	Absorbance, %
Unoxidized fibrinogen	0,40 ± 0,02	277,8 ± 5,8	100
Oxidized Fibrinogen	0,29 ± 0,03	264,8 ± 18,2	73
Unoxidized fibrinogen+DHQ	0,48 ± 0,02	278,8 ± 15,3	100
Oxidized fibrinogen+DHQ	0,42 ± 0,03	270,8 ± 25,2	88

Note. Data are presented as mean values and standard errors of the mean (M ± s.e.m.).

To assess the functional activity of fibrinogen after ozone oxidation in the absence and in the presence of DHQ, we controlled the ability of fibrinogen to form a fibrin clot after addition of thrombin. The results of this study are presented in Table 2.

Table 2. The time of a fibrin clot formation after addition of thrombin to the unoxidized fibrinogen, oxidized fibrinogen, and oxidized mixture of fibrinogen and ihydroquercetin (fibrinogen+DHQ)

T, Unoxidized fibrinogen, s	T, Oxidized fibrinogen, s	T, Oxidized fibrinogen+DHQ (2.5×10^{-4} M), s	T, Oxidized fibrinogen+DHQ (0.8×10^{-4} M), s
59 ± 1	–	64 ± 1	86 ± 2

Note. Data are presented as mean values and standard errors of the mean (M ± s.e.m.).

The table above shows that fibrinogen, oxidized in the absence of dihydroquercetin, completely lost its functional activity, whereas fibrinogen, oxidized in the presence of dihydroquercetin, retains its functional properties.

Conclusion

1. In the presence of dihydroquercetin ozone oxidation of fibrinogen is negligible (absorption peak of fibrinogen is decreased by 12% in the presence of dihydroquercetin, whereas in the case of oxidation without dihydroquercetin absorption peak of fibrinogen is decreased by 27%).

2. In the presence of dihydroquercetin fibrinogen retains its functional properties after ozone oxidation, whereas in the case of ozone oxidation of fibrinogen without dihydroquercetin, the functional activity of fibrinogen is completely lost.

REFERENCES

[1] Shacter, E., Williams, J.A., Levine, R.F. Oxidative modification of fibrinogen inhibits thrombin-catalyzed clot formation. *Fee Radic. Biol. Med.* Vol. 18, 815-831, (1995).

[2] Belisario, M.A., Di Domenico, C., Pelagalli, A., Della Morte,R., Staiano, N. Metal-ion catalyzed oxidation affects fibrinogen activity on platelet aggregation and adhesion *Biochimie.*,Vol. 79, 449-455, (1997).

[3] Roitman E. V., Azizova O. A., Morozov Yu. A., Aseichev A. V. Effect of oxidized fibrinogens on blood coagulation. *Bul. of exp. biology and medicine.* Vol. 138, 527-530, (2004).

[4] Rosenfeld M. A., Leonova V. B., Shchegolikhin A. N., Razumovskii S. D., Konstantinova M. L., Bychkova A. V. and Kovarskii A. L. Oxidized modification of fragments D and E from fibrinogen induced by ozone. *Biochemistry* (Moscow). Vol. 75, 1285-1293, (2010).

[5] Azizova OA, Maksyanina EV, Romanov YA, Aseichev AV, Scheglovitova ON. Fibrinogen and its oxidized form induce interleukin-8 [correction of interleukin-2] production in cultured endothelial cells of human vessels. *Bul. of exp. biology and medicine.* Vol. 137, 406-409, (2004).

[6] Shcheglovitova ON, Azizova OA, Romanov YA, Aseichev AV, Litvina MM, Polosukhina ER, Mironchenkova EV. Oxidized forms of fibrinogen induce expression of cell adhesion molecules by cultured endothelial cells from human blood vessels. *Bul. of exp. biology and medicine.* Vol. 142, 277-281, (2006).

[7] Aseychev AV, Azizova OA, Shulenina LV, Piryazev AP. Effect of oxidized fibrinogen on aggregation of activated platelets and neutrophils. *Bul. of exp. biology and medicine.* Vol. 147, 284-290, (2009).

[8] Men'shchikova E.B., Lankin V.Z., Zenkov N.K.et al. Oxidative stress. *Prooxidants and antioxidants.* M.: Slovo, 556, (2006).

[9] Lushchak V. I. Free radical oxidation of proteins and its relationship with functional state of organisms. *Biochemistry* (Moscow). Vol. 72, 995-1017, (2007).

[10] Ivanova I.P., Knyazev D.I., Kudryavtseva Y.V., Chuprov A.D., Trofimova S.V. Oxidation of lipids and proteins in lens and blood plasma of rats in ageing. *Modern Technologies in Medicine.* Vol. 3, 16-20, (2011).

[11] Pandey KB, Mishra N, Rizvi SI. Protein oxidation biomarkers in plasma of type 2 diabetic patients. *Clinical Biochemistry.* Vol. 43, Issues 4-5, Pages 508-511, March, (2010).

[12] Stadtman, E.R. Protein oxidation and aging *Free Radic. Res.* Vol. 40, 1250-1258, (2006).

[13] Rosenfeld M. A., Leonova V. B., Konstantinova M. L. and Razumovskii S. D. Mechanism of enzymatic cross-linking of fibrinogen molecules. *Izv. Ros. Akad Nauk, Biol. Ser.,* No. 6, 578-584 (2008).

[14] Dijkgraaf, L.C., Zaardeneta, G, Corddewener, F.W., Liems, R.S., Schmitz, J.P., de Bont, L.G., Milan, S.B. Crosslinking of fibrinogen and fibrinonectin by free radicals: a possible initial step in adhesion formation in osteoarthitis of the temporomandibular joint. *J. Oral Maxillofac Surg.* Vol. 61, 101-111, (2003).

[15] Razumovskii S.D., Zaikov G.E. Ozone and its Reactions with Organic Compounds. *Nauka,* Moscow (1974).

In: Biology and Communication in the Information Age
Editor: Rafiqul Islam
ISBN: 978-1-62808-252-4

Chapter 6

IMPROVED PARTICLE SWARM OPTIMIZATION FOR FUZZY RULE EXTRACTION AND MEMBERSHIP FUNCTION TUNING

***C. Rani*[*1], *P. Ganesh Kumar*[2] *and S. N. Deepa*[3]**
[1]Department of Information Technology
[2]Department of Information Technology
[3]Department of Electrical and Electronics Engineering
[1,2,3]Anna University of Technology Coimbatore, TamilNadu, India

ABSTRACT

This paper proposes an Improved Particle Swarm Optimization (IPSO) for extracting knowledge from data for the fuzzy rule based classifier model. In the standard PSO, if the swarm uses a small inertia weight or constriction coefficient, then its convergence is premature. Further, if the global best is found early in the searching process, then it may get struck in poor local minima. To overcome this issue, the concept of selection from Genetic Algorithm (GA) is incorporated in addition to standard velocity and position updating of PSO. The performance of the proposed approach is evaluated using ten publicly available bench mark data sets. From the simulation study, it is found that the proposed approach improves the convergence and generates a compact and comprehensible fuzzy rule based system with high classification accuracy for all the data sets. Statistical analysis of the test result shows the superiority of the proposed method over other approaches reported in the literature.

Keywords: Fuzzy Logic, If-Then rules, Membership function, Particle Swarm Optimization and Genetic Operation

---

[*] E-mail:merrylin_c@yahoo.com.

1. Introduction

Data Mining [1] has attracted a great deal of attention in the information industry due to wide availability of huge amounts of data and the imminent need for turning such data into useful knowledge. Classification is a popular data mining tasks that builds models describing the important data classes and provide a better understanding of the data at large. A number of algorithms have been proposed for solving the classification problem. Statistical approaches like Fisher's Linear Discriminant Analaysis [2], Naïve Baye's Approach [3], and K-Nearest Neighbor [4] are memory resistant, deal with small sized data sets and are not scalable. Machine Learning Approaches like Artificial Neural Network [5] and Support Vector Machine [6] are theory rich approach for data classification but the results produced by them are very hard and difficult to interpret. They are popularly called as "Black Box" method as they focus only on classification performance and do not provide any knowledge about the problem.

Rule Based Classification (RBC) is one kind of approach that uses a set of If-Then rules for data classification. As rules are good ways of representing knowledge, RBC seems to be an appropriate approach for classification and understanding of data. Decision Tree [7] is used to construct a rule based classifier model. Even though the rules produced by them contain biologically meaningful terms, it is a sensitive type of classifier. Small disturbances in the training sample lead to large differences in the tree structure. Symbolic machine learning approach [8] can be used to extract human understandable rules from neural networks but they can not completely bring out the hidden information in the data. Moreover they are lacking in robustness with respect to the noisy and missing data.

Data Classification using the knowledge extracted from the data is a complex problem that involves a decision-making process. This decision making process has lot of uncertainties since the real world data are vague in nature and are very hard to predict as the boundaries between the output class labels are not well defined. With the ability of Fuzzy Logic [9] that deals with uncertain situation and vagueness, it seems to be an appropriate approach for classification using the knowledge extracted from data. Typical fuzzy classifiers consist of interpretable if-then rules with fuzzy antecedents and class labels in the consequent part. The antecedents (if-parts) of the rules partition the input space into a number of fuzzy regions by fuzzy sets, while the consequents (then-parts) described the output of the classifier in these regions.

An important issue in the design of Fuzzy Rule Based System (FRBS) [10] is the formation of fuzzy if-then rules and the membership functions. In general, the rules and membership functions are formed from the experience of the human experts. But, for classification problems with large number of input variables, the possible number of rules increases exponentially, which makes it difficult for experts to define a complete rule set. Data-driven approaches [11] have been proposed for developing the FRBS from numerical data without the knowledge of domain experts. But this approach is very weak in self-learning and determining the required number of fuzzy if-then rules.

The design of FRBS [12] can be formulated as a search problem in high dimensional space where each point represents a rule set, membership function and the corresponding system behavior. Given some performance criteria, the performance of the system forms a hyper surface in the space. Developing the optimal fuzzy system is equivalent to finding the optimal location of this hyper surface. This makes Evolutionary Algorithms (EA) or Swarm Intelligence (SI) based optimization algorithms as better candidates for searching the hyper surface than conventional optimization methods. A popular swarm-intelligence-based algorithm is the Particle Swarm Optimization (PSO) [13] that models the social behavior of bird flocking or fish schooling.

The standard PSO has been subjected to empirical and theoretical investigations by several researchers [14]. In many occasions, the convergence is premature, especially if the swarm uses a small inertia weight or constriction coefficient. As the global best found early in the searching process, PSO may get struck in a poor local minima. Now a day, the research trend is to merge or combine the PSO with other techniques, especially the other evolutionary computation techniques.

In this paper, an Improved Particle Swarm Optimization (IPSO) algorithm is proposed that includes the concept tournament selection from Genetic Algorithm (GA) [15] for optimal design of fuzzy classifier that can evolve the rule set along with the number of rules and the membership function simultaneously.

The performance of the proposed IPSO approach has been tested using ten bench mark datasets available in the UCI machine learning repository [16].

The rest of the paper is organized as follows. Section 2 briefly introduces the task of classification solved using fuzzy if-then rules. In Section 3, standard PSO algorithm is presented. Details of the proposed IPSO and its implementation issues are discussed in Section 4 and 5 respectively. Simulations conducted using ten benchmark data sets are discussed and the results are reported in Section 6. Concluding remarks are given in Section 7.

2. Data Classification Using Fuzzy If-Then Rules

Classification is a supervised learning technique that takes labeled data samples and generates a model (classifier) that classifies new data samples into different predefined groups or classes. This classification problem can be solved by using fuzzy logic with interpretable if-then rules and membership function. The following subsections present the various components of a fuzzy classifier system [9]

2.1. Fuzzy Sets

Fuzzy set theory generalizes classical set theory to allow partial membership with a smooth boundary. The degree of membership in a set is expressed by a number between 0 and 1. 0 means entirely not in the set, 1 means completely in the set, and a number in between means partially in the set. Mathematically, a fuzzy set A in the universe of discourse X is defined to be a set of ordered pairs,

$$A = \{(x, \mu_A(x)) \mid x \in X\} \quad (1)$$

where $\mu_A(x)$ is called the membership function of x in A. Triangular and Trapezoidal are the most commonly used membership function.

2.2. Fuzzy If-Then Rules

Fuzzy classifiers consist of interpretable if-then rules representing the input features and the output class of the form:

$$R_j : \text{if } x_{p1} \text{ is } A_{ji} \text{ and } \ldots \text{ and } x_{pn} \text{ is } A_{jn} \text{ then class } C_j \tag{2}$$

where A_{ji} ,...... A_{jn} are antecedent fuzzy sets of the input variable x_{p1} .... x_{pn} and C_j is one of the output class label. A collection of such rules are used as a knowledge base by the classifier upon which qualitative reasoning is performed to derive conclusion.

2.3. Fuzzy Inference System

With input-output relationship expressed as a collection of fuzzy if-then rules, in which the input uses linguistic variables and the consequent have class labels, qualitative reasoning is performed to infer the results. In this paper, Mamdani inferencing system with product t-norm and max t-conorm is used. Here the set of input variable is matched against the if part of each if-then rule, and the response of each rule is obtained through fuzzy implication operation. The response of each rule is weighted according to the extent to which each rule fires. The responses of all the fuzzy rules for a particular output class are combined to obtain the confidence with which the input is classified to the corresponding output class.

In general, the rules and the membership functions used by the fuzzy logic for data classification are formed from the experience of the human experts which is a difficult and time consuming task. In this paper, a hybrid Particle Swarm Optimization approach is proposed to extract knowledge in the form of if-then rules from the data and tuning of membership function for effective data classification.

3. Review of Particle Swarm Optimization

Particle Swarm Optimization (PSO) [13] is a generalized search and optimization technique inspired by the behavior of social insects. Figure 1 shows the various steps in applying PSO for any optimization problem.

PSO maintains a population of particle's position that represents candidate solutions to the problem. Each position of particle in the population has an adaptable velocity, according to which the particle moves in the search space. Each position in the population is evaluated to give some measure of its fitness to the problem using the objective function. During each iteration, new population is created according to the following two equations:

$$v_i^{g+1} = w_i v_i^g + c_1 rand \times (p_{best} - s_i^g) + c_2 rand \times (g_{best} - s_i^g) \tag{3}$$

$$s_i^{g+1} = s_i^g + v_i^{g+1} \tag{4}$$

where = velocity of particle i at generation g, v_i^{g+1} = velocity of particle i at generation g+1,

s_i^g = position of particle i at generation g, s_i^{g+1} = position of particle i at generation $g+1$, = inertia weight, = self confidence factor, = swarm confidence factor, = particle's individual best, g_{best} = global best at generation g. The equation (3) updates the velocity and equation (4) updates the position of the respective particle at each iteration until the convergence criterion is satisfied.

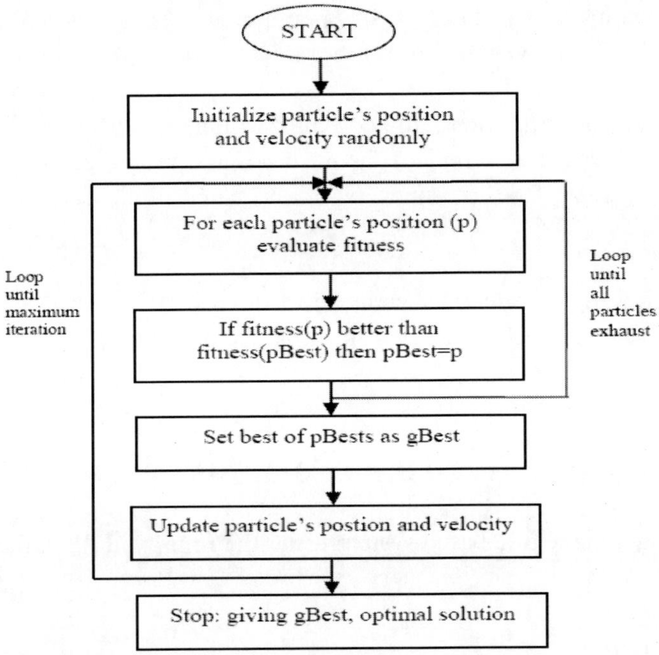

Figure 1. Flowchart of PSO.

4. Improved Particle Swarm Optimization

In a standard PSO, velocity and position updation are the two major operations used to update the solution space iteratively. In general PSO will produce improved performance, if there is a balance between exploration and exploitation in its searching process This exploration and exploitation tradeoff is influenced by modifying and tuning parameters namely current motion, inertia weight, cognitive and social coefficients. If the swarm uses a small inertia weight or constriction coefficient, then its convergence is premature. Further, if the global best is found early in the searching process, then it may get struck in poor local minima. To overcome this difficulty, the proposed PSO incorporates the concept of selection from Genetic Algorithms [15].

4.1. Tournament Selection

Selection is one of the genetic operators that serve to redirect the search towards those positions in the search space that have shown a relative advantage over others recently visited. While there is a form of selection implicit in particle swarm, it is extremely weak. Hence this paper proposes to incorporate a most common form of tournament selection used in genetic algorithm. The tournament selection is performed as follows:

1. The fitness of each particle, based on its current position, is compared to the fitness of *k* other particles and scores a point for each with worse fitness.
2. The population is then sorted using this score with the individuals having the highest scores appearing at the head of the population. The fitness of the personal best positions for the individuals are not considered during the scoring or sorting of the particles.
3. Once the population is sorted, the current positions and velocities of the best half of the population are used to replace the positions and velocities of the worst half of the population leaving the personal best associated with each of the individuals unmodified.

With the addition of the selection method just described, the proposed IPSO will have more exploitative search mechanism and it finds better optima more consistently than standard PSO.

5. Improved PSO Implementation

While designing a fuzzy rule based system using the Improved PSO; the following issues are to be addressed:

- Representation
- Fitness function formation

5.1. Representation

The first important consideration in IPSO is the representation strategy to be followed. A fuzzy system is specified only when the rule set and the membership function associated with each fuzzy set are specified. A typical particle's position in the swarm will look like the following:

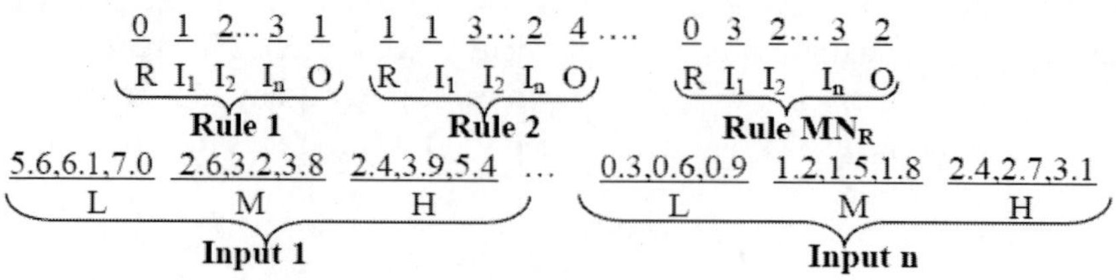

Figure 2. Representation of single particle's position in the swarm.

As shown in figure 2, each rule in the rule set has three sections: rule selection ('R'), representation for the input variables (antecedent – I_1, I_2… I_n) and the representation for the output classes (consequent – 'O'). Rule selection may take either '1' to select the rule otherwise '0'. Each antecedent part may take an integer value ranges from 1 to 3 such that '1' represents "low", '2' represents "medium", '3' represents "high". The output class is also represented by an integer value whose range depends on the number of class label. The term 'MN_R' denotes maximum number of rules.

To represent membership function, the range of each input variable is partitioned into three areas (Low, Medium and High) determined by fuzzy sets as shown in figure 3.

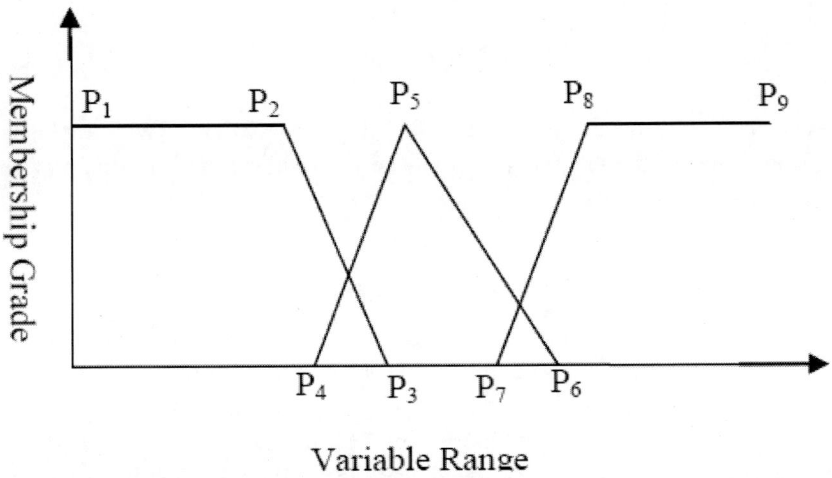

Figure 3. Fuzzy Space.

A total of nine membership points (P_1, P_2, P_3, P_4, P_5, P_6, P_7, P_8, P_9) are required for representing each input variable as a fuzzy set. In that nine points, first and last points (P_1 and P_9) are fixed that are the minimum and maximum of the input variable. The remaining seven membership points are evolved between the dynamic range such that P_2 has $[P_1,P_9]$, P_3 has $[P_2,P_9]$, P_4 has $[P_2,P_3]$, P_5 has $[P_4,P_9]$, P_6 has $[P_5,P_9]$, P_7 has $[P_5,P_6]$ and P_8 has $[P_7,P_9]$ as limits.

5.2. Fitness Function

The next important consideration following the representation is the choice of fitness function. Evaluation of the individuals in the population is accomplished by calculating the objective function value for the problem using the parameter set. The result of the objective function calculation is used to calculate the fitness value of the individual. In the classification problem under consideration, there are two objectives; one is to maximize the correctly classified data and the other one is to minimize the number of rules. These two objectives are conflicting objectives. This is overcome by reformulating the first objective of maximizing the correctly classified data as minimizing the difference between total number of samples and the correctly classified data.

Given the total number of samples ('S') and the maximum number of rules ('MN_R'), the task is to find out the difference between 'S' and the correctly classified data ('Cc') for the selected number of rules('SN_R') of every IPSO run. During the IPSO run, the objective is to find out the minimum of the above said value. This is mathematically represented as,

$$Min\ \ f = (S - Cc) + (k \times SN_R) \tag{5}$$

where 'k' is a constant introduced to amplify 'SN_R' whose value is usually small. In this paper, the value of 'k' is taken as 5.

In general, PSO searches for a solution with maximum fitness function value. Hence, the minimization objective function given by (5) is transformed to a fitness function to be maximized as

$$\text{Fitness} = \frac{K}{f} \tag{6}$$

where 'K' is another constant used to amplify ($1/f$), the value of which is usually small; so that the fitness value of the chromosome will be in a wider range. In this paper, the value of 'K' is taken as 1000.

6. Simulation Result

This section presents the details of the simulation carried out on ten benchmark datasets available in the UCI machine learning repository [16]. The details of the dataset are given in table 1. The proposed approach is developed using Matlab R2009 and executed in a PC with Pentium IV processor with 2.40 GHz speed and 256 MB of RAM. To demonstrate the effectives of the proposed approach, experiments are also conducted using simple RGA and standard PSO. The details of the implementation and the performance of the classifier for the Wine data are presented here. The wine data contains the chemical analysis of 178 wines with 13 input variables and 3 output classes.

As shown in figure 3, each input variable of the wine data is represented by three fuzzy sets namely, "low", "medium" and "high". Trapezoidal membership function is used to represent the "low" and "high" fuzzy sets. Triangular membership function is used to represent the "medium" fuzzy set. Seven points are required to represent an input variable and hence a total of ninety one (13×7=91) membership function points are needed. Each rule for a wine data set requires fifteen integer numbers (1 for rule selection + 13 for input variables + 1 for output class).

A maximum of six rules are included and hence a total of ninety integer numbers (6×15 =90) are needed to represent the complete rule set. First the learning ability of the proposed IPSO is examined by using all the 178 samples of wine data as training patterns. The proposed approach is run with 30 independent trials with different values of random seeds and control parameters. The optimal results are obtained with the following setting:

Max. Iteration: 5000
Swarm Size: 30
W_{min} - W_{max}: 0.4 – 0.9
C_1 and C_2: 2 and 2
Tournament Size: 2

Table 1. Details of dataset used

Area	Dataset Name	# Attributes	Attribute Types	# Classes	Total # Samples	Class wise # Samples
Life Science	Breast Cancer Wisconsin (Original)	10	All are Integer	2	699	458,241
	Pima Indians Diabetes	8	Integer (1,2,3,4,5,8) Real (6,7)	2	768	500,268
	Iris	4	All are Real	3	150	50,50,50
	Ecoli	8	All are Real	8	336	143,77,52, 35,20,5,2,2
	Yeast	8	All are Real	10	1484	463, 429, 244, 163, 51, 44, 37, 30, 20, 5
Physical Science	Magic Gamma Telescope	11	All are Real	2	19020	12332, 6688
	Wine	13	Integer(1,6,13) Real(2,3,4,5,7,8,9, 10,11,12)	3	178	59, 71, 48
	Glass Identification	9	All are Real	6	214	70,17,76, 13,9,29
Financial	Credit Approval	15	Categorical (1,4,5,6,7,9,10,12,13) Integer (3,11,14,15) Real (2,8)	2	690	307,383
Computer	Page Blocks Classification	10	Integer (1,2,3,8,9,10) Real (4,5,6,7)	5	5473	4913, 329, 28, 88, 115

A fuzzy system with three rules and 98.13% classification accuracy is evolved by the proposed IPSO. The three rules evolved by the proposed IPSO are given below:

1. If alcohol and color intensity are very low and non-flavanoid phenols are very high then it is wine 1.
2. If magnesium and proline are medium and malic acid and color intensity are high then it is wine 2.
3. If magnesium and ash is low and alcohol and color intensity are high then it is wine 3.

During the course of run, the ranges of each membership function points are evolved and tuned simultaneously. After 1000 iterations, the points of membership function tuned by the proposed IPSO have begun to distribute uniformly over the range of the variables with halfway overlap between them. The optimal membership function obtained using the proposed IPSO for the input variable ‘Magnesium’ of wine dataset is shown in figure 4. The convergence behavior of the proposed IPSO is shown in figure 5.

From the figure 5, it is found that the fitness value increases sharply for the first eight hundred generations. After that the proposed IPSO has drastic improvement in its fitness value up to thousand generations and above that it reaches the optimal value. Table 2 compares the learning ability of the proposed IPSO with RGA and PSO for wine data.

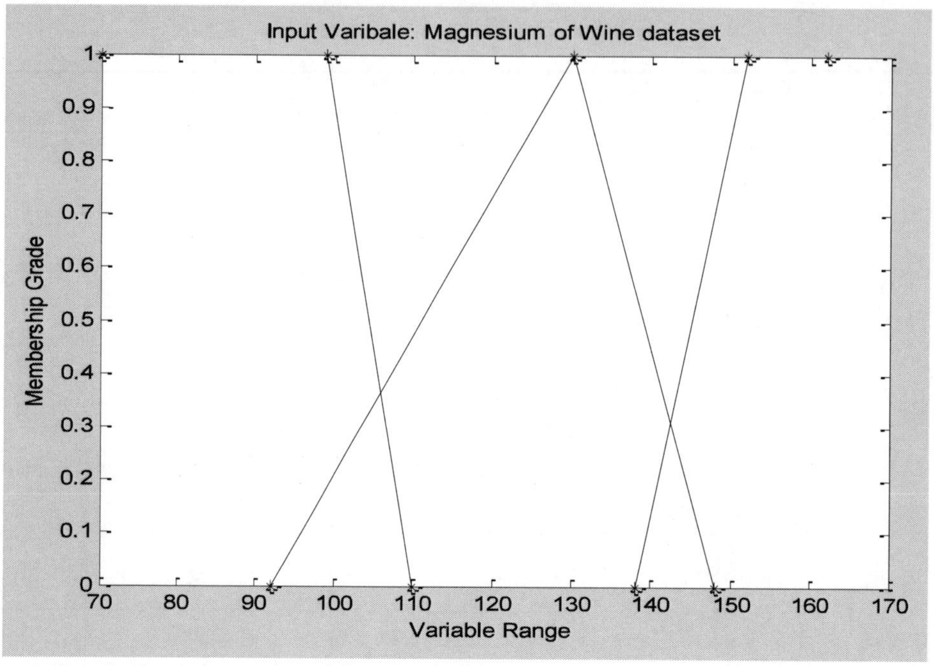

Figure 4. Optimal Membership.

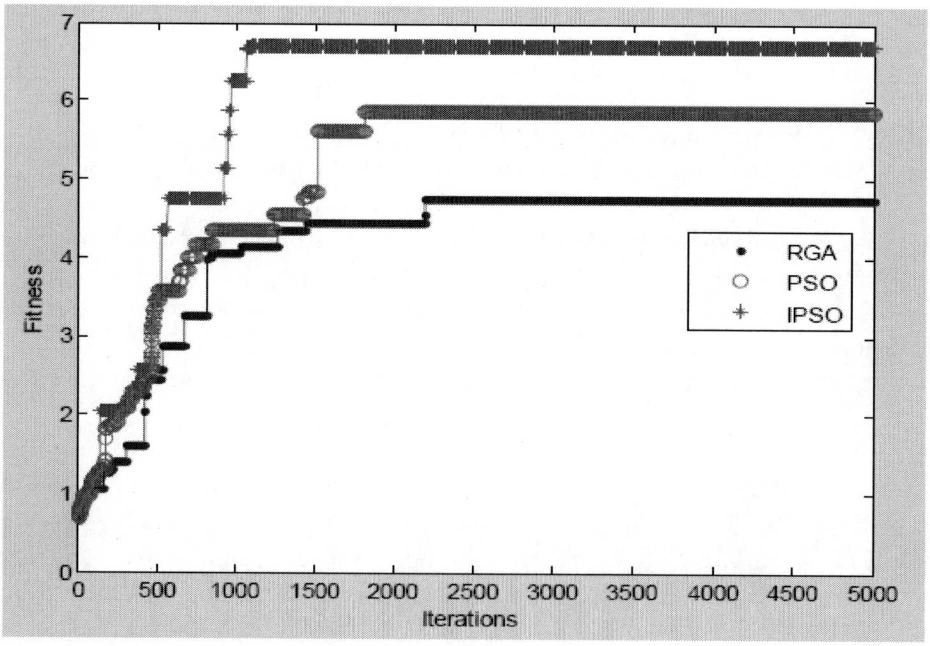

Figure 5. Convergence of proposed IPSO during learning.

From table 2, it is inferred that the proposed IPSO classifies 178 data correctly in 1042 generations showing very good learning ability when compared with RGA and PSO. It is clearly seen from table 2, that the proposed IPSO works well in fine tuning the solution variable thereby results in faster convergence than RGA and PSO. Table 3 shows the performance comparison of proposed IPSO with RGA and PSO during learning for the other data sets.

Table 2. Learning ability of proposed IPSO for wine data

Performance Parameter	RGA	PSO	IPSO
No. of Iterations	2485	1758	1042
Correctly Classified data (Average)	171.2	174.8	177.9
No. of Rules(Average)	3.9	3.6	2.9
Accuracy in percentage	94.62%	96.57%	98.13%
CPU Time in Seconds	1238.92	886.79	667.48

Table 3. Learning ability of proposed IPSO for the other data sets

DATA SETS	Simple RGA			Standard PSO			Proposed IPSO		
	*nCc	*CA	SN_R	*nCc	*CA	SN_R	nCc	CA	SN_R
Breast Cancer	665.2	95.13	7.2	672.3	96.28	6.4	697.1	99.73	3.1
Pima Indians Diabetes	736.3	95.83	8.4	742.1	96.15	6.7	761.7	99.38	3.3
Iris	149.7	100	3.4	149.9	100	3.1	150.0	100	2.9
Ecoli	311.4	92.55	13.1	318.4	94.49	11.9	331.2	98.19	8.5
Yeast	1432.1	96.49	16.5	1461.3	98.56	14.8	1482.1	99.69	10.7
Magic Gamma Telescope	18806.3	98.87	8.6	18825.4	98.72	6.3	19012.8	99.54	4.6
Glass Identification	197.4	92.05	10.8	202.3	94.91	8.5	211.9	98.92	6.2
Credit Approval	671.5	97.24	8.5	678.6	98.65	5.2	685.3	99.76	3.6
Page Blocks Classification	5231.2	95.57	9.6	5248.5	95.82	7.6	5466.2	99.11	5.8

*nCc=number of Correctly classified data (Average) *CA = Classification Accuracy in percentage.

From this table 3, it is understood that the incorporation of the elitism and tournament selection avoids the problem of premature convergence and local minima when compared with RGA and PSO for all the data sets.

Next, the generalization ability of the proposed system is examined by using standard Leave-One-Out Cross-Validation (LOOCV) [17] method. The LOOCV procedure works as by dividing all samples into *K* subsets randomly, where *K* is the total number of samples.

Then K-1 subsets are used to train the model and the remaining K^{th} sample is used for testing. This procedure is repeated for *K* times such that each sample is given a chance for testing the performance. The LOOCV accuracy is calculated using

$$LOOCV\ \ Accuracy = \frac{nCc}{K} \tag{7}$$

For wine data set, all the 178 samples were divided into 177 samples for finding the optimal membership function and the rule set and a single sample for testing the performance of the designed fuzzy system. This procedure is iterated 178 times so that each sample is used for evaluating the performance of obtained membership function and rule set. Table 4 shows percent of the samples that are correctly classified, incorrectly classified and unclassified by each method in the LOOCV evaluation for all the data sets.

Table 4. Comparison of LOOCV accuracy

Data Sets	Correct in %			Incorrect in %			Unclassified in %		
	RGA	PSO	IPSO	RGA	PSO	IPSO	RGA	PSO	IPSO
Wine	94.62	96.57	98.13	2.56	1.21	0	0.82	0.29	0
Breast Cancer	95.13	96.28	97.73	3.78	2.53	0.03	1.09	1.19	0.24
Pima Indians Diabetes	95.83	96.15	96.38	2.31	1.78	0.01	1.86	2.07	0.61
Iris	99.8	99.8	99.8	0.1	0.1	0.1	0.1	0.1	0.1
Ecoli	92.55	94.49	96.19	4.67	3.27	0.11	2.78	2.24	1.7
Yeast	96.49	98.56	97.69	1.24	0.78	0	2.27	0.66	0.31
Magic Gamma Telescope	94.87	96.72	96.54	0.58	0.23	0	0.55	1.05	0.46
Glass Identification	92.05	94.91	95.92	6.72	3.52	0.34	1.23	1.57	0.74
Credit Approval	95.24	96.65	96.76	1.62	0.16	0.05	1.14	1.19	0.19
Page Blocks Classification	95.57	95.82	96.11	2.98	2.04	0	1.45	2.14	0.99

Table 5. Statistical comparison of proposed IPSO

DATA SETS	RGA	PSO	IPSO
Wine	0.571(3)	0.595(2)	0.743(1)
Breast Cancer	0.638(3)	0.692(2)	0.897(1)
Pima Indians Diabetes	0.412(3)	0.454(2)	0.753(1)
Iris	0.399(1)	0.399(1)	0.399(1)
Ecoli	0.568(3)	0.623(2)	0.934(1)
Yeast	0.634(3)	0.712(2)	0.862(1)
Magic Gamma Telescope	0.426(3)	0.564(2)	0.598(1)
Glass Identification	0.749(3)	0.887(2)	0.912(1)
Credit Approval	0.369(3)	0.498(2)	0.654(1)
Page Blocks Classification	0.325(3)	0.348(2)	0.434(1)

Average Rank 2.8 1.9 1

From this table it is found that proposed IPSO has almost zero incorrectly classified samples for all the data sets which clearly depict the superiority of the proposed approach than the RGA and PSO in designing a fuzzy rule based system for data classification. Statistical evaluation of experimental results is considered an essential part of validation of new machine learning methods.

Table 6. Performance comparison of proposed IPSO

Data set	Approaches	Classification accuracy
Wine	Castillo et al., 2001 [19]	96.76%
	Setnes et al., 2000 [20]	98.31%
	Proposed IPSO	100%
Breast Cancer	Pena-Reyes et al., 1999 [21]	97.51%
	Ganeshkumar et al., 2008 [22]	98.2%
	Proposed IPSO	97.73%
Iris	Ishibuchi et al., 2004 [23]	96.4%
	Devaraj et al., 2010 [24]	99.7%
	Proposed IPSO	99.8%
Ecoli	Horton et al., 1996 [25]	81.1%
	Proposed IPSO	96.19%
Yeast	Horton et al., 1996 [25]	54.9%
	Proposed IPSO	97.69%
Glass Identification	Quinlain, 1987 [26]	63.2%
	Ishibuchi et al., 1999 [27]	64.4%
	Proposed IPSO	95.92%
Credit Approval	Quinlain, 1987 [26]	85.8%
	Ishibuchi et al., 1999 [27]	86.7%
	Proposed IPSO	96.76%

For comparing the performance of the proposed algorithm with other algorithms on multiple data sets, Receiver Operating Characteristic (ROC) analysis [18] is performed. ROC graphs are useful technique for organizing classifiers and visualizing their performance. Table 5 gives the value of Area under ROC Curve (AUC) for proposed RGA and Simple RGA.

From this statistical analysis, it is found that proposed IPSO has the ability to provide a richer measure of classification performance with compact rule set than RGA and PSO in designing the fuzzy rule based system for all the ten data sets.Table 6 gives the comparison of results obtained by the proposed approach with the already published results for some popular data sets. From this table, it is clear that the incorporation of genetic operator like elitism and tournament selection into PSO improved the performance of fuzzy classifier and outperforms the results of existing classifier in the literature.

CONCLUSION

Data Classification using the knowledge extracted from the data is a complex data mining problem. This paper has proposed a hybrid form of particle swarm optimization for classification rule discovery and tuning of membership function for fuzzy logic based classifier model. Concept of elitism and tournament selection from genetic algorithm is incorporated in addition to standard velocity and position updating operations of PSO. The performance of the proposed IPSO is tested using ten publicly available bench mark data sets. From the simulation result it is observed that incorporation of elitism operation rapidly increase the performance of PSO and prevents losing the best found solution. Further the addition of the selection operation provides more exploitative search mechanism and it finds better optima more consistently than RGA and PSO. For all the data sets, the proposed approach generated a compact fuzzy system with high classification accuracy than the RGA, PSO and the other approaches reported in the literature.

REFERENCES

[1] J.Han, M.Kamber, *Data Mining Concepts and Techniques*, Morgan Kaufmann publisher, (2006).

[2] R.J.Durrant, A. Kaban, Compressed fisher linear discriminant analysis: classification of randomly projected data, *proceedings of the 16th ACM SIGKDD international conference on Knowledge discovery and data mining*, (2010), 1119-1128.

[3] B. Krishnapuram, A.J.Hartemink, L.Carin, M.A.Figueiredo, A bayesian approach to joint feature selection and classifier design, *IEEE Trans Pattern Anal Mach Intell* 26(9) (2004),1105-1111.

[4] C.C.Holmes, N.M. Adams, A probabilistic nearest neighbor method for statistical pattern recognition, *Journal of the Royal Statistical Society, 64(2) (2002), 295-306.*

[5] G.P.Zhang, Neural Networks for data classification: A Survey, IEEE Trans on System Man and Cybernetics-Part C: *Application and Reviews,* 30(4) (2000), 451-462.

[6] S.Osowski, L.T.Hoai, T.Arkiewicz, Support vector machine-based expert system for reliable heartbeat recognition. *IEEE Trans Biomed Eng*, 51 (2004), 582-89.

[7] D.P.Bhukya, S.Ramachandram, Decision Tree Induction: An Approach for Data Classification Using AVL-Tree, *International Journal of Computer and Electrical Engineering*, 2(4) (2010), 1793-8163.

[8] Z.Zhou, Y.Jiang and S.Chen, Extracting symbolic rules from trained neural network ensembles, *AI Commun*. 16(1) (2003), 3-15.

[9] T.Ross, Fuzzy Logic with Engineering Application, *Tata McGraw Hill inc*, (1995).

[10] E. G. Mansoori, M. J. Zolghadri, S. D. Katebi, Using distribution of data to enhance performance of fuzzy classification systems, *Iranian Journal of Fuzzy Systems*, 4(1) (2007), 21-36.

[11] S.Abe, M.S. Lan, "A Method for fuzzy rule extraction directly from numerical data and its application to pattern classification," *IEEE Trans. Fuzzy Syst.,* vol. 3, pp. 18-28, Feb, 1995.

[12] D.Devaraj, P.GaneshKumar, Design of Optimal Fuzzy Classifier Using Enhanced Genetic Algorithm" *Journal of Information, Intelligence and Knowledge,* 1(1), 103-116, 2009.

[13] J.Kennedy, R.C.Eberhart, Y.Shi, Swarm Intelligence. *Morgan Kaufman Publishers*, San Francisco, (2001).

[14] Y.Valle, G.K.Venayagamoorthy, S. Mohagheghi, J.Hernandez, R.G.Harley, Particle Swarm Optimization: Basic Concepts, Variants and Applications in Power Systems, *IEEE Transaction on Evolutionary Computation*, 12(2) (2008.), 171-195.

[15] S.N.Sivanandam, S.N.Deepa, Introduction to Genetic Algorithms, *Springer*, (2007).

[16] D.J.Newman, S.Hettich, C.L.Blake, UCI Repository of machine learning databases, University of California, Irvine, (1998)

[17] Kohavi, Ron. A study of cross-validation and bootstrap for accuracy estimation and model selection, *Proceedings of the Fourteenth International Joint Conference on Artificial Intelligence.* 2(12) (1995), 1137–1143.

[18] T. Fawcett, An Introduction to ROC analysis, *Pat. Recog. Letters*, 27 (2006) 861-874.

[19] L.Castillo, A.Gonzalez, P.Perez, Including a simplicity criterion in the selection of the best rule in a genetic fuzzy learning algorithm, Fuzzy Sets and Systems. 120(2) (2001), 309-321.

[20] M. Setnes, H. Roubos, GA-Fuzzy Modeling and Classification: Complexity and Performance, *IEEE Transaction on Fuzzy Systems.* 8(5) (2000), 509-522.

[21] C.A.Pena-Reyes, M. Sipper, A fuzzy-genetic approach to breast cancer diagnosis. *Artificial Intelligence in Medicine*, 17 (1999), 131-155.

[22] P.GaneshKumar, D.Devaraj, Improved Genetic Fuzzy System for Breast Cancer Diagnosis, *Int. Journal of Systemics, Cybernetics and Informatics,* (2008), 24-28.

[23] H.Ishibuchi, T.Yamamoto, Fuzzy rule selection by multi-objective genetic local search algorithms and rule evaluation measures in data mining, *Fuzzy Sets System*. 141 (2004), 59–88.

[24] D.Devaraj, P.Ganeshkumar, Mixed Genetic Algorithm approach for Fuzzy Classifier Design, *Int. Journal of Comput. Intelligence and Applications*, 9(1) (2010), 49-67.

[25] P.Horton, K.Nakai, A Probablistic Classification System for Predicting the Cellular Localization Sites of Proteins, *Intelligent Sys. in Molecular Biology*, (1996),109-115.

[26] J.R.Quinlain, Simplifying decision trees. *Int. Jrnl. of Man-Mach. Stud.* 27 (1987), 221-234.

[27] H. Ishibuchi, T.Nakashima, T.Murata, Performance Evaluation of Fuzzy Classifier Systems for Multidimensional Pattern Classification Problems, *IEEE Transaction on System, Man and Cybernetics-Part B*, 29(5) (1999), 601-617.

In: Biology and Communication in the Information Age
Editor: Rafiqul Islam
ISBN: 978-1-62808-252-4

Chapter 7

Tissue Specificity Study of Bioregulators Isolated from Liver and Lungs of Mammals

D. I. Malcev[1*], V. P. Yamskova[2] and I. A. Yamskov[1]
[1]Nesmeyanov Institute of Organoelement Compounds,
Russian Academy of Sciences, Moscow, Russia
[2]Kol'tsov Institute of Developmental Biology,
Russian Academy of Sciences, Moscow, Russia

Abstract

The tissue specificity of a new bioregulators group, isolated from different mammalian tissues, is studied in this article. It was shown that protein-modulators which are serum albumin isoforms are responsible for tissue specificity of MHTBs membranotropic action. The tissue specificity of biological action of MHTBs-containing fractions isolated from liver and lung mammalian tissues are demonstrated on the model of roller organotypic culturing of newt liver tissue *in vitro*.

Keywords: Bioregulators, liver, lung, small intestine, protein-modulators, roller, organotypic culturing

Introduction

In the earlier studies, membranotropic homeostatic tissue-specific bioregulators (MHTBs) are discovered in different mammalian tissues including liver and lung tissues [1, 2]. This group bioregulators are localized in extracellular tissue space and display the membranotropic action – they change the elastic-plastic properties and permeability of cytoplasmic membrane [2, 3]. MHTBs stimulate recovering and reparing processes *in vivo*, provide a protective action on cells under culturing *in vitro* (increase their viability, contribute to the maintenance of tissue structure entirety and intercellular interactions) [4, 5]. It was

* E-mail: mal-dima@yandex.ru.

shown that activity of MHTBs is characterized by the tissue specificity, but the absence of species specificity [6]. This type of bioregulators activity is maintained both for their action at ultra low doses and at lower concentrations. Molecular mechanisms which on based MHTBs tissue-specific action are still unknown. It should be mentioned that MHTBs have a complex competition: they contain peptides which exhibit the membranotropic action and protein-modulators which influence on biological action of these peptides. It was shown that interaction between peptides and protein-modulators is carried out according to the ligand-receptor interaction in presence of Ca^{2+} ions. Protein-modulators have different effects on specific activity of peptides. For example, in eye tissues complex of peptide and protein-modulators is active at ultra low doses. Modulator of serum MHTB inhibites peptide action.

Aim

The aim of this work was to study the tissue specificity of bioregulators, isolated from liver and lung of mammalian tissues.

Background

A new group of bioregulators was discovered in different animal tissues. These bioregulators affect biological processes, such as cell migration, adhesion, and proliferation when used at ultra low doses corresponding to concentration 10^{-8}-10^{-15} mg/ml. Bioregulators of this group stimulate reparative processes in pathologically altered tissues, facilitating the repair of their disturbed structure. The physicochemical properties of this group bioregulators are similar, for example, they are presence in the form of nanoparticles in solutions of different concentrations.

Experimental Part

The study was performed on male and female Wistar rats weighing 200-250 g maintained in a vivarium of N. K. Kol'tsov Institute of Developmental Biology, Russian Academy of Sciences.

Protein Extraction

Using the experimental approach developed earlier for this group bioregulators we obtained liver and lung tissue extracts [7]. Before extraction we carried out perfusion of organs with the normal seline through the portal vein *in situ* to disposal most of the blood in liver and lung tissues. Then we cut tissues into small fragments and extracted with the normal seline for 2h at 4-8°C. Thus, we have two extracts – extract of liver and lung tissues, which then we studied using adhesion method.

Method for Determination of Bioregulators Membranotropic Activity

The membranotropic activity of extractions obtained from liver and lung tissues of mammals and also of these extractions fractions was determined by adhesion method using experimental models of multiple organotypic culturing of liver and lung tissues. The study was performed on male F1 CBA/C57Bl mice weighing 18-20 g [8].

Determination of Protein Concentration

Protein concentration in the tissue extracts was determined spectrophotometrically.

Method of Roller Organotypic Culturing of Newt Liver Tissue *In Vitro*

For obtaining organotypic liver tissue cultures, newts were decapitated under urethane anesthesia, the liver was surgically removed, 3x3x3 fragments were taken from the central lobe, transferred into a nutrient medium for amphibian, and cultured at 21-23 ^{0}C in dark-glass vials. For each experiment, liver fragments from one newt were used for different experimental series and control. Tissue fragments (n=5) were placed on a roller and rotation velocity was set at 35 rpm.

SDS-Electrophoresis of Protein in Polyacrylamide Gel

Electrophoresis in 15% polyacrylamide gel (PAAG) was performed according to Laemmli [9]. Coomassie G-250 was used for gels staining.

Fingerprinting Method

Protein hydrolysis was performed at 37°C for 18 h, then 7 μl of 0,5% TFA in 10% acetonitrile were added and intimately mixed. This solution was used for MALDI-TOF mass-spectrometric analysis. Identification of the protein was implemented using the Mascot program (www. matrixscience.com).

Mass-spectrometric research was performed by Orekhovich Institute of Biomedical Chemistry of RAMS (Moscow).

Results and Discussion

According to method, which had been earlier developed for MHTBs, two extractions were obtained (each volume of 200 ml). Protein concentration for liver extract was 1,5 mg/ml and for lung extraction – 0,7 mg/ml.

The research of membranotropic activity for both of extracts was performed on multiple organotypic cultures of liver and lung mouse tissues *in vitro*. It was shown that both of extracts influences on elastic-plastic properties cytoplasmic membrane of liver and lung cells in tissue-specific way (Table 1). Membranotropic activity for both of extracts characterized by polymodal dose dependence and was manifested at ultra low doses.

The research of specific activity using the model of roller organotypic culturing of newt liver tissue *in vitro* for fractions, obtained from liver and lung of mammals demonstrated that these fractions action is characterized by the presence of tissue specificity (Table 2). Only in experimental series with addition of bull and rat liver extract supernatant fractions at ultra low doses the area of pigmented cells was larger than in control.

Table 1. Membranotropic activity for fractions, obtained from extracts of liver and lung tissues of rats

Fractions	The value of membranotropic activity for fractions	
	liver tissue	lung tissue
Extract of liver	141,0 ± 7,0*	104,2 ± 5,1
Extract of lung	108,3 ± 5,2	162,0 ± 8,1*
Supernatant of liver extract	146,6 ± 7,2*	145,3 ± 4,2*
Supernatant of lung extract	135,4 ± 4,5*	152,6 ± 6,2*
Pellet of liver extract	105,5 ± 7,1	106,6 ± 4,7
Pellet of lung extract	107,4 ± 8,2	109,4 ± 8,1
Pellet of liver extract + Supernatant of liver extract	165,5 ± 7,8*	102,5 ± 3,5
Pellet of lung extract + Supernatant of lung extract	104,6 ± 2,1	170,6 ± 6,5*
Pellet of liver extract + Supernatant of lung extract	154,6 ± 5,2*	165,3 ± 1,5*
Pellet of lung extract + Supernatant of liver extract	157,6 ± 4,7*	153,4 ± 4,6*

* - $p<0,05$ compared to the control.

Data obtained in the study of biological activity corresponded to results of this bioregulators group investigation, which demonstrated that after tissue extracts salting out by addition of ammonium sulfate supernatant fractions don´t demonstrate tissue-specificity of membranotropic activity.

Nevertheless, in experiments on organ tissue cultures *in vitro* and *in vivo* in which the bioregulators effect on the state of the tissue was studied, we observed the tissue-specific character of their biological action [6].

Apparently, it is connected with the complex constitution of MHTBs in which certain components are responsible for different type of these bioregulators biological action. Indeed, if supernatant, obtained after salting out, was joined with the appropriate pellet then tissue specificity of membranotropic action is recovered. But after interaction between supernatants and protein pellets, obtained from another tissue extracts, we didn´t observe tissue specificity of membranotropic action recovery (Table 1).

These results indicate that in pellets there are substances which are responsible for tissue specificity of MHTBs biological action. For instance, for MHTBs, isolated from blood serum and pigment epithelium eye bull such substances are albumin isoforms with the values of molecular mass 67 854 ± 2 and 66 283 ± 2 Da.

Table 2. Investigation of specificity activity for fractions, obtained from liver and lung of mammals on model of roller organotypic culturing of newt liver

Fraction	The area of pigment cells (% of the total area of histological section)
Control	6,0 ± 2,6%
Supernatant of rat liver extract	12,1 ± 3,4% *
Supernatant of rat liver bull	10,3 ± 2,6% *
Supernatant of rat lung extract	7,4 ± 2,2%
Native liver	10,5 ± 2,7%

* - *p<0,05* compared to the control.

The albumin molecular mass in pellet, obtained from extract of rat lung was determined by this method and the value was 68 674 ± 2 Da. The result obtained indicates the presence of rat serum albumin isoform in this pellet.

Table 3. Fingerprinting of protein with molecular mass 66 000 Da, isolated after PAAG of pellets fraction

	Homologous protein	M.W.	Concordance	Description
1	gi|55628	68674	254	Unknown protein
2	gi|124028612	68686	254	Serum albumin precursor
3	gi|158138568	68714	236	Serum albumin precursor
4	gi|149033753	51521	132	Serum albumin isoform

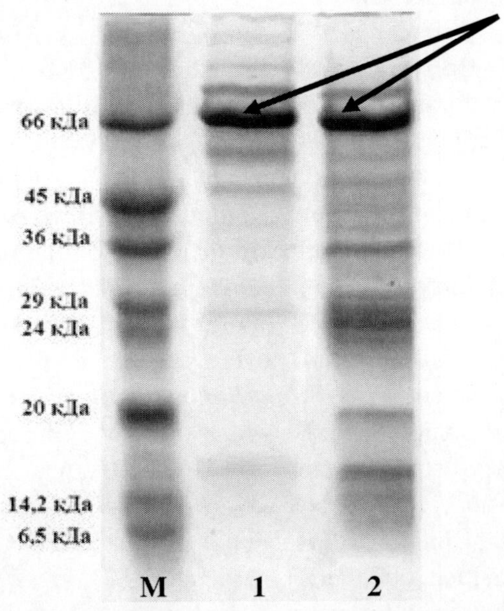

Figure 1. SDS-electrophoresis of pellet of rat lung extract (1) and pellet of rat liver extract (2). M – marker proteins. Proteins which were analyzed by fingerprinting shown by the arrows.

Study of pellets, obtained after liver and lung extracts salting out, by methods of PAAG and MALDI-TOF mass-spectrometry demonstrated that protein with molecular mass 65 915 ± 2 Da is the major component (Figure 1).

These results correspond to fingerprinting data which demonstrated the presence in pellets, obtained after liver and lung extracts salting out of protein with high homology to serum albumin of rat (Table 3).

So, in liver and lung tissue there are albumines, which belong to superfamily of serum albumines and like MHTBs, isolated from blood serum and bull eye pigment epithelium they can perform function of protein-modulators in these MHTBs tissues [2, 4, 5, 10].

Conclusion

Thus, we can conclude that liver extracts show tissue specificity of membranotropic activity. It was established that supernatants, obtained from studied tissue extracts didn´t show tissue-specific membranotropic action, but at the same time they show tissue-specific action on model of roller organotypic culturing of newt liver. It was also shown that protein-modulators belonging to superfamily of serum albumines are responsible for these bioregulators displaying of membranotropic action tissue specificity.

References

[1] Yamskov I.A., Yamskova V.P., Danilenko A.N., Klemenkova Z.S., Antipov B.G., Chernikov F.R., Gusynina, M.M., Rybakova E.Yu. Experimental evidence of the physic-chemical factors role in mechanism biological action of ultra low doses // *Russian Journal of Chemistry.* 1999. V. 43. № 5. p. 34–39.

[2] Yamskova V.P., Skripnikova V.S., Molyavka A.A., Iliina A.P., Krasnov M.S., Margasyuk D.V., Borisenko A.V., Berezin B.B., Kuznetsova E.S., Buryak A.K., Yamskov I.A. Structural – functional characteristic of a new bioregulator isolated from bovine pigmented epithelium tissue // *Biochemistry*. 2009. V. 74. № 9. p. 1195 - 1203.

[3] Yamskova V.P., Nechaeva N.V., Tumanova N.B., Yurovitsky Yu.G., Novikova T.E., Fateeva T.I., Gvazava I.G. Investigation of the effect of macromolecular adhesion factors isolated from serum and liver of mammals on the rate of protein synthesis and ATP content in hepatocytes // *Biology Bulletin*. 1994. № 2. p. 190-196.

[4] Margasyuk D.V., Krasnov M.S., Yamskov I.A., Yamskova V.P. The role of the regulatory protein, isolated from bovine cornea, in the activation of regeneration cellular sources of the cornea *in vitro* // *Biology Bulletin*. 2008. № 6. p. 736-745.

[5] Margasyuk D.V., Krasnov M.S., Blagodatskikh I.V., Grigoryan E.N., Yamskova V.P., Yamskov I.A. “Regulatory Protein from Bovine Cornea: Localization and Biological Activity”, pp. 49-56 // In the book «*Biochemical Physics Frontal Research*», Ed. by Varfolomeev S.D., Burlakova E.B., Popov A.A., Zaikov G.E. Hauppauge NY. Nova Science Publishers Inc. 2007. P.47-60 Marastoni S, Ligresti G, Lorenzon E, Colombatti A., Mongiat M. Extracellular matrix: a matter of life and death // *Connect.Tissue Res.* 2008. V.49. p. 203-206.

[6] Yamskova V.P., Krasnov M.S., Rybakova E.Yu., Vecherkin V.V., Borisenko A.V., Yamskov I.A."Analysis of regulatory proteins from bovine blood serum that display biological activity at ultra low doses: 2. Tissue localization and role in wound healing", P. 68-77 // In the book "*Biochemical Physics Frontal Research*", Ed. by Varfolomeev S.D., Burlakova E.B., Popov A.A., Zaikov G.E. Hauppauge NY. Nova Science Publishers Inc. 2007. p.160.

[7] Yamskova V.P., Tumanova N.B., Loginov A.S. A comparative study of the liver extracts of mice C57Bl and CBA action on hepatocytes adhesion // *Bulletin of Experimental Biology and Medicine*. 1990. № 3. p. 303-306.

[8] Yamskova V.P., Reznikova M.M. Low molecular weight polypeptide of warm-blooded serum: effects on cell adhesion and proliferation // *Journal of general Biology*, 1991, V. 52. № 2. p. 181-191.

[9] Laemmli U.K. Cleavage of structural proteins during the assembly of the head of bacteriophage T4 // Nature. 1970. V.227. p. 680-685.

[10] Yamskova V.P., Rybakova, E.Yu., Vinogradov A.A., Vecherkin V.V., Yamskov I.A. The study of adhesive glycoprotein protein-inactivator from the blood serum of mammals // *Applied Biochemistry and Microbiology*, 2004. V. 40. № 4. p. 407-413.

In: Biology and Communication in the Information Age
Editor: Rafiqul Islam
ISBN: 978-1-62808-252-4

Chapter 8

MARKETING RESEARCH, GOALS, PROBLEMS AND OBJECTIVES

***Abdalnaser A. Jaradat*[1] *and Kadri S. AL-Shakri**[2]**
[1]Ajloun National, Faculty of Business Administration and Information technology Department of Marketing, Ajloun Jordan
[2]Ajloun National, Faculty of Business Administration and Information technology Department of Management Information Systems, Ajloun Jordan

ABSTRACT

The present work concerns the general conceptions of marketing, marketing research, its goals, problems, categories, objectives. Plus in the given work are mentioned the ways of marketing program evaluation, the ways to cut the impact of reduced spending, different steps important to take in order to help the consumer feel comfortable. A special attention is pained to the basics of marketing, different points of views on them. The present work may also help to navigate away from the marketing and business mistakes that will cost money, time and future growth potential.

Keywords: Marketing, Marketing research, Marketing program evaluation

1. INTRODUCTION

Managers need information in order to introduce products and services that create value in the mind of the customer. But the perception of value is a subjective one, and what customer's value this year may be quite different from what they value next year. As such, the attributes that create value cannot simply be deduced from common knowledge. Rather, data must be collected and analyzed.

* Corresponding Author, E-mail: dr.kadri70@yahoo.com.

The goal of marketing research is to provide the facts and direction that managers need to make their more important marketing decisions ,John et. al (2008). Other literature sources, Ross Brennan et al, (2009), Sidney J. Levy, (2002).

To maximize the benefit of marketing research, those who use it need to understand the research process and its limitations.

Marketing Research vs. Market Research

These terms often are used interchangeably, but technically there is a difference. Market research deals specifically with the gathering of information about a market's size and trends. Marketing research covers a wider range of activities. While it may involve market research, marketing research is a more general systematic process that can be applied to a variety of marketing problems.

The Value of Information

Information can be useful, but what determines its real value to the organization? In general, the value of information is determined by; The ability and willingness to act on the information, The accuracy of the information, The level of indecisiveness that would exist without the information, The amount of variation in the possible results, The level of risk aversion, The reaction of competitors to any decision improved by the information, and the cost of the information in terms of time and money.

The Marketing Research Process

Once the need for marketing research has been established, most marketing research projects involve these steps; Define the problem, Determine research design, Identify data types and sources, Design data collection forms and questionnaires, Determine sample plan and size, Collect the data, Analyze and interpret the data, and Prepare the research report, Jenny Poolton, Hossam S. Ismail, Iain R. Reid, Ivan C. Arokiam (2008).

2. Problem Definition

The decision problem faced by management must be translated into a market research problem in the form of questions that define the information that is required to make the decision and how this information can be obtained. Thus, the decision problem is translated into a research problem. For example, a decision problem may be whether to launch a new product. The corresponding research problem might be to assess whether the market would accept the new product. The objective of the research should be defined clearly. To ensure that the true decision problem is addressed, it is useful for the researcher to outline possible scenarios of the research results and then for the decision maker to formulate plans of action under each scenario. The use of such scenarios can ensure that the purpose of the research is agreed upon before it commences.

3. Research Design

Marketing research can classify into one of three categories; Exploratory research, Descriptive research, and Causal research. These classifications are made according to the objective of the research. In some cases the research will fall into one of these categories, but in other cases different phases of the same research project will fall into different categories.

Exploratory research has the goal of formulating problems more precisely, clarifying concepts, gathering explanations, gaining insight, eliminating impractical ideas, and forming hypotheses. Exploratory research can be performed using a literature search, surveying certain people about their experiences, focus groups, and case studies. When surveying people, exploratory research studies would not try to acquire a representative sample, but rather, seek to interview those who are knowledgeable and who might be able to provide insight concerning the relationship among variables. Case studies can include contrasting situations or benchmarking against an organization known for its excellence. Exploratory research may develop hypotheses, but it does not seek to test them. Exploratory research is characterized by its flexibility, McCauley, Joseph l. (2004).

Descriptive research is more rigid than exploratory research and seeks to describe users of a product, determine the proportion of the population that uses a product, or predict future demand for a product. As opposed to exploratory research, descriptive research should define questions, people surveyed, and the method of analysis prior to beginning data collection. In other words, the who, what, where, when, why, and how aspects of the research should be defined. Such preparation allows one the opportunity to make any required changes before the costly process of data collection has begun, Ronald L. Rardin (1997).

There are two basic types of descriptive research: longitudinal studies and cross-sectional studies. Longitudinal studies are time series analyses that make repeated measurements of the same individuals, thus allowing one to monitor behavior such as brand-switching. However, longitudinal studies are not necessarily representative since many people may refuse to participate because of the commitment required. Cross-sectional studies sample the population to make measurements at a specific point in time. A special type of cross-sectional analysis is a cohort analysis, which tracks an aggregate of individuals who experience the same event within the same time interval over time. Cohort analyses are useful for long-term forecasting of product demand.

Causal research seeks to find cause and affect relationships between variables. It accomplishes this goal through laboratory and field experiments.

4. Data Types and Sources

Secondary Data

Before going through the time and expense of collecting primary data, one should check for secondary data that previously may have been collected for other purposes but that can be used in the immediate study. Secondary data may be internal to the firm, such as sales invoices and warranty cards, or may be external to the firm such as published data or commercially available data. The government census is a valuable source of secondary data.

Secondary data has the advantage of saving time and reducing data gathering costs. The disadvantages are that the data may not fit the problem perfectly and that the accuracy may be more difficult to verify for secondary data than for primary data.

Some secondary data is republished by organizations other than the original source. Because errors can occur and important explanations may be missing in republished data, one should obtain secondary data directly from its source. One also should consider who the source is and whether the results may be biased.

There are several criteria that one should use to evaluate secondary data, these are; Whether the data is useful in the research study, How current the data is and whether it applies to time period of interest, Errors and accuracy - whether the data is dependable and can be verified, and many others.

Primary Data

Often, secondary data must be supplemented by primary data originated specifically for the study at hand. Some common types of primary data are; Demographic and socioeconomic characteristics, Psychological and lifestyle characteristics, Attitudes and opinions, Awareness and knowledge - for example, brand awareness, Intentions - for example, purchase intentions. While useful, intentions are not a reliable indication of actual future behavior, and Motivation - a person's motives are more stable than his/her behavior, so motive is a better predictor of future behavior than is past behavior, McCauley, Joseph l. (2004).

Primary data can be obtained by communication or by observation. Communication involves questioning respondents either verbally or in writing. This method is versatile, since one need only to ask for the information; however, the response may not be accurate. Communication usually is quicker and cheaper than observation. Observation involves the recording of actions and is performed by either a person or some mechanical or electronic device. Observation is less versatile than communication since some attributes of a person may not be readily observable, such as attitudes, awareness, knowledge, intentions, and motivation. Observation also might take longer since observers may have to wait for appropriate events to occur, though observation using scanner data might be quicker and more cost effective. Observation typically is more accurate than communication.

Personal interviews have an interviewer bias that mail-in questionnaires do not have. For example, in a personal interview the respondent's perception of the interviewer may affect the responses.

5. Questionnaire Design and Measurement Scales

The questionnaire is an important tool for gathering primary data. Poorly constructed questions can result in large errors and invalidate the research data, so significant effort should be put into the questionnaire design. The questionnaire should be tested thoroughly prior to conducting the survey.

Attributes can be measured on nominal, ordinal, interval, and ratio scales: Nominal numbers are simply identifiers, with the only permissible mathematical use being for counting. Example: social security numbers. Ordinal scales are used for ranking. The interval between the numbers conveys no meaning. Median and mode calculations can be performed on ordinal numbers. Example: class ranking. Interval scales maintain an equal interval between numbers. These scales can be used for ranking and for measuring the interval between two numbers. Since the zero point is arbitrary, ratios cannot be taken between numbers on an interval scale; however, mean, median, and mode are all valid. Example: temperature scale. Ratio scales are referenced to absolute zero values, so ratios between numbers on the scale are meaningful. In addition to mean, median, and mode, geometric averages also are valid. Example: weight

6. Validity and Reliability

The validity of a test is the extent to which differences in scores reflect differences in the measured characteristic. Predictive validity is a measure of the usefulness of a measuring instrument as a predictor. Proof of predictive validity is determined by the correlation between results and actual behavior. Construct validity is the extent to which a measuring instrument measures what it intends to measure, Nargundkar, (2008).

Reliability is the extent to which a measurement is repeatable with the same results. A measurement may be reliable and not valid. However, if a measurement is valid, then it also is reliable and if it is not reliable, then it cannot be valid. One way to show reliability is to show stability by repeating the test with the same results.

7. Attitude Measurement

Many of the questions in a marketing research survey are designed to measure attitudes. Attitudes are a person's general evaluation of something. Customer attitude is an important factor for the following reasons; Attitude helps to explain how ready one is to do something, Attitudes do not change much over time, Attitudes produce consistency in behavior, Attitudes can be related to preferences, Mehdi Mohammadi et al (2010).

Attitudes can be measured using the following procedures; Self-reporting - subjects are asked directly about their attitudes, Self-reporting is the most common technique used to measure attitude.

Observation of behavior - assuming that one's behavior is a result of one's attitudes, attitudes can be inferred by observing behavior. For example, one's attitude about an issue can be inferred by whether he/she signs a petition related to it.

Performance of objective tasks - assumes that one's performance depends on attitude. For example, the subject can be asked to memorize the arguments of both sides of an issue. He/she is more likely to do a better job on the arguments that favor his/her stance.

Physiological reactions - subject's response to a stimuli is measured using electronic or mechanical means. While the intensity can be measured, it is difficult to know if the attitude is positive or negative.

Multiple measures - a mixture of techniques can be used to validate the findings, especially worthwhile when self-reporting is used.

There are several types of attitude rating scales. Equal-appearing interval scaling - a set of statements are assembled. These statements are selected according to their position on an interval scale of favorableness. Statements are chosen that has a small degree of dispersion. Respondents then are asked to indicate with which statements they agree. Likert method of summated ratings - a statement is made and the respondents indicate their degree of agreement or disagreement on a five point scale (Strongly Disagree, Disagree, Neither Agree Nor Disagree, Agree, Strongly Agree). Semantic differential scale - a scale is constructed using phrases describing attributes of the product to anchor each end. For example, the left end may state, "Hours are inconvenient" and the right end may state, "Hours are convenient". The respondent then marks one of the seven blanks between the statements to indicate his/her opinion about the attribute. Stapel Scale - similar to the semantic differential scale except that 1) points on the scale are identified by numbers, 2) only one statement is used and if the respondent disagrees a negative number should marked, and 3) there are 10 positions instead of seven. This scale does not require that bipolar adjectives be developed and it can be administered by telephone.

8. Sampling Plan and Data Collection

The sampling frame is the pool from which the interviewees are chosen. The telephone books often are used as a sampling frame, but have some shortcomings. Telephone books exclude those households that do not have telephones and those households with unlisted numbers. Since a certain percentage of the numbers listed in a phone book are out of service, there are many people who have just moved who are not sampled. Such sampling biases can be overcome by using random digit dialing. Mall intercepts represent another sampling frame, though there are many people who do not shop at malls and those who shop more often will be over-represented unless their answers are weighted in inverse proportion to their frequency of mall shopping, Ekrem Erdem, (2010).

In designing the research study, one should consider the potential errors. Two sources of errors are random sampling error and non-sampling error. Sampling errors are those due to the fact that there is a non-zero confidence interval of the results because of the sample size being less than the population being studied. Non-sampling errors are those caused by faulty coding, untruthful responses, respondent fatigue, etc.

There is a tradeoff between sample size and cost. The larger the sample size, the smaller the sampling error but the higher the cost. After a certain point the smaller sampling error cannot be justified by the additional cost.

While a larger sample size may reduce sampling error, it actually may increase the total error. There are two reasons for this effect. First, a larger sample size may reduce the ability to follow up on non-responses. Second, even if there are a sufficient number of interviewers for follow-ups, a larger number of interviewers may result in a less uniform interview process.

In addition to the intrinsic sampling error, the actual data collection process will introduce additional errors. These errors are called non-sampling errors. Some non-sampling errors may be intentional on the part of the interviewer, who may introduce a bias by leading the respondent to provide a certain response. The interviewer also may introduce unintentional errors, for example, due to not having a clear understanding of the interview process or due to fatigue.

Respondents also may introduce errors. A respondent may introduce intentional errors by lying or simply by not responding to a question. A respondent may introduce unintentional errors by not understanding the question, guessing, not paying close attention, and being fatigued or distracted. Such non-sampling errors can be reduced through quality control techniques.

9. Data Analysis

Before analysis can be performed, raw data must be transformed into the right format. First, it must be edited so that errors can be corrected or omitted. The data must then be coded; this procedure converts the edited raw data into numbers or symbols. A codebook is created to document how the data was coded. Finally, the data is tabulated to count the number of samples falling into various categories. Simple tabulations count the occurrences of each variable independently of the other variables. Cross tabulations, also known as contingency tables or cross tabs, treats two or more variables simultaneously. However, since the variables are in a two-dimensional table, cross tabbing more than two variables is difficult to visualize since more than two dimensions would be required. Cross tabulation can be performed for nominal and ordinal variables.

Cross tabulation is the most commonly utilized data analysis method in marketing research. Many studies take the analysis no further than cross tabulation. This technique divides the sample into sub-groups to show how the dependent variable varies from one subgroup to another. A third variable can be introduced to uncover a relationship that initially was not evident.

Conjoint Analysis

The conjoint analysis is a powerful technique for determining consumer preferences for product attributes.

Hypothesis Testing

A basic fact about testing hypotheses is that a hypothesis may be rejected but that the hypothesis never can be unconditionally accepted until all possible evidence is evaluated. In the case of sampled data, the information set cannot be complete. So if a test using such data does not reject a hypothesis, the conclusion is not necessarily that the hypothesis should be accepted.

The null hypothesis in an experiment is the hypothesis that the independent variable has no effect on the dependent variable. The null hypothesis is expressed as H0. This hypothesis is assumed to be true unless proven otherwise. The alternative to the null hypothesis is the hypothesis that the independent variable does have an effect on the dependent variable. This hypothesis is known as the alternative, research, or experimental hypothesis and is expressed as H1. This alternative hypothesis states that the relationship observed between the variables cannot be explained by chance alone.

There are two types of errors in evaluating hypotheses: Type I error: occurs when one rejects the null hypothesis and accepts the alternative, when in fact the null hypothesis is true. Type II error: occurs when one accepts the null hypothesis when in fact the null hypothesis is false.

Tests of Statistical Significance

The chi-square ($\chi 2$) goodness-of-fit test is used to determine whether a set of proportions have specified numerical values. It often is used to analyze bivariate cross-tabulated data. Some examples of situations that are well-suited for this test are; A manufacturer of packaged products test markets a new product and wants to know if sales of the new product will be in the same relative proportion of package sizes as sales of existing products, A company's sales revenue comes from Product A (50%), Product B (30%), and Product C (20%). The firm wants to know whether recent fluctuations in these proportions are random or whether they represent a real shift in sales.

The chi-square test is performed by defining k categories and observing the number of cases falling into each category. Knowing the expected number of cases falling in each category, one can define chi-squared as:

$$X_0^2 = \sum_{i=1}^{k} \frac{(O_i - E_i)^2}{E_i}$$

where; O_i: the number of observed cases in category i,

E_i : the number of expected cases in category i,

k : the number of categories,

Before calculating the chi-square value, one needs to determine the expected frequency for each cell. This is done by dividing the number of samples by the number of cells in the table.

To use the output of the chi-square function, one uses a chi-square table. To do so, one needs to know the number of degrees of freedom (df). For chi-square applied to cross-tabulated data, the number of degrees of freedom is equal to (Number of columns - 1) (Number of rows - 1).

This is equal to the number of categories minus one. The conventional critical level of 0.05 normally is used. If the calculated output value from the function is greater than the chi-square look-up table value, the null hypothesis is rejected.

ANOVA

Another test of significance is the Analysis of Variance (ANOVA) test. The primary purpose of ANOVA is to test for differences between multiple means. Whereas the t-test can be used to compare two means, ANOVA is needed to compare three or more means. If multiple t-tests were applied, the probability of a TYPE I error (rejecting a true null hypothesis) increases as the number of comparisons increases.

One-way ANOVA examines whether multiple means differ. The test is called an F-test. ANOVA calculates the ratio of the variation between groups to the variation within groups (the F ratio). While ANOVA was designed for comparing several means, it also can be used to compare two means. Two-way ANOVA allows for a second independent variable and addresses interaction. To run a one-way ANOVA, use the following steps, Identify the independent and dependent variables. Describe the variation by breaking it into three parts - the total variation, the portion that is within groups, and the portion that is between groups (or among groups for more than two groups).

The total variation (SStotal) is the sum of the squares of the differences between each value and the grand mean of all the values in all the groups. The in-group variation (SSwithin) is the sum of the squares of the differences in each element's value and the group mean. The variation between group means (SSbetween) is the total variation minus the in-group variation (SStotal - SSwithin), Measure the difference between each group's mean and the grand mean, Perform a significance test on the differences.

Results Interpretation

This F-test assumes that the group variances are approximately equal and that the observations are independent. It also assumes normally distributed data; however, since this is a test on means the Central Limit Theorem holds as long as the sample size is not too small.

ANOVA is efficient for analyzing data using relatively few observations and can be used with categorical variables. Note that regression can perform a similar analysis to that of ANOVA.

Cluster Analysis

Market segmentation usually is based not on one factor but on multiple factors. Initially, each variable represents its own cluster. The challenge is to find a way to combine variables so that relatively homogenous clusters can be formed. Such clusters should be internally homogenous and externally heterogeneous. Cluster analysis is one way to accomplish this goal. Rather than being a statistical test, it is more of a collection of algorithms for grouping objects, or in the case of marketing research, grouping people. Cluster analysis is useful in the exploratory phase of research when there are no a-priori hypotheses.

CONCLUSION

The format of the marketing research report varies with the needs of the organization. The report often contains the following sections; Authorization letter for the research, Table of Contents, List of illustrations, Executive summary, Research objectives, Methodology, Results, Limitations, Conclusions and recommendations, Appendices containing copies of the questionnaires, etc,. Marketing research by itself does not arrive at marketing decisions, nor does it guarantee that the organization will be successful in marketing its products. However, when conducted in a systematic, analytical, and objective manner, marketing research can reduce the uncertainty in the decision-making process and increase the probability and magnitude of success.

To make a conclusion we may say that the marketing research covers a wide range of activities. The goal of marketing research is to provide the facts and direction that managers need to make their more important marketing decisions. Many of the questions in a marketing research survey are designed to measure attitudes which are a person's general evaluation of something. Marketing research by itself does not arrive at marketing decisions, nor does it guarantee that the organization will be successful in marketing its products.

As the statistics shows, consumers are buying less, looking for deals, or switching to different brands, product categories, or stores. So it is important to follow seven key steps to minimize the impact of reduced spending: to stay focused, to enlist trusted partners, to value experience and judgment, to seize opportunities overseas, to go online with a dash of skepticism, not to cut across the board, to keep an eye on the new consumer

REFERENCES

Ekrem Erdem, Onur Gozbasi, M. Fatih Ilgun, Saban Nazlioglu (2010), Stock market and economic growth nexus in emerging markets: cointegration and causality analysis,*International Journal of Business Forecasting and Marketing Int elligence* - Vol. 1, No.3/4 pp. 262 - 274.

Jenny Poolton, Hossam S. Ismail, Iain R. Reid, Ivan C. Arokiam (2008), Implementation of an agile marketing approach for the manufacturing-based SME , *International Journal of Entrepreneurship and Small Business* (IJESB), Vol. 6, No. 2.

John A. Saunders, Veronica Wong, Gary Armstrong, and Philip Kotler, (2008). *Principles of marketing,* 11th Edition, Prentice Hall.

McCauley, Joseph l. (2004), Thermodynamic analogies in economics and finance: instability of markets. Published in: Physica A 329 (2003): pp. 199-212.

Mehdi Mohammadi, Ali Azadeh, Morteza Saberi, Amir Azaron, (2010), Genetic algorithm-based clustering ensemble: determination number of clusters, *International Journal of Business Forecasting and Marketing Intelligence,* Vol. 1, No.3/4 pp. 201 - 216.

Nargundkar, (2008), Marketing Research: Text And Cases, Tata McGraw-Hill Education, ISBN: 9780070220874.

Ronald L. Rardin (1997), Optimization in Operations Research, Prentice Hall, ISBN-10: 0023984155, ISBN-13: 978-0023984150.

Ross Brennan, Philip Kotler, Michael Harker, and Gary Armstrong, (2009). Marketing: An Introduction, 6th Edition, Prentice Hall.

Sidney J. Levy, (2002) "Revisiting the marketing domain", *European Journal of Marketing*, Vol. 36 Iss: 3, pp.299 – 304.

In: Biology and Communication in the Information Age
Editor: Rafiqul Islam

ISBN: 978-1-62808-252-4

Chapter 9

STUDY OF NEW PLANT BIOREGULATORS

O. G. Kulikova*[*1], A. P. Il'ina, M. S. Krasnov[1], D. V. Margasyuk[1], V. P. Yamskova[2] and I. A. Yamskov[1]

[1]Nesmeyanov Institute of Organoelement Compounds,
Russian Academy of Sciences, Moscow, Russia
[2]Kol'tsov Institute of Developmental Biology,
Russian Academy of Sciences, Moscow, Russia

ABSTRACT

In different plants there are new group of bioregulators that can classified into the group of regulatory molecules that function as 'tuners' of organ and tissue homeostasis. These bioregulators with physical, chemical properties and biological activity similar to those of MHTBs which had been found earlier in different animal tissues. It was also shown that these bioregulators exhibited the biological effect characteristic of these plants.

Keywords: Plant, bioregulators, structure, wound healing

INTRODUCTION

Plants are one of the most important sources for different biological active substances isolating. But search and study of biologically active compounds that determine medicinal properties of plants is still a relevant problem. Thus, searching of new biological active substances in plants is very important problem of modern biology. In this aspect, we can note a new group of membranotropic homeostatic tissue specific bioregulators (MHTBs) earlier determined in different tissues of mammals.

* E-mail: koulikova_olga@mail.ru.

AIM

The aim of this work was to identify bioregulators of new group in plant tissues, study their structure, physical and chemical properties, their membranotropic and specific activity, and also bioregulators location in plant tissues.

BACKGROUND

A new group of bioregulators was discovered in different animal tissues in our earlier studies. At ultralow doses corresponding to concentrations 10^{-8}–10^{-15}mg/ml these bioregulators affect biological processes, such as adhesion, proliferation and cell migration [1-6].

It was established that bioregulators are located in the extracellular space and that their activity is characterized by the presence of tissue specificity and the absence of species specificity. Bioregulators of this group stimulate reparative processes in pathologically altered tissues, facilitating the repair of their disturbed structure [1-3]. In essence, they can be classified into the group of regulatory molecules that function as "tuners" of organ and tissue homeostasis.

EXPERIMENTAL PART

Plantago major L. and *Anethum graveolens L.* leaves, *Allium cepa L.* and *Allium sativum L.* bulbs were the main objects for extraction of bioregulators. Bioregulators isolated from *Aloe arborescens L.* leaves were used in comparative experiments for biological activity study. All of these plants were received from Tsitsin`s Botanical garden (Russian Academy of Sciences). Ammonium sulphate, ammonium nitrate, potassium nitrate, hydrate of calcium chloride, sodium azide (chemically pure), and bidistilled water (16 Mom) were used.

Protein Extraction and Purification

Plants were preliminary sliced into fragments 1-2 cm. Extraction was carried out for 5-6 h at 8-10 ^{0}C in solution containing NH_4NO_3 2.06 x 10^{-2} M, KNO_3 1.88 x 10^{-2} M, $CaCl_2$ x $2H_2O$ 3 x10^{-2} M, $MgSO_4$ x $7H_2O$ 1.5 x 10^{-3}M, KH_2PO_4 1.25 x 10^{-3} M. The plant extracts were filtered through tissue and centrifuged at 3 000 g for 30 min. Pellets were wasted.

Then dry ammonium sulfate was added until saturation (780 g/l) keeping pH 7.5-8.0. The protein mixtures were kept at 4 ^{0}C for 2 weeks. The protein pellets were separated. Supernatants obtained were dialyzed against water until total removal of sulfate ammonium, filtered and concentrated using rotor vacuum evaporator at 37-40 ^{0}C.

Biotesting Method

Biotesting method was developed earlier as adhesiometric method for MHTBs identification in different fractions on each step of purification. The study was performed on male F1 CBA/C57Bl mice weighing 18-20 g [7].

Determination of Protein Concentration

Protein concentrations in fractions of plant bioregulators were determined spectrophotometrically.

Protien Separation by Chromatography

Supernatant obtained after salting out onion extract was analyzed using a high pressure chromatograph (Kontron, US) equipped with a Jupiter C_5 hydrophobic column (2x150 mm) (Phenomex, US). Elution was conducted in a gradient of concentration of acetonitrile (0-70%) in 0.1 % trifluoroacetic acid (pH 2.2) with velocity 0.3 ml/min for 70 min. Detection was carried out at 210 nm.

Method of Dynamic Laser Desorption

The solutions with bioregulators were preliminary purified from the dust by filtration thought the membranes Durapore with pore diameter 0.45 μm. Hydrodynamic radius of particles in the aqueous solution of supernatants was investigated. The measurement was carried out on a PhotoCor Complex apparatus (FotoKor, Russia).

MALDI-TOF Mass-Spectrometry

Analysis of peptides molecular weight was carried out using UltraFlex 2 time-of-flight mass-spectrometer (Bruker Daltonic, Germany). Samples for analysis were obtained by drying followed by dilution in 70% acetonitrile with 0.1% trifluoroacetic acide. α-cyano-4-hydroxycinnamic acid was used as the matrix.

Determination of C-Terminal Amino Acid Sequence

Peptide (600 pmol) was solved in 10 μl of sodium citrate buffer (50 mM, pH 6.0) and added with carboxypeptidase Y (Sigma Aldrich, United States) in the ratio enzyme-peptide of 1:50 on mass. Aliquots 1 μl were taken after 1, 2, 5, 10, 15, 30, 45, 60, 90, and 120 min of incubation at 37°C, mixed with 1.5 μl matrix solution and placed on the plate for further mass-spectrometry of reaction products.

Isoelectric Focusing (IEF)

Isoelectric focusing was performed in a sucrose density gradient in an LKB 440 column (Sweden) at 4°C and 500–2000V for 96h. Protein fractions were detected spectrophotometrically at 280 nm.

Electrophoresis of Proteins in Polyacrylamide Gel

PAGE was performed according to Laemmli [8]. The mixture containing molecular_weight protein markers (14.4–94 kDa; Helicon, Russia) was supplemented with 3.5 kDa oxidized B-chain of bovine pancreatic insulin (Sigma, Germany); final dilution, 1 mg/ml).

Immunohistochemical Staining with FITC-Labeled Secondary Antibodies

To study expression of the slightly acidic protein isolated from the onion bulbs, rabbit antiserum was preliminarily obtained, which was then used as primary antibodies to this protein at a dilution of 1:10. We used the technique of indirect immunostaining with antirabbit antibodies at a dilution of 1:35 using secondary antibodies (IgG, fill-length molecule; Sigma, United States) labeled with fluorescein isothiocyanate (FITC). The cross-sections stained with the secondary antibodies alone as well as the cross-sections stained with the antiserum obtained from a nonimmunized rabbit served as a control. The location of the protein was studied and images were obtained with an Olympus microscope (Japan).

Investigation of Wound-Healing Effect of the Bioregulator Isolated from the Aloe Leaves on an Experimental Model of Mouse Skin Wound in Vivo

Experimental mice were divided into three groups, five animals in each group. A piece of skin (approximately 1 cm in diameter) was excised from the back of each mouse under ester anesthesia. Traumatized mice of the control group remained untreated. Mice of another control group were daily treated after wounding with 100 μl of saline applied on the wound area. Mice of the experimental group were daily treated after wounding with 100 μl of the fraction isolated from the aloe leaves. On day 11 after wounding, all animals were withdrawn from the experiment in an ester atmosphere. Tissue in the wound area was accurately excised, and histological cross-sections 7 mm thick embedded in paraffin were prepared. Tissues were fixed in Bouin fluid; cross-sections were stained with hematoxylin/eosin by the standard procedure. The histological samples were analyzed by light microscopy under an Olympus microscope (Japan).

Results and Discussion

Using the experimental approach investigated for isolating MHTBs of mammals we obtained plant bioregulators. Similar to MHTBs bioregulators isolated from plant tissues remained in the saturated salt solution. It was demonstrated when supernatants and pellets obtained after salting out plant extracts were investigated. Figure 1 shows the diagram illustrating the membranotropic activity of the supernatant obtained from garlic extract. Biologically active supernatants then were separated by different methods: isoelectric focusing (pH 3.5-10.0), HPLC, PAGE. Fractions of bioregulators obtained were identified by biotesting, MALDI-TOF mass-spectrometry and dynamic laser desorption. The specific activity of some bioregulators was studied in certain experiments.

IEF separation of supernatant showed that plant bioregulators are fraction with pH<3.0 similar to MHTBs. This fraction exhibited the membranotropic activity.

Figure 2 shows the separation of onion extract supernatant using reversed phase HPLC. The main fraction eluted at 9% concentration of acetonitrile. This fraction also exhibited the membranotropic activity.

By method of MALDI-TOF mass-spectrometric analysis we determined that peptide is the major component of HPLC-fraction. The molecular weight of this peptide was 4036±2 Da (Figure 3). The enzymatic hydrolysis by carboxypeptidase Y of the peptide followed by MALDI-TOF mass-spectrometry analysis of reaction mixture has determined C-terminal amino acid sequence of 18 residues: GlyPheGlyGluGlyAlaTyrTrpGlyAlaValAla AlaGlyTrpGluGlyArg. Using the comparative analysis of peptides in the Uniprot data base we didn`t find any peptides with similar amino acid sequence.

Investigation of plant extracts supernatants by method of dynamic laser desorption demonstrated the presence of nanoparticles in aqueous solution of bioregulators. This is one of the characteristic for mammals MHTBs.

For plant bioregulators the presence of great nanoparticles with the size 88-110 nm was shown (Tab.1). Results obtained showed that in spite of low-level protein concentration (0.1 mg/ml) in fraction containing plant bioregulators similar to MHTBs there are great nanoparticles.

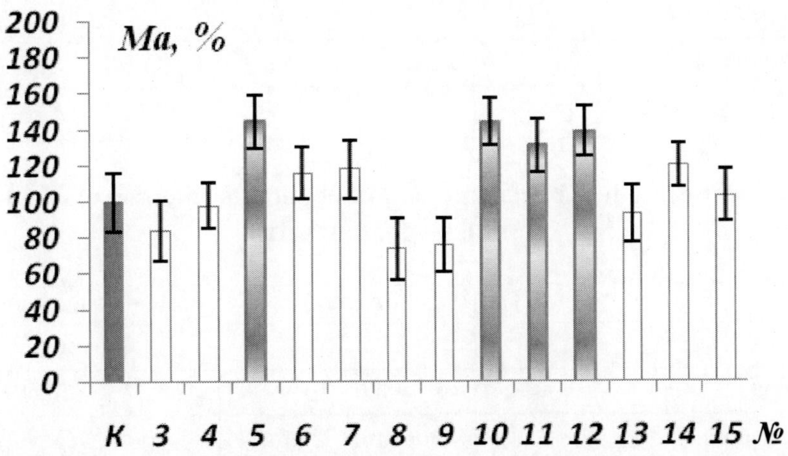

Figure 1. Membranotropic action of supernatant garlic extract ($p< 0.05$).The axis of ordinates– effect value Ma (%); the axis of abscissa – the degree of tenfold sequ ential dilutions of the fraction; K - control; Initial concentration 0.1 mg/ml.

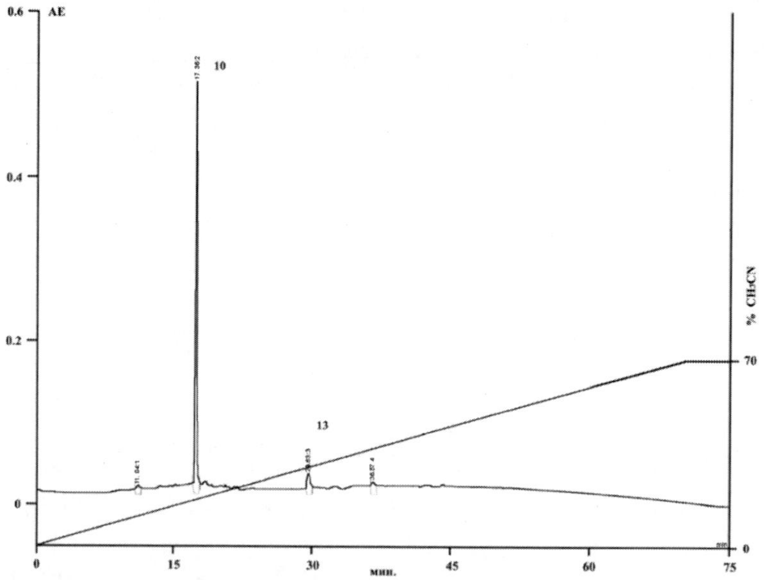

Figure 2. Separation of onion extract supernatant using reversed phase HPLC equipped with Jupiter C5 column (2 × 150 mm). Elution was carried out in a concentration gradient of acetonitrile (0–70%) in 0.1% trifluoroacetic acid (pH 2.2) and velocity of 0.3 ml/min for 70 min. Detection at 210 nm.

Table 1. The results of plant extract supernatants investigation by method of dynamic laser desorption (protein concentration 0,1 mg/ml)

Plant	The value of hydrodynamic radius (nm)
Plantago major L.	110 ± 5,5
Allium cepa L.	99,2 ± 4,9
Allium sativum L.	101,7 ± 5,1
Anethum graveolens L.	88 ± 4,4

Table 2. Identification of plant supernatant peptides be method of MALDI-TOF mass-spectrometry

Plant	M (m/z)	Protein concentration (mg/ml)
Plantago major L.	2374, 2963	0,064
Allium cepa L.	4036, 8079	2,69
Allium sativum L.	3061, 4367, 8743	0,79
Anethum graveolens L	1100, 1391, 2229	1,37

In this study we can underline other important similarity of plant bioregulators and MHTBs: the presence in their structure low-molecular peptides. This fact was demonstrated using MALDI-TOF mass-spectrometry (Table 2).

Data of MALDI-TOF mass-spectrometric analysis proved out by the results of PAGE (Figure 4). For example, IEF investigation of *Plantago major L.* fraction showed that peptide is the main component demonstrated membranotropic activity at ultra low doses characterized for MHTBs.

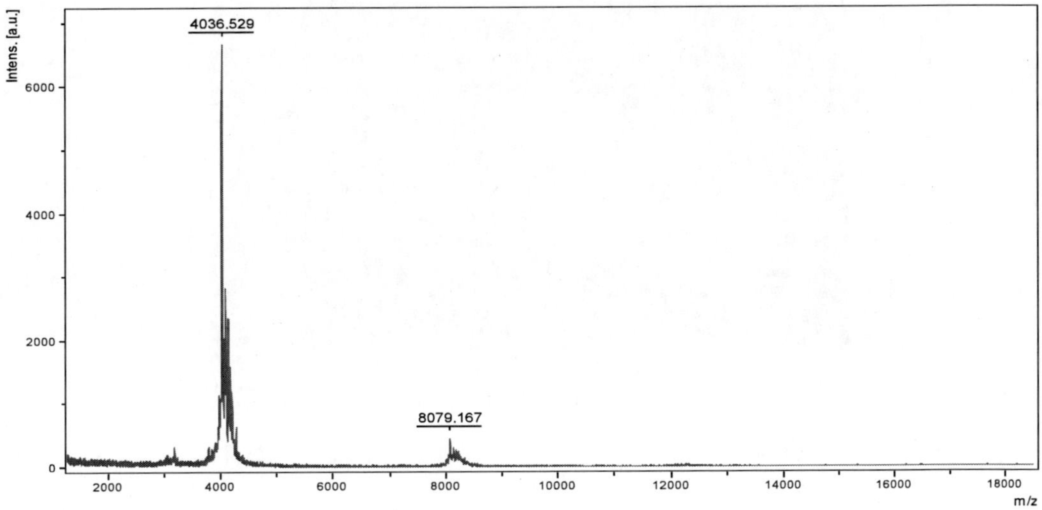

Figure 3. MALDI-TOF mass-spectrum of supernatant fraction after HPLC.

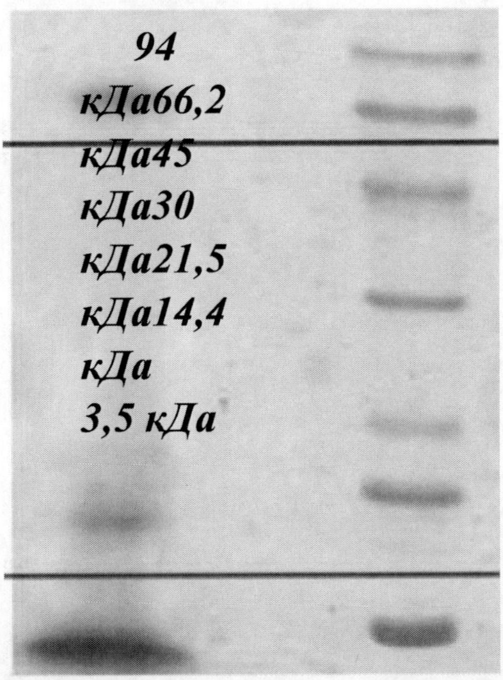

Figure 4. SDS-electrophoresis in of IEF-fraction isolated from Plantago major L. in polyacrylamide gel (pH < 3.0). Markers protein: oxidized B chain of insulin from the pancreas of bull – 3.2 kDa, α-lactalbumin – 14.4 kDa, soybean trypsin inhibitor – 21.5 kDa, carbonic anhydrase – 31 kDa, ovalbumin – 45 kDa, bovine serum albumine – 66.2 kDa, phosphorylase b – 94.6 kDa.

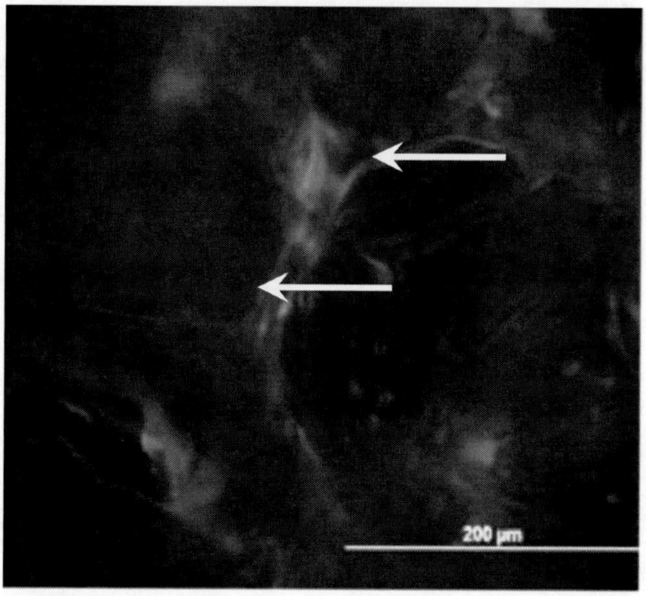

Figure 5. Bioregulator location in the extracellular space of Allium cepa L. bulbs (shown by the arrows).

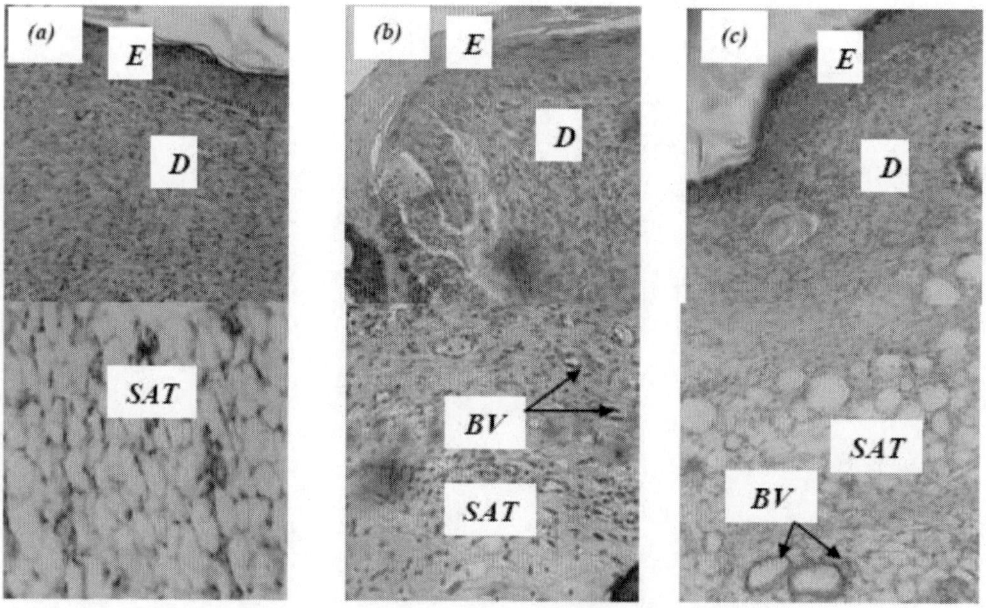

Figure 6. Healing wounds on mouse skin on 7 days after operation: a – without wound treatment, control; b- saline wound treatment; c – bioregulator wound trentment. Designations: E – epidermis; D – derma; SAT – subcutaneous adipose tissue; BV – blood vessels.

Using the modern immunogistochemical method with the FITC-conjugated secondary antibodies we studied the location of bioregulator isolated from *Allium cepa L.* bulbs in bulbs tissue. It was shown that this bioregulator localized in the extracellular space (Figure 5). Data obtained are consistent with the data on the localization of MHTBs of mammals.

To study the specific activity of bioregulator isolated from *Aloe arborescens L.* we used the experimental model of mouse skin wound *in vivo* (Figure 6). In all experimental animals on day 11 after operation we observed a nearly complete reepithelization of the wound surface. In the control group of mice (untreated wound) there was formation of a fibrous scar. We observed the parallel location to the epidermis of collagen fibers in the form of solid cords (Figure 6a). In the group of mice with saline treatment of wound there was also the formation of a connective-tissue scar and a chronic inflammation focus in the center of wound (Figure 6b). In experimental group of mice with plant bioregulator treatment the recovery of subepidermal layers was observed (Figure 6c). The epithelium and derma recovered almost completely, in the latter, blood vessels were observed, fibers were arranged more loosely than in control.

Thus, it was shown that bioregulator isolated from *Aloe arborescens L.* stimulate wound healing in mice *in vivo* when used at ultra low doses. So, in this study we demonstrated that bioregulator discovered in the *Aloe arborescens L.* exhibited the biological effect characteristic of this plant.

Conclusion

So, we can conclude that bioregulators with physical, chemical properties and biological activity similar to those of MHTBs which had been found earlier in different animal tissues were discovered in plant tissues.

Acknowledgments

This work was supported by the Russian Foundation for Basic Research, project no.10-04-00706-a.

References

[1] Krasnov M.S., Gurmizov E.P., Yamskova V.P., Yamskov I.A. Analysis of a Regulatory Peptide from the Bovine Eye Lens: Physicochemical Properties and Effect on Cataract Development in vitro and in vivo // In the book "Biochemical Physics Frontal Research". Ed. by S.D. Varfolomeev, E.B. Burlakova, A.A. Popov and G.E. Zaikov. HauppaugeNY. Nova Science Publishers Inc. 2007. p. 21-34.

[2] Nazarova P.A., Yamskova V.P., Krasnov M.S., Filatova A.G., Yamskov I.A. "Regulatory proteins biologically active in ultralow doses from mammalian glands and their secretions", pp. 78-88 // In the book "*Biochemical Physics Frontal Research*", Ed. by Varfolomeev S.D., Burlakova E.B., Popov A.A. and Zaikov G.E., Hauppauge NY, Nova Science Publishers Inc. 2007. p. 73-82.

[3] Borisenko A.V., Yamskova V.P., Krasnov M.S., Blagodatskikh I.V., Vecherkin V.V., Yamskov I.A. "Regulatory proteins from the mammalian liver that display biological activity at ultra low doses" pp. 35-45 // In the book "*Biochemical Physics Frontal*

Research", Ed. by Varfolomeev S.D., Burlakova E.B., Popov A.A. and Zaikov G.E., Hauppauge NY, Nova Science Publishers Inc. 2007. P. 35-46.

[4] *Yamskova V.P., Rybakova E.Yu., Vecherkin V.V., Berezin B.B., Filatova A.G., Blagodatskikh I.V. and Yamskov I.A.* "Analysis of regulatory proteins from bovine blood serum that display biological activity at ultra low doses: 1. Isolation, purification and physicochemical properties.", pp. 61-70 // In the book "*Biochemical Physics Frontal Research*", Ed. by S.D. Varfolomeev, E.B. Burlakova, A.A. Popov and G.E. Zaikov, Hauppauge NY, Nova Science Publishers Inc. 2007. p. 127.

[5] Yamskova V.P., RybakovaE.Yu., Vecherkin V.V., Berezin B.B., Filatova A.G., Blagodatskikh I.V., YamskovI.A. Analysis of Regulatory Proteins from Bovine Blood Serum that Display Biological Activity at Ultra Low Doses: 1. Isolation, Purification and Physicochemical Properties // In the book "*Biochemical Physics Frontal Research*". Ed. by S.D. Varfolomeev, E.B. Burlakova, A.A. Popov and G.E. Zaikov. HauppaugeNY. Nova Science Publishers Inc. 2007. p. 61-70.

[6] Margasyuk D.V., Krasnov M.S., Blagodatskikh I.V., Grigoryan E.N., Yamskova V.P., Yamskov I.A. "Regulatory Protein from Bovine Cornea: Localization and Biological Activity", pp. 49-56 // In the book «*Biochemical Physics Frontal Research*», Ed. by Varfolomeev S.D., Burlakova E.B., Popov A.A. and Zaikov G.E., Hauppauge NY, Nova Science Publishers Inc. 2007. P. 47-60. Yamskova V.P., Reznikova M.M. Low molecular weight polypeptide of warm-blooded serum: effects on cell adhesion and proliferation // *Journal of general Biology*. 1991. V. 52. № 2. p. 181-191.

[7] Laemmli U.K. Cleavage of structural proteins during the assembly of the head of bacteriophage T4 // *Nature*. 1970. V.227. p. 680-685.

In: Biology and Communication in the Information Age
Editor: Rafiqul Islam
ISBN: 978-1-62808-252-4

Chapter 10

ARCHITECTURE OF THE SOFTWARE FOR THE COMPLEX PSYCHOPHYSIOLOGICAL MONITORING

G. Kilessa[1*], N. Khorseva[2], P. Grigoryev[3] and A. Glivenko[3]

[1]Tavrida National University named after V.I.Vernadsky, Simferopol, Ukraine
[2]Emanuel Institute of Biochemical Physics of Russian Academy of Sciences, Moscow, Russian Federation
[3]Crimea State Medical University named after S. I. Georgievsky, Simferopol, Ukraine

ABSTRACT

The software product intended for the integrated monitoring of the human health status by means of express-evaluation techniques of emotional state, central and autonomic nervous system, and system of hemodynamics. It can be used by a wide range of users without age restrictions is presented and described here. The designed software complements our earlier information system "Universal monitoring of ecological health."

Keywords: Information system of diagnostics, software, psychophysiological monitoring, website

INTRODUCTION

Development of the software allowing to obtain valid and reliable results of complex human state with minimal tools has been the main task of our work. This product can be used in applied psychology, hygiene, medicine and human ecology.

* E-mail: kilessa@gmail.com.

Aim and Background

Design and implementation of information systems (IS) for the complex psycho-physiological monitoring becomes an important goal of the day. The authors have developed a client-server IS "Universal monitoring of ecological health" [2, 3, 4, 5]. This system is oriented to the global network, however, the customers who have no access to Internet, or it has time limits, are not able to use all available tools of the IS. In this connection, we have designed a local version of the software related to the early developed information technology, which has an option of local testing and presentation of results.

Results and Discussion

We have developed software (SW) for a PC running Windows 98/2000/XP/Vista/7 for comprehensive monitoring of the human health state by means of express-evaluation of emotional state, central and autonomic nervous system, and system of hemodynamic. It can be used by a wide range of users without age restrictions. To obtain a diagnostic data it will require only personal computer and tonometer for blood pressure and heart rate measurement.

The program is written in programming language C #, for software platform. NET Framework 2.0 for operating system (OS) Windows. For the selection of .Net version the criterion was the following: compatibility with the earlier versions of Windows. The latest version .NET Framework was 4.0, which supports operating systems Windows XP (Home\Professional Edition)\\Windows Vista\Windows 7, while version 2.0 is also supported by Windows 98.

At the first stage it has been developed a general algorithm for operating software. The block diagram is shown in Figure 1.

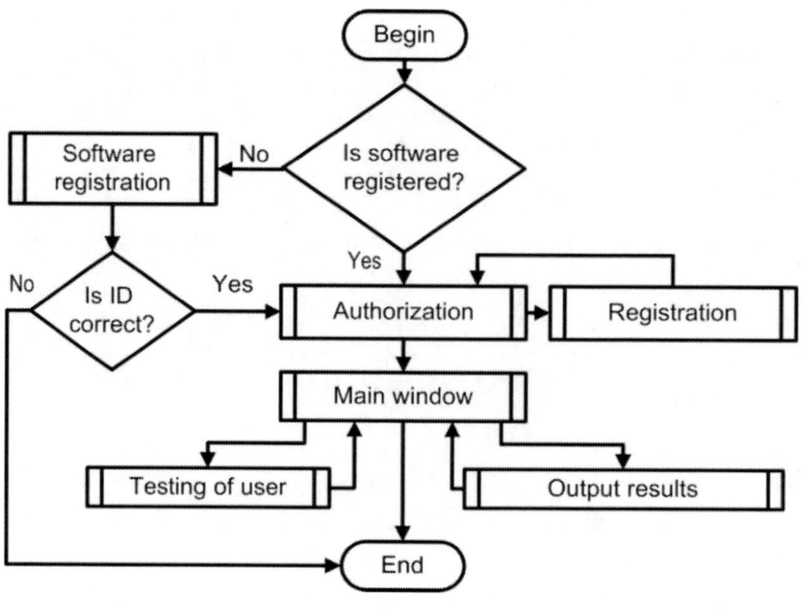

Figure 1. General Algorithm of operating software.

As we can see on Figure 1, after start the program identifies if the software is authorized. In absence of registration the user is asked to enter an unique serial number received along with the purchase of the product. The registration has to be performed only once, and after successful authorization the user can create his/her account by filling up the following fields: username and password to enter the program, name, age, gender, e-mail. The main menu allows to start test and check the results after login. Before testing, the user should create a list of preferred tests which he/she is going to take. This software also has an option for monitoring and rehabilitation of patients (after stroke, cerebral infantile paralysis, and some other neurological disorders).

There are following tests are available in the software:

- auditory-motor response to determine the parameters of reaction to the audible stimulus;
- visual-motor response, to determine the parameters of reaction to the visual stimulus;
- corrècture test (determination of voluntary attention stability);
- test of fine motor skills of fingers (chaotic ten-finger tapping, RF Patent № 2314743), allows to determine violations of the fine motor function, to characterize the functional status and efficiency of the motor zones of the cerebral cortex, interhemispheric interaction, lateral preferences;
- methodology for studying dynamics of psycho-emotional state (activity, mood, anxiety, aggression);
- hemodynamic data - blood pressure and pulse rate, the fields are filled by the user after measuring of BP by tonometer;
- mechanical memory, to estimate the level of mechanical memory (mechanical memory capacity);
- semantic memory for evaluation of semantic memory level (logical memory factor).

The combination of these techniques will allow to conduct a comprehensive express-diagnosis of the main functional systems of the human organism. After the last test, the user can see the results representing the overall data, as well as conclusions and recommendations based on analysis of data including his/her age, sexual norms, and individual standards (obtained through the repeated testing of the subject).

Results can be represented in two modes: "one-day" and "days-long". In one-day mode the data is displayed in a single session of the selected test. In "days-long" mode the results are presented as a interval of dates, and show dynamics of the studied parameter. The report can be saved in pdf-format. Figure 2 shows a diagram of classes responsible for tested users and displays the results. We have identified the common parameters (such as the date of test) and declared them in the abstract base class «TestBase». Each technique is represented by a class inherited from «TestBase». We made the same for the data viewing, each test has its own reprezentator, which deals with visualization of the data according to the method.

The calculated parameters for each test are represented in separate classes inherited from the base «TestResult» (which declares all general characteristics) and interface «IResultIO», which describes options determined in each class-results (for example, «Save» , «Load»).

The developed software product is a local tool, and data can be saved on the user's computer. The mechanism of data transfer from user to specialist in electronic form is an important part of the program. The software has an option of export of the test results for a selected period of time in csv-format (text format for presentation of tabular data).

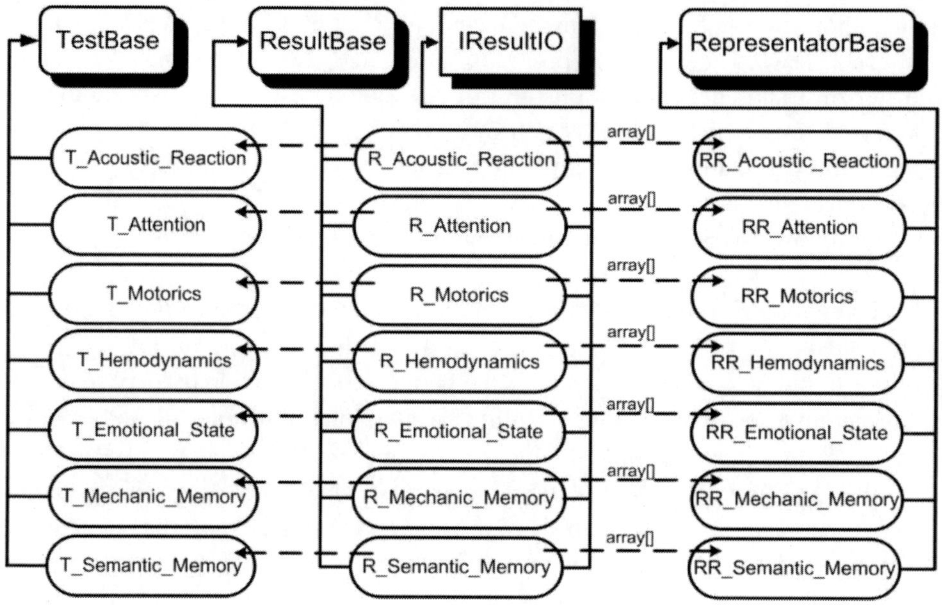

Figure 2. Diagram of classes responsible for blocks Test and Output (solid lines – inheritance; dotted lines – class as a field).

The user can send the obtained data to the specialist by e-mail, or save on flash memory \ optical disk cant. Professionals can use an extended version of the software product with data visualization, where the data source supports csv-file, but not a local data base.

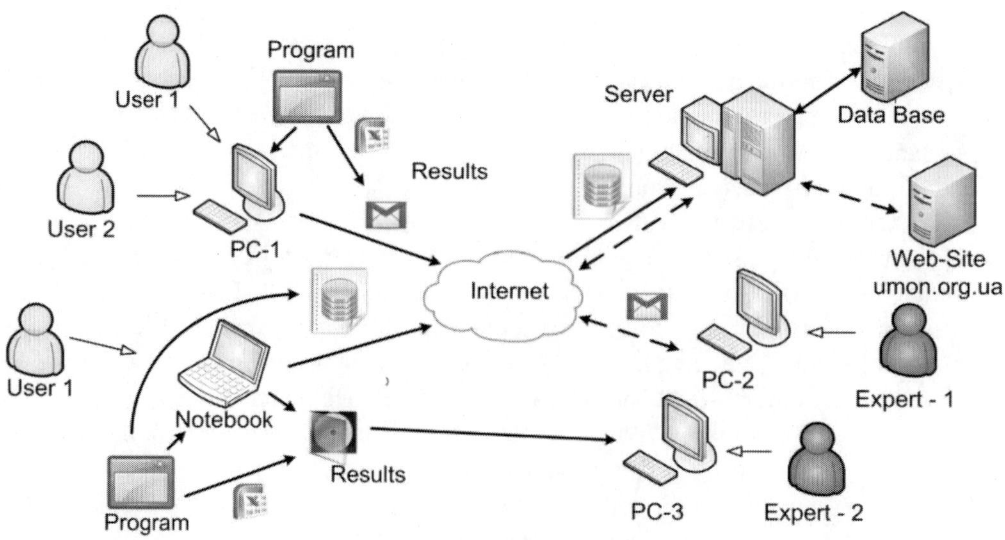

Figure 3. Structure of the software operating in the framework of IT «Universal monitoring of ecological health» and mechanism of interaction with experts.

Designed software product is inbuilt into the earlier information technology "Universal monitoring of ecological health." The program has an option of sending data to the site www.umon.org.ua.

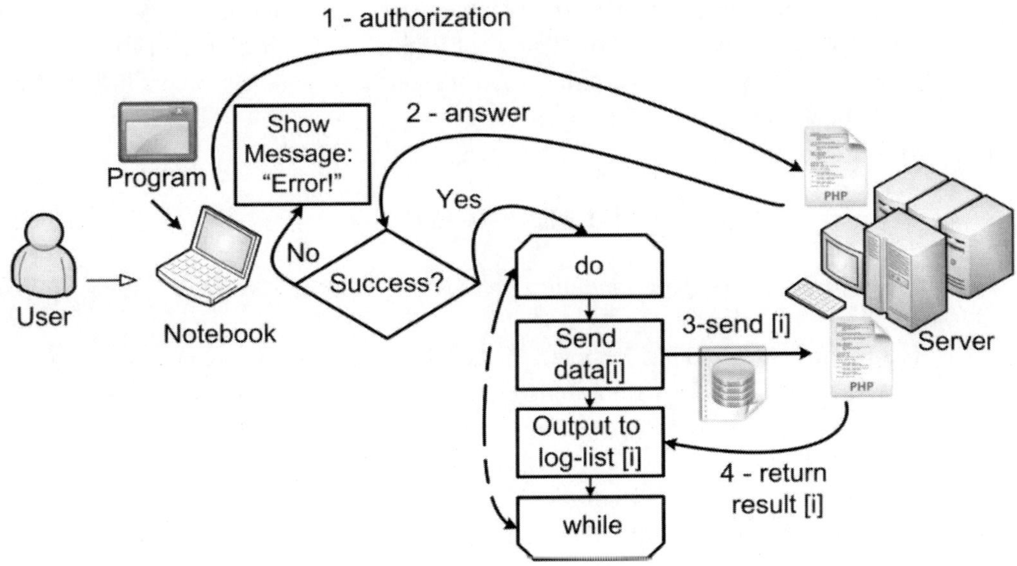

Figure 4. Algorithm of data transfer on server.

Algorithm of data transfer from the local program to the server is shown on Figure 4. Due to the user's registration details on the website (email and password) may differ from those in the local program (login and password), in menu "Settings" user can adjust his email address and password according to the registration details used for website. This data is used to send results of the first stage - authentication. Server verifies result and sends it to a local program (Figure 4, step №2 - answer). If the server denies access to database, the program displays an error message. If registration was successful the program forms an array that should be send. Then the cycle (selected dates and tests): at each step it sends [i]-th element of the formed array (Figure 4, step №3 - send [i]). After receiving the data server provides transaction to add data into the database and returns the result to the local program (Figure 4, step №4 - return result).

Thus, developed software product enables a comprehensive monitoring of the human health status as well as presentation of results as graphs, charts, and numerical values. This software can be used in scientific research and in clinical practice such as applied psychology, speech therapy and rehabilitation.

Conclusion

1. The software designed for PC running Windows 98/2000/XP/Vista/7 for comprehensive monitoring of the human health state by means of express-evaluation of emotional state, central and autonomic nervous system, and system of hemodynamics.

2. Developed software product provides testing, results visualization in a user-friendly diagrams and tables, and saves reports in pdf-format. Besides that, it has an option to export the user's local database to file csv-format (text format for representing tabular data).
3. The software program integrates and complements the earlier developed information system "Universal monitoring of ecological health", providing transfer of test results to the website: www.umon.org.ua for further analysis, conclusions and recommendations.

REFERENCES

[1] B. Stroustrup. *The C++ Programming Language*. – Moscow: «Binom-Press», 2008. – 1104 p.

[2] Grigoryev P.E., Kilessa G.V., Khorseva N.I., Khramov V.V., Olenchuk A.V., Asanov E.E. Information system for determining the dependence of the human condition on the effects of cosmophysical factors // *Clinical Informatics and Telemedicine*. – 2010. – V.6. № 7. – P. 83-89.

[3] Kilessa G.V., Asanov E.E., Grigoryev P.E., Khorseva N.I. Information technology of monitoring the human condition for determining its dependence on cosmophysical factors // *Proceedings of the X-th international conference for young scientists "Biochemical physics"*. – Moscow, RAS, 2010. – P. 94-98.

[4] Olenchuk A.V., Asanov E.E., Kilessa G.V., Grigoryev P.E., Khorseva N.I. The server part of an information system for determining the dependence of the human condition on cosmophysical factors // *Proceedings of the X-th international conference for young scientists "Biochemical physics"*. – Moscow, RAS, 2010. – P. 98-100.

[5] Kilessa G. V., Schmidt M. V., Grigoryev P.E., Khorseva N.I. Client software of information system for determining the dependence of the human condition on cosmophysical factors // *Proceedings of the X-th international conference for young scientists "Biochemical physics"*. – Moscow, RAS, 2010. – P. 100-103.

[6] William S. Davis, David C. Yen. *The Information System Consultant's Handbook. Systems Analysis and Design*. CRC Press, 1998 – 710 p.

In: Biology and Communication in the Information Age
Editor: Rafiqul Islam
ISBN: 978-1-62808-252-4

Chapter 11

HIGH-GAIN KOCH FRACTAL LOOP ANTENNA AND ITS ARRAY FOR RADAR APPLICATIONS

Mukh Ram **[1*]**, ***R. K. Yadav*** **[2]** ***and R. L. Yadava*** **[1]**

[1]Department of Electronics and Communication Engineering,
I.T.S. Engineering College, Gr. Noida, U.P. India
[1]Department of Electronics and Communication Engineering,
Galgotia's College of Engineering and Technology, Gr. Noida, U.P. India

ABSTRACT

In this paper a loop antenna is designed at 2 GHz and its performances is compared with fractal loop antennas.

It has been seen that as the order of iteration of the loop antenna increases so does the values of gain, standing wave ratio and reflection coefficient. Standing wave ratio reaches the coveted value of 1 and reflection ratio reaches -10dB which are required for proper matching.

It is also found that the fractal loop antenna exhibits improved impedance and Standing Wave Ratio (SWR) performances in a reduced physical area as compared to non-fractal Euclidean geometries. Therefore it is very suitable for direction finding in Radars applications.

Keywords: Loop antennas, fractal antennas, Euclidean geometries and Radars

I. INTRODUCTION

The use of wireless receivers and Bellini-Tosi loop antennas for direction finding (DF) was developed by the Marconi Company before the World-War and by 1916; secret chains of DF stations were in use for tracking Zeppelins, submarines and surface craft [1].

* E-mail: mram2679@yahoo.co.in.

The properties of circular and square loop antennas as probes for measuring the magnetic field was investigated by H. White de ESIDE, has shown that large errors may occur when a single loop is used, unless its diameter is less than 0.01 λ [2].

Due to loop antenna's size and directivity, it is used for transmitting and receiving signals, mainly on the MF and HF bands. The loop antennas are very directional with the pickup pattern shaped like a figure eight which also allows signals on opposite sides to be received.

Whereas in off sides of the loop antenna signal, will decrease or be nulled out. The nulling feature allows one to remove a local station on a frequency and pick up another on the same frequency by removing the local signals. In spite of various advantages, there are certain disadvantages with the loop antennas; low radiation resistance and very high quality factor (Q). The reason behind low resistance is that single turn small loop antennas are made of very thick wire, or a tubular conductor. While high Q causes a narrow bandwidth that limits the carrier and its sidebands.

In addition, large size of loop antennas (LP) limits its applications as radio antennas [3] and as the frequency increases, additional lobes appear causing principal lobe to point more in the direction normal to the loop [4]. An alternate approach to remove such limitations is to replace LP with fractal shape loop antennas.

The concept of "Fractals" was first invented by Benoit Mandelbrot in 1975, and fractal antenna whose dimensions are not complete numbers has been used as replacement of broadband Microstrip Antenna (MSA) by Gutton H. [5]. However most fractals have some sort of self similarity at arbitrarily small scales [6], and each repetition is on different scale [7]. Basically there are Koch, Minkowski and Sierpinski types of fractal geometries. However Koch fractal antennas are mostly preferred due to their simple geometry, low profile and easy design. Therefore in present paper authors first designed conventional loop antenna and then 1st and 2nd second iterated fractal Koch loop antennas. The various parameters; Radiation patterns, SWR, Reflection coefficients, have been calculated with NEC2 (version 2.0).

II. Loop Antenna Designs

Loop Antenna 0-Iteration

This section contains design of conventional loop antenna of diameter 2 mm and total length of 0.16 m. The antenna was fed at middle point of one the sides of a wire length with coaxial feed of 50 ohms impedance.

Loop Antenna 1 and 2 Iterations

In this case a single iteration of the loop antenna is carried out where in equal portions of the loop antennas are removed symmetrically from the four ends. During the second course of iteration the antenna is further split into smaller portions which reduces the effective area and increases the perimeter (Figure 1).

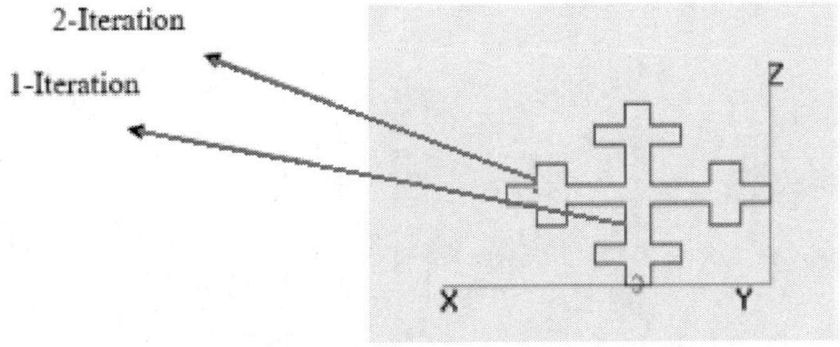

Figure 1. 1st and 2nd iterated fractal loop antennas.

III. Loop Antenna Array

Fractal Loop Antenna Array: Array of Iteration-1(three elements)

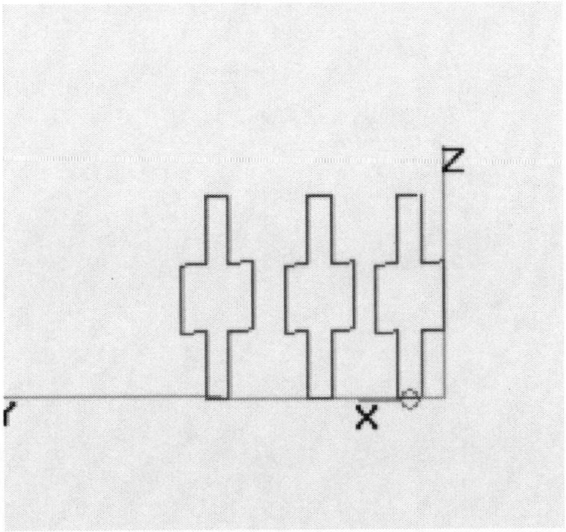

Figure 2. Array of 1st iterated fractal loop antenna

In Figure 2 loop fractal antennas are placed in close proximity to each other to form an antenna array. This process improves the gain, SWR and the reflection coefficient of the antenna system. The gain improves, the SWR comes closer to the value of 1 and the reflection coefficient approaches to -10 dB.

Array 2: Fractal Loop Antenna Array: Array of Iteration-2(three elements)

As above, three 2nd iterated fractal loop antennas are placed in close proximity to each other to form an antenna array. This process improves the gain, SWR and the reflection coefficient of the antenna. In this design successive antennas are placed $\lambda/4$ distance to each other (Figure 3). The inverse fractal loop antenna and loop antenna array are shown in Figures 4 and 5 respectively.

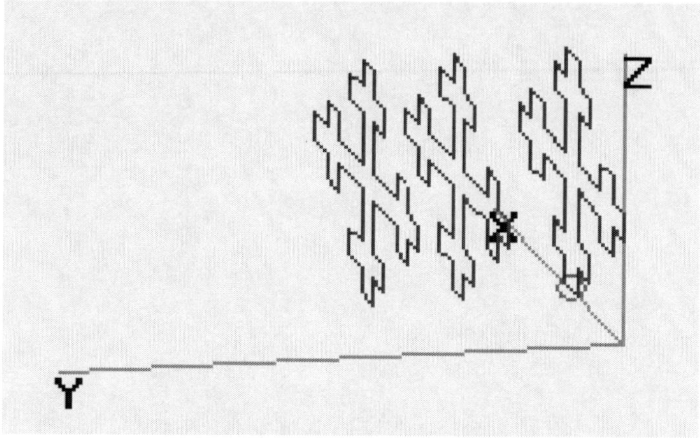

Figure 3. Array of 2nd iterated loop antenna.

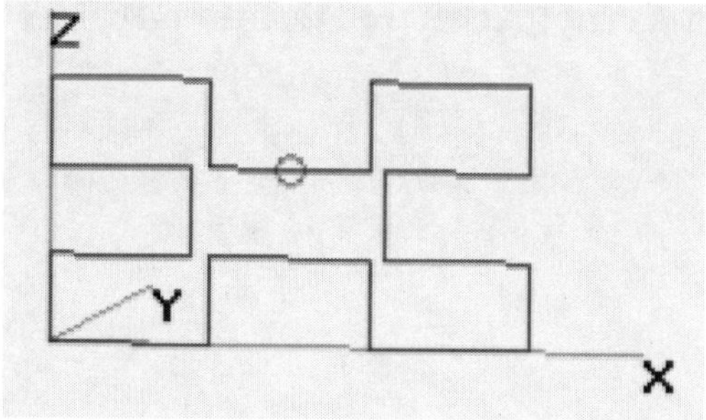

Figure 4. Inverse fractal antennas.

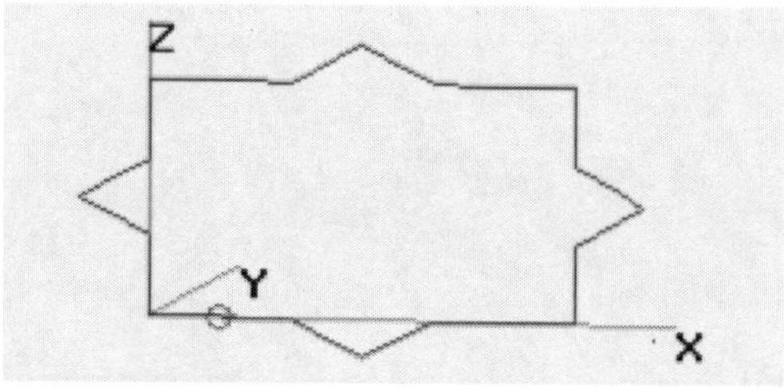

Figure 5. An element of Inverse fractal loop array.

IV. Radiation Pattern and Gain

Using NEC2 (version 2.0), proposed antennas and its arrays has been analyzed. The obtained results are shown in figure 6-19. The figre6 shows variation of normalized power density versus elevation angle ϕ of a conventional loop antenna. The radiation pattern assumes an elliptical path with the maximum radiations along the XY plane and comparatively lesser radiations along the Z direction. The antenna offers a maximum gain of 3.32 dB and a minimum of -0.6 dB.

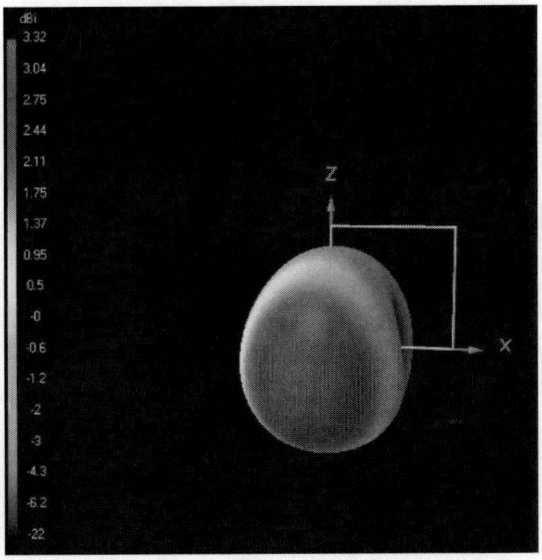

Figure 6 3-D Radiation pattern of a conventional loop antenna at 2.00 GHz.

The radiation pattern is circular in shape. However, the first iteration brings about a change in the value of gain, the SWR and reflection coefficient. The maximum gain of the antenna is 2.78 dB and minimum is 0.79 dB. This is shown in the Figure 7. After second iteration the maximum gain of the antenna is found to be 2.65 dB and minimum value is 1.29 dB which can clearly be seen from the Figure 8. The maximum and minimum values of gains of the antenna array are 7.67 dB and are -19 dB respectively. These values are much greater than the single loop antenna; i.e. one antenna array can increase gain by 2 dB.

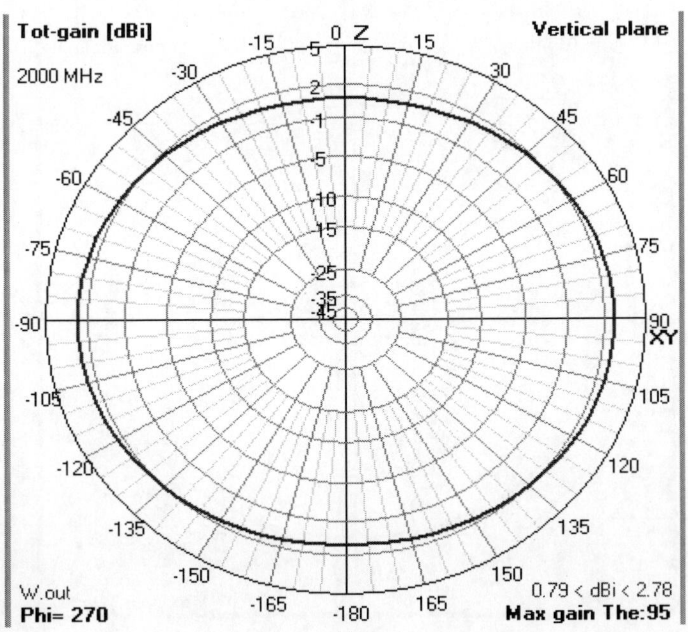

(a)

Figure 7. (Continued).

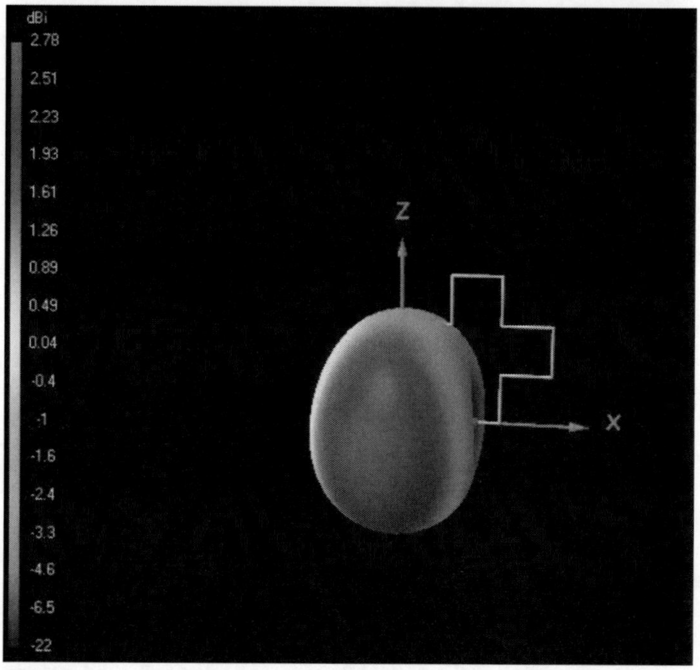

(b)

Figure 7. (a) 2-D radiation pattern of 1-st iteration loop antennas (b) 3-D radiation pattern.

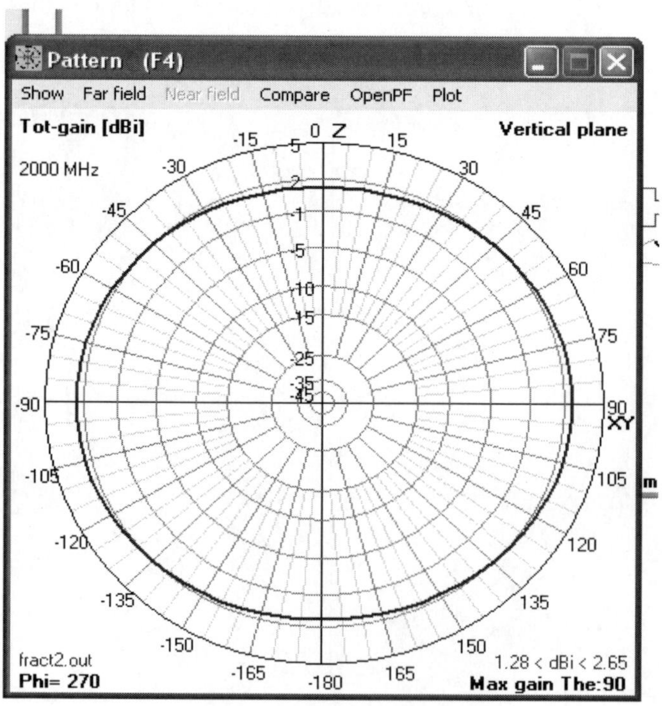

(a)

Figure 8. (Continued).

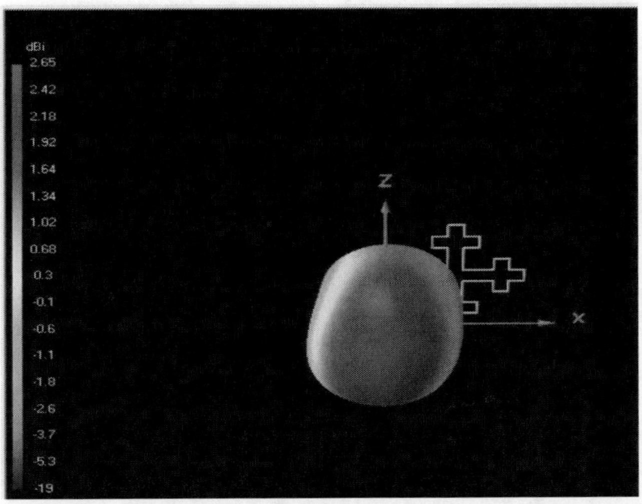

(b)

Figure 8. (a) 2D Radiation pattern of 2nd iteration loop antennas (b) 3-D radiation pattern at 2.00GHz.

So; as per gain requirement we can increase number of loop antennas in the array (Figure 9). The radiation pattern of antenna arrays of second iterated three elements fractal loop antenna is shown in Figure 10. The radiation pattern of inverse fractal antenna becomes 4 lobed wherein the gain and intensity along the XY direction is higher than that of along the Z direction. The pattern assumes an elliptical shape with maximum gain of 5 dB and minimum is 0.2dB (Figure 11). Whereas the gain of the fractal array antenna varies between -2.4 dB and 3.56 dB. The radiation patterns are elliptical in shape as shown in Figure 12.

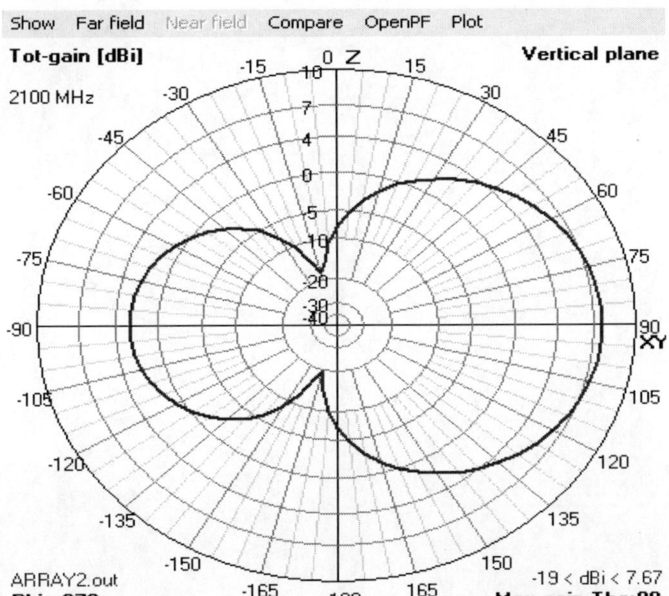

(a)

Figure 9. (Continued).

(b)

Figure 9. (a) 2-D Radiation pattern of fractal loop antenna array (b) 3-D radiation pattern at 2.10 GHz.

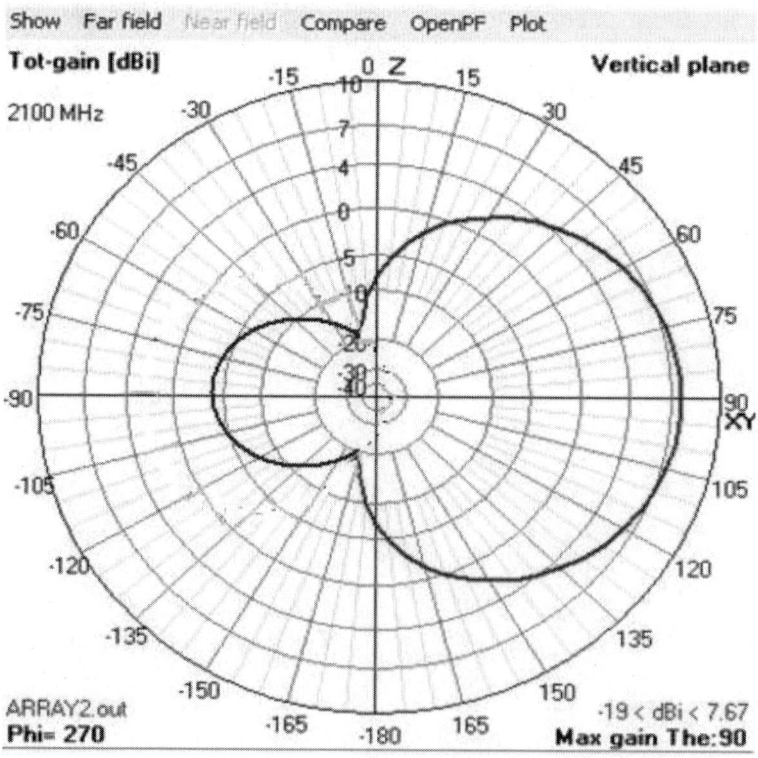

(a)

Figure 10. (Continued).

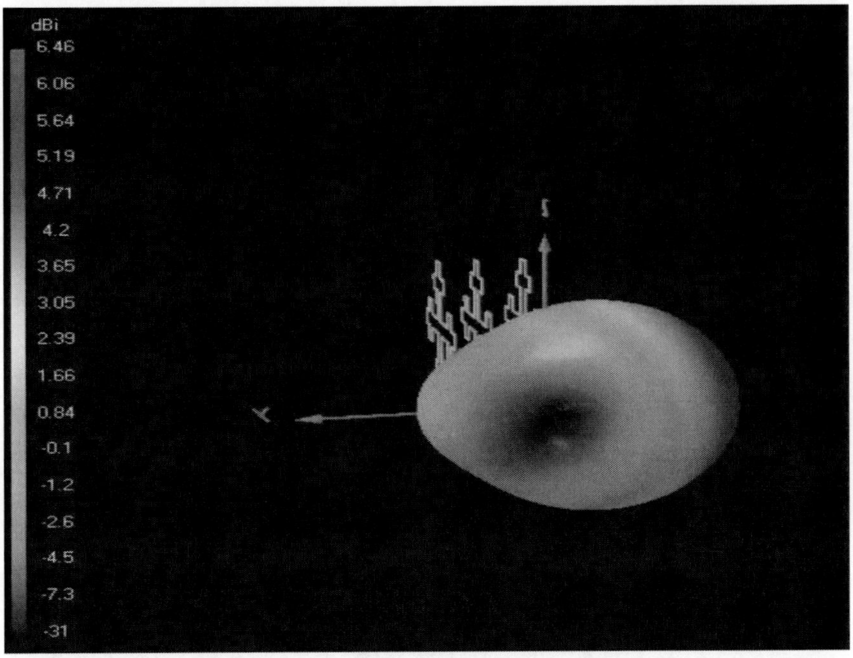

(b)

Figure 10. (a) 2-D pattern of antennas array (b) 3-D. Pattern antennas array at 2.10 GHz.

(a)

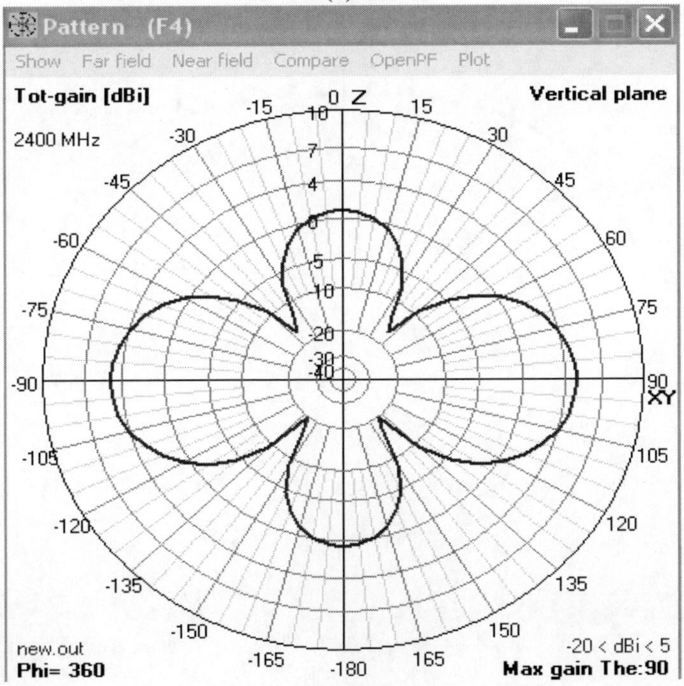

(a)

Figure 11. (Continued).

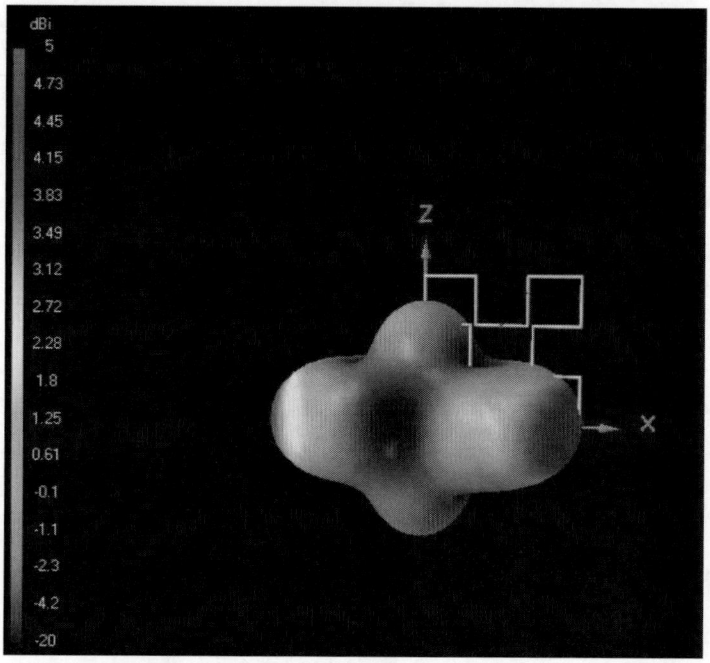

(b)

Figure 11. (a) 2-D Radiation pattern of proposed fractal loop antenna (b) 3-D radiation pattern at 2.10 GHz.

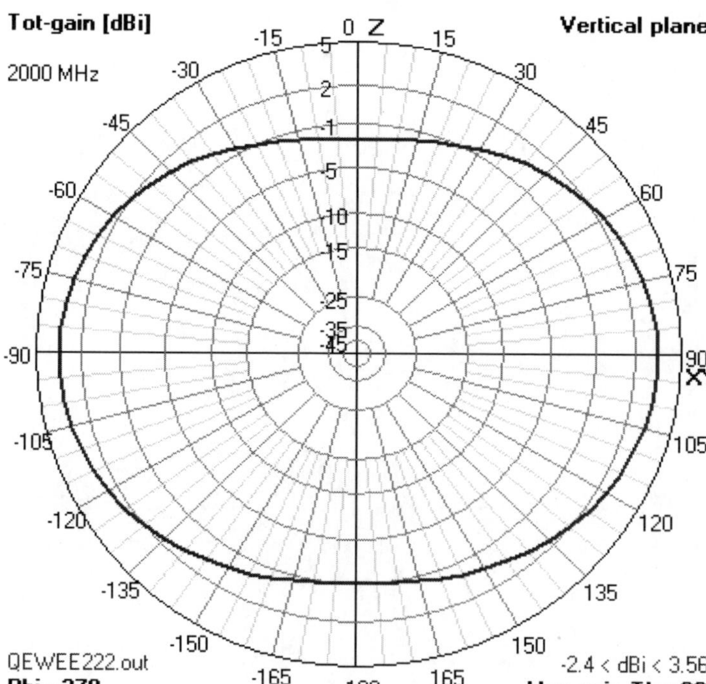

(a)

Figure 12. (Continued).

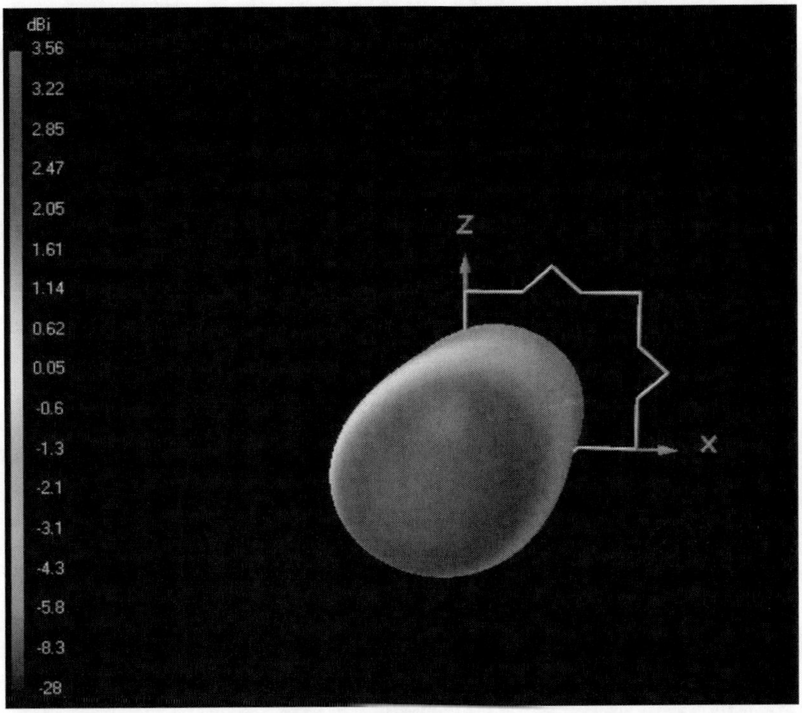

(b)

Figure 12. (a) 2-D Radiation pattern of proposed fractal loop antenna (b) 3-D radiation pattern.

V. VARIATION OF SWR AND REFLECTION COEFFICIENT

The variation of the values of SWR and reflection coefficient of basic loop antenna, shown in Figure 13, indicates that their values are 1.77699 and -11.063 dB, which are very close to ideal values i.e. 1 and -10 dB respectively. A single iteration brings about a change in the value of gain, the SWR and reflection coefficient data. The values of standing wave ratio are close to the coveted value of reflection coefficient i.e. -25dB at 2126 MHz (Figure 14). However, from the figure 15, it is clear that on carrying out second iteration the values of both the SWR as well as the reflection coefficient improves. The new value of SWR becomes 1.16586 whereas reflection coefficient; -22.318 dB at 1931 MHz.

However the values of SWR and reflection coefficient, for first iterated three element fractal loop array antennas are found to be 1.77699 and -10.063 dB respectively at 2050 MHz, figure 16. From the figure 17, it is clear that on carrying out iteration the values of both the SWR as well as the reflection coefficient further improves. The new value of SWR and reflection coefficient become 1.4489 and -14.736 dB at 1850 MHz for 2nd -iterated three element fractal loop array antennas. However from Figure 18 it is clear that the new SWR becomes 1.2738, which is further closer to 1 and reflection coefficient becomes -18.386dB at 2580 MHz. Similar variations are observed at a frequency of 1950 MHz as shown in figure 19. The obtained results have also been summarized in Table 1.

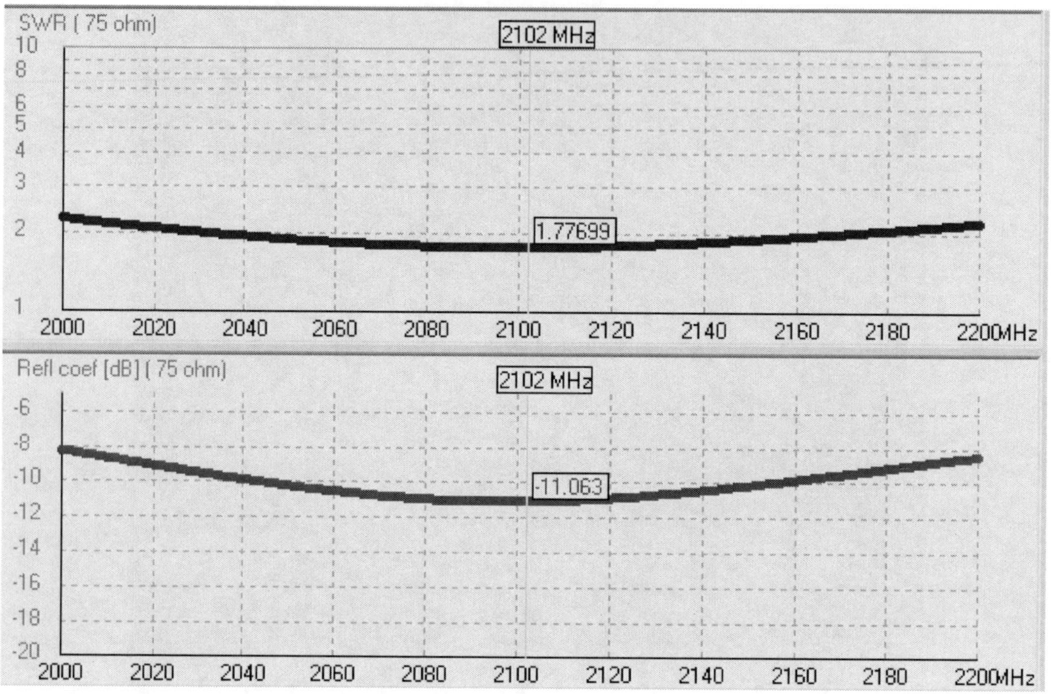

Figure 13. SWR and reflection coefficient of the basic loop antennas at 2.00 GHz.

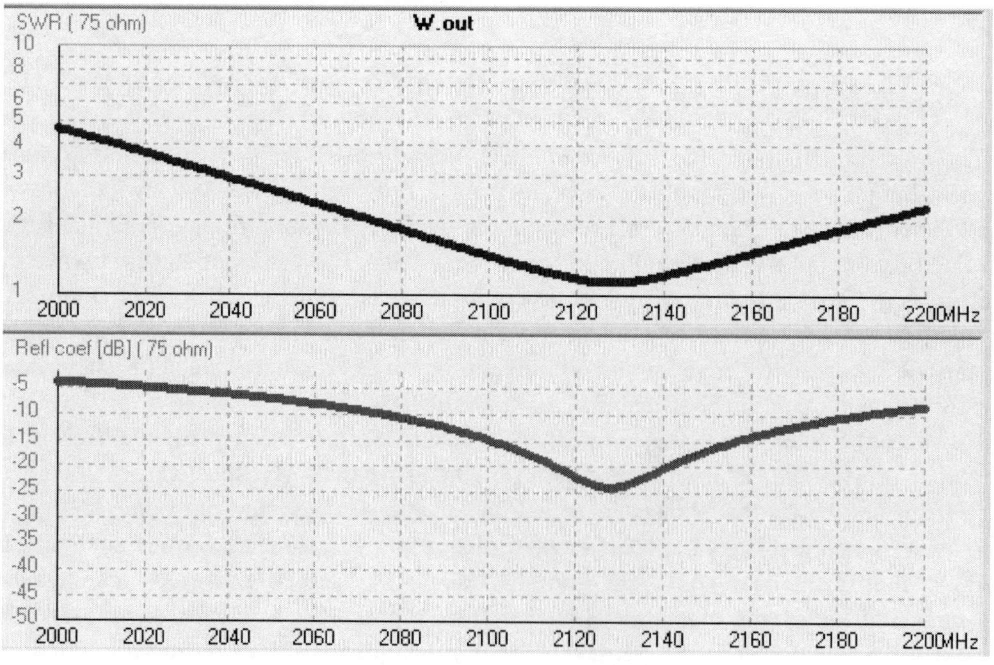

Figure 14. SWR and reflection coefficient of the 1st iteration fractal loop antennas at 2.00 GHz.

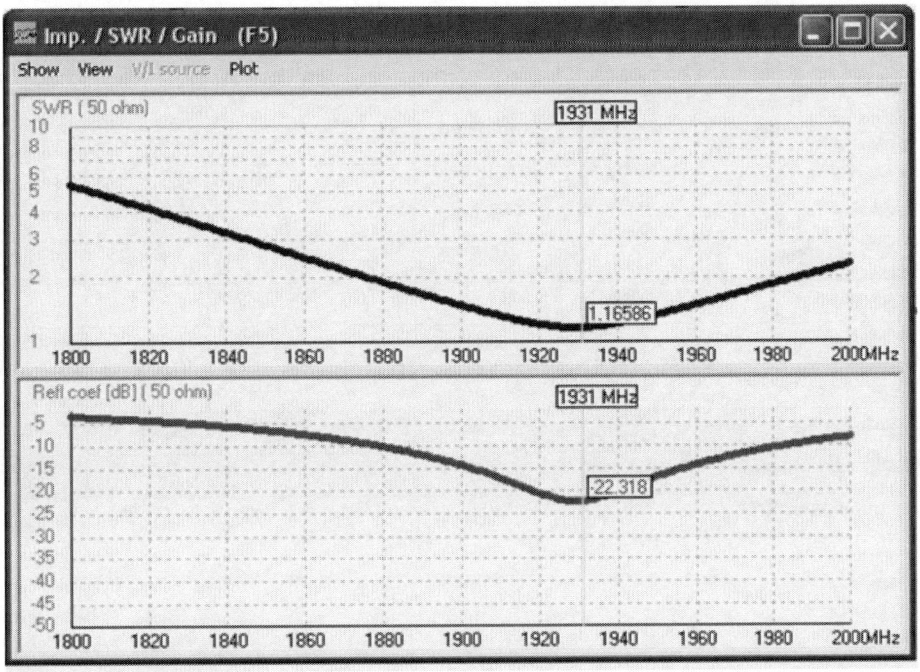

Figure 15. SWR and reflection coefficient of 2nd iterated fractal loop antennas at 2.00 GHz.

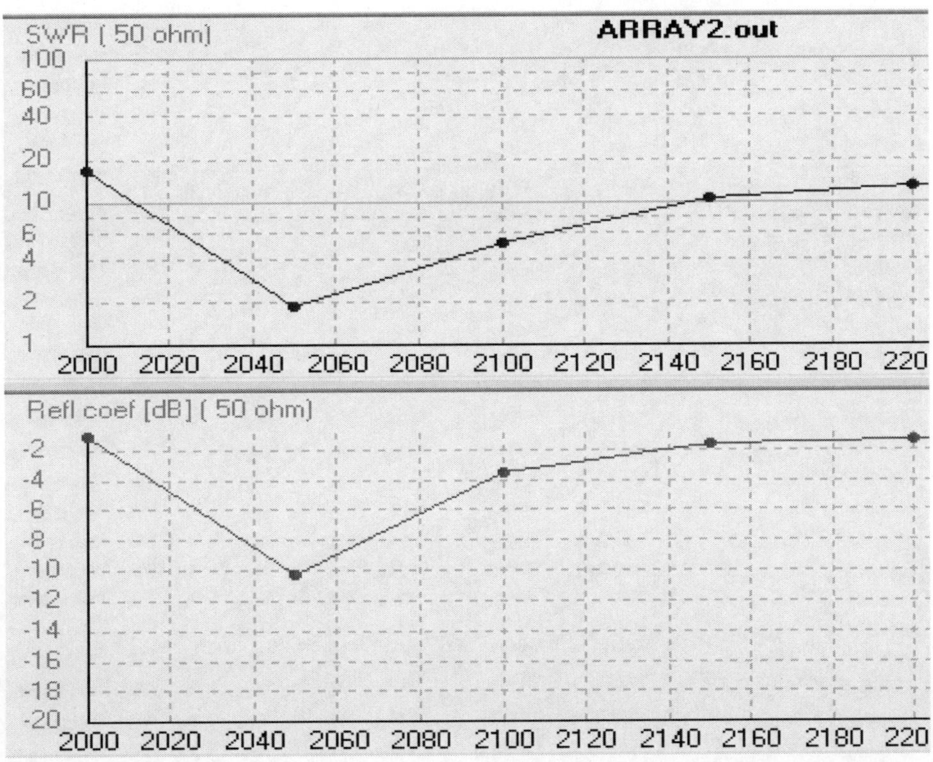

Figure 16. SWR and reflection coefficient of 1-iterated three element fractal loop array antennas at 2.1GHz.

Table 1. Parameters of various proposed fractal antennas

Proposed antennas types	Designed Frequency (GHz)	Simulated Frequency (MHz)	Gain (dB)		SWR	S_{11} (dB)
			Max	Min		
Loop antenna	2	2000	3.32	-0.6	1.77699	-11.063
1-Iteration	2	2126	2.78	0.79	1.25	-25
2-iteration	2	1931	2.56	1.29	1.16586	-22.318
1-Iteration array	2.1	2050	7.67	-19	1.77699	-10.063
2-Iteration array	2.1	1850	8.45	-11	1.4489	-14.736
Inverse Fractal antenna	2.1	2580	5	0.2	1.2738	-18.386
Loop antenna array	2	1950	3.56	2.4	1.69041	-11.814

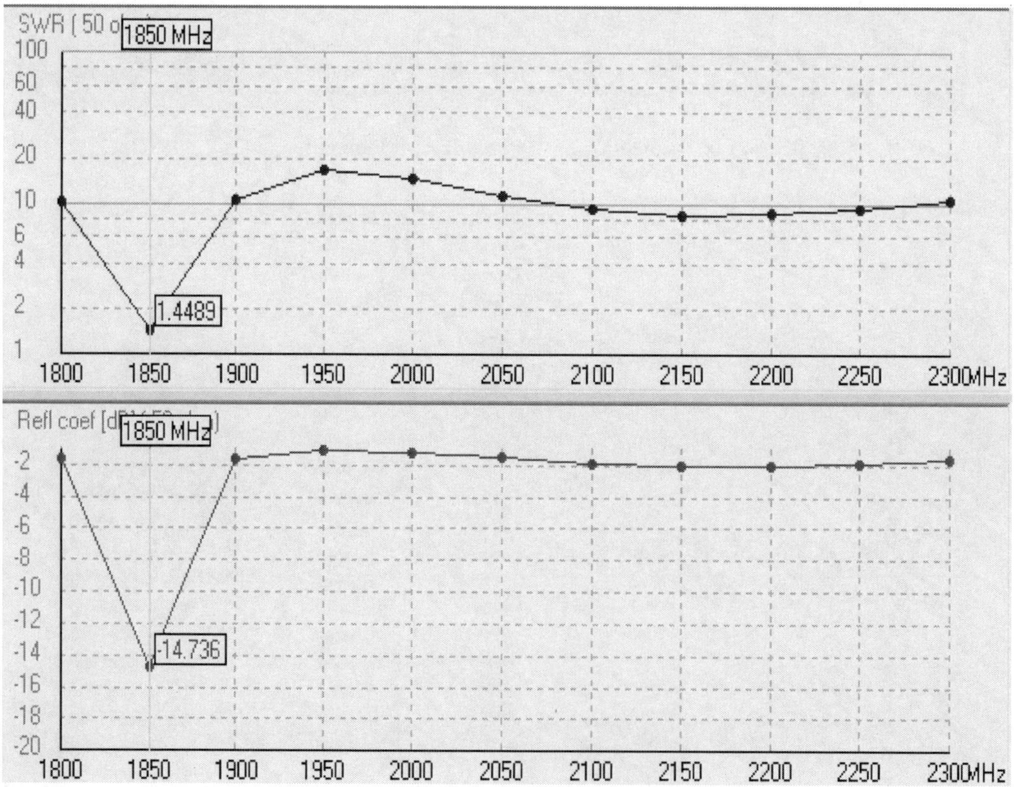

Figure 17. Variation of VSWR and reflection coefficient of three elements array at 2.1 GHz.

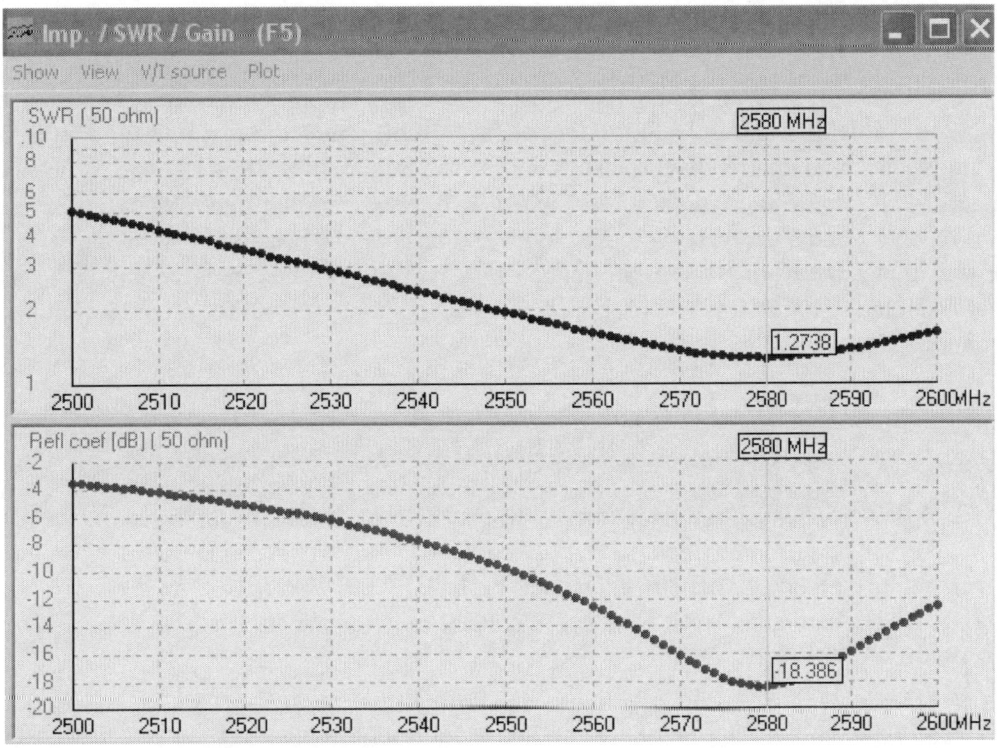

Figure 18. SWR and Reflection coefficient.

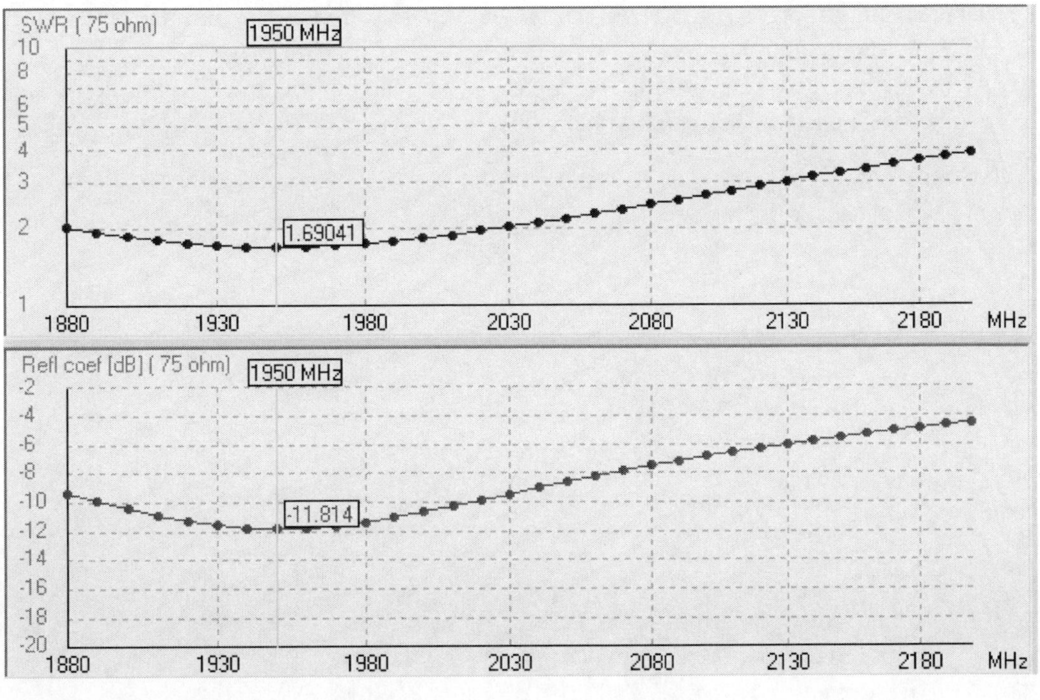

Figure 19. SWR and Reflection coefficient.

CONCLUSION

From the above study, it is concluded that SWR of the proposed antenna systems reaches the coveted value of 1 and reflection ratio reaches -10 dB which are required for appropriate matching, hence maximum power transformation. The minimum SWR = 1.16586, is observed in case of 2nd-iterated at frequency 1931 MHz. It is also seen that due to self-similar scale factor of the fractal geometry, fractal loop antennas have been shown to demonstrate improved impedance and SWR performance in a reduced physical area compared to non-fractal Euclidean geometries. The proposed antennas may very useful for direction finding in radar applications.

REFERENCES

[1] R. J. Eassom, R. L. J. Awcock and B. M.Sosin "HF Transmitters and receivers for naval radio", *Years of radio, Conference Publication 411, IEEE*, 5-7 September 1995.

[2] H. White de Eside, "The loop antenna as probe", PP.291-296 *IEEE Trans.*1964.

[3] N. H. Shepherd and W. G. Chaney, "Personal radio antennas," pp.23-31, IEEE *Trans.* 1995.

[4] D. Foster, "Loop antennas with uniform current", *Proceedings of the I.R.E*. Nov.1944.

[5] H. Gutton, and G. Baissinot, "Flat aerial for ultra-high frequencies", *French Patent no.* 703 113, 1955.

[6] J. Wojciech, and Krzysztofik, "Modified Sierpinski fractal monopole for ISM –bands handset applications", *IEEE Transactions on Antennas and Propagation*, vol. 57 No.3, pp. 606, March, 2009.

[7] S. R. Best, "The fractal loop antenna: A comparison of fractal and non-fractal geometries", pp.146-149, *IEEE* Trans. 2001.

In: Biology and Communication in the Information Age
Editor: Rafiqul Islam
ISBN: 978-1-62808-252-4

Chapter 12

A SURVEY ON XOR BASED ERASURE CODES FOR CANONICAL STORAGE SYSTEMS

Jun Feng, Yu Chen and Douglas H. Summerville
Department of Electrical and Computer Engineering,
State University of New York at Binghamton, Binghamton, NY, US

ABSTRACT

Reliable data storage has been recognized as one of the main dimensions of information technology. Traditionally, the storage systems achieve higher fault tolerance and reliability by introducing redundancy. Mirroring technology is a simple and reliable method but would result in low storage efficiency. In recent years, researchers have revealed that many XOR (exclusive-OR) erasure codes can be applied to storage systems. This paper presents a survey on state-of-the-art development in XOR based erasure codes for the construction of data storage technologies, and discusses their characteristics. In this paper, we have divided XOR erasure codes into two categories: namely Specialized Tolerance Codes and more powerful General Tolerance Codes. The characteristics and encoding/decoding algorithms are described in detail. We hope this survey will inspire more active research in this area.

Keywords: Mass storage, Design, Algorithms, Documentation, XOR Operation, erasure code, data security, fault tolerance, distributed storage System

1. INTRODUCTION

Canonical storage systems introduce additional information to support fault tolerance and improve reliability (Pinheiro et al. 2007; Schroeder and Gibson 2007) so that if one or more disks fail they can be recovered by accessing data from the remaining disks. To date, there are two different categories of fault-tolerant techniques: mirroring technology and erasure-coding based technologies.

The principle of mirroring technology is straightforward. By duplicating entire sets of data, such as disks, files or blocks, on multiple physical devices, the system can continue to provide reliable services in the event of a device or storage malfunction. However since such reliability is achieved by accessing more storage resources, the cost of adding such devices can quickly become excessive.

In contrast, erasure-coding technologies can provide a tradeoff by offering the same reliability but at a lower cost (Hellerstein et al. 1994; Kempf 2007; Plank 2005; Rabin 1989; Reed and Solomon 1960). Erasure coding adds redundant data to each record, so that the original data can be retrieved from a subset of that record. Thus, this technology has become very popular for canonical storage and has been adopted in many practical systems.

In canonical storage systems, XOR (exclusive OR) based erasure codes are preferable for high performance (Bindel et al. 2000; Huang et al. 2007b). For this reason, many XOR based erasure codes have been proposed in recent years (Plank 2005; Blaum and Roth 1993; Cassuto and Bruck 2006).

In this survey, we will introduce some XOR based erasure codes that are being reported for storage systems. We put them in two categories according to fault tolerance namely: Specialized Tolerance Codes (STCs) and more powerful General Tolerance Codes (GTCs). "Specialized" means that there is an upper bound for fault tolerance. When the number of failed disks exceeds the bound, failed disks cannot be recovered even if more parity disks are added. In contrast, "General" means failed disks can be recovered with enough parity disks. There are other applications for codes besides canonical storage systems such as LDPC and network codes. In particular, LDPC's are used in wide-area content-distribution frameworks, and network codes are used when data "lives" on a collection of distributed storage nodes. Since they have been surveyed by other researchers (MacKay 2005; Thul 2003; Liva 2006; Brack 2007; Dimakis et al. 2011), we will not discuss them in detail.

The rest of this paper is organized as follows: Section 2 specifies the terms and introduces the basic coding and decoding ideas to be used in this paper. Then, according to their categories, Section 3 describes STCs and Section 4 discusses GTCs. Section 5 summarizes the characteristics of all the algorithms and provides some insight. Section 6 summarizes the paper.

2. Terms and Basic Ideas

In order to help our readers understand the XOR based codes, we will first present the definition of terms to be used, give a brief introduction to RAID (Redundant Array of Independent Disks) due to their popularity in storage systems, and then discuss the concepts used to build the algorithms, and include illustrations and examples. Please note that we use " $\oplus$ " to stand for the XOR operation.

2.1. Definitions of Terms

Device: Any type of related storage, each of which can fail independently from each other. It can be a node in a distributed storage or a disk in storage systems.

Symbol: The fundamental unit which is usually a block or a disk in one device. It also could be a byte or a bit in one code. In the following section, we assume each symbol has only one bit so there is no confusion. Some are called "data symbols" because they store "data". Others store the XOR result and are called "parity symbols".

Chain: A set of data and parity symbols where the parity symbol is generated by XORing the data symbols. For example, if $c_0 = d_0 \oplus d_1 \oplus d_2$, the set is $\{d_0, d_1, d_2, c_0\}$. It is called "Parity chain c_0".

Optimal Code: Suppose there are n input data symbols and $n+r$ output. If any n of $n+r$ symbols can reconstruct the original n data symbols, the code is optimal. Optimal Code is also known as a Maximum Distance Separable (MDS) Code.

Update Complexity: The average number of parity symbols affected by a data symbol.

Storage Efficiency: The ratio of data symbols to the total of all symbols, and normally it is the ratio of $n/(n+r)$. It is also called the "rate" of the code.

Horizontal Code: Horizontal Code devices hold data or parity.

Vertical Code: Vertical Code devices hold both data and parity (Hafner 2005a0.

In-degree: the number of data symbols that are involved in computing a parity symbol.

Syndrome: In this paper, when there are some symbols that failed in one chain, we XOR the remaining symbols together. The result is called "syndrome". For example, suppose the chain is $\{d_0, d_1, d_2, c_0\}$. If d_1 is failed, the "syndrome" is $d_0 \oplus d_2 \oplus c_0$.

In order to avoid confusion and inconsistency, for horizontal codes, "n" stands for the number of information devices, and "r" is the number of parity devices. For vertical codes, "n" stands for the number of occurrences of data and "r" is the number of parity symbols in one device. Also "d" indicates the data symbol and "c" corresponds to the parity symbol. And normally, "i" is used to specify the row and "j" is used to specify the column. For convenience, we use d_i and d_{ij} for data symbols in different situations to avoid confusion. Also, we use c_i and c_{ij} for parity symbols.

2.2. A Brief Introduction to RAID

Because RAID 2, 3, and 4 are not commonly used in industry, we do not discuss them in this paper. In this section, an introduction of RAID systems 0, 1, 5, and 6 are presented. They are more commonly used in industry, and RAID (Patterson et al. 1988) is also used to improve the reliability and availability of data.

RAID-0 distributes the data into different disks. It can improve both read and write performance, and the storage efficiency is 100%. But it cannot provide fault tolerance. Any disk failure is unrecoverable.

RAID-1 is also called "mirror" technology. The data in one disk is copied to multiple disks. Suppose there are $1+r$ disks in total. It can provide r fault tolerance, but the storage efficiency is only $1/(1+r)$.

RAID-5 backs up data information with parity symbols and it can tolerate only one disk failure. It has gained popularity because of the low cost associated with redundancy.

In recent years there has been an increasing need for a more fault-tolerant RAID in large scale storage systems. Such a requirement has led to the RAID-6 system. RAID-6 is different from other RAID methods. It has no specific algorithms (Wikipedia 2011). It uses independent data disks with two independent distributed parity schemes (IBM 2009). Any algorithm that meets the requirement in IBM (2009) can be applied on any RAID-6 system.

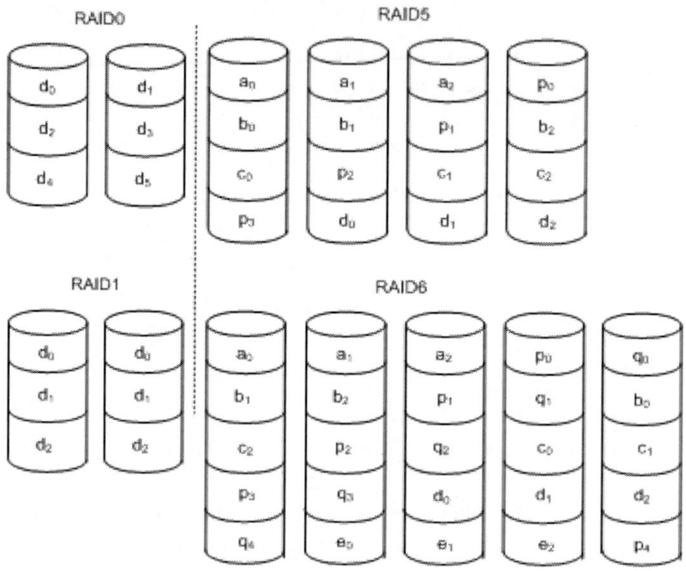

Figure 1. Illustration of RAID systems.

In this paper, we do not focus on how to distribute parity symbols along disks However it is worth mentioning that parity symbols can be moved to one device in a RAID-5 system. The device is called a "*P*" device. For a RAID-6 system, parity symbols can be moved to two parity devices, they are called "*P*" and "*Q*" devices. We redraw the two systems in Figure 2.

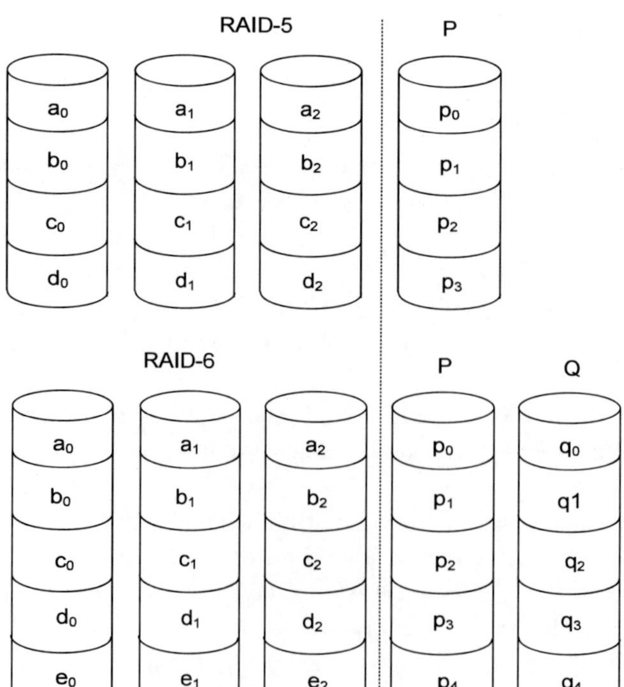

Figure 2. RAID-5 and RAID-6 systems.

Thus a RAID-5 system can be extended into a RAID-6 system by adding another parity disk to tolerate two disk-failures. The Reed-Solomon code can be implemented on a RAID-6 system (Lin et al. 2004). Plank gives a tutorial about the algorithm (Lin et al. 2004; Plank 1999). Some XOR based algorithms can be found in reference (Blaum et al. 1995; Corbett et al. 2004; Plank 2008) and we will introduce them in Section 3.

2.3. Examples and Basic Ideas of Chain Based Codes

There are two ways to build XOR based codes. One is based on the application of matrices, which use multiplication and an inverse operation to code and decode symbols. We do not give an example here because the principle is simple. The other is based on chain operation. The following is an example that illustrates the basic principles of chain based erasure codes.

Suppose there are seven devices, five for data and two for parity. Then n=5 and r=2, which can be arranged as a 5 rows × 7 columns array. Each column stands for one device. Data symbols are labeled as d_{ij}. The symbols in the P or Q device can be labeled as p_i and q_i, as shown in Figure 3.

d_{00}	d_{01}	d_{02}	d_{03}	d_{04}	p_0	q_0
d_{10}	d_{11}	d_{12}	d_{13}	d_{14}	p_1	q_1
d_{20}	d_{21}	d_{22}	d_{23}	d_{24}	p_2	q_2
d_{30}	d_{31}	d_{32}	d_{33}	d_{34}	p_3	q_3
d_{40}	d_{41}	d_{42}	d_{43}	d_{44}	p_4	q_4

(a). Horizontal code.

d_{00}	d_{01}	d_{02}	d_{03}	d_{04}
d_{10}	d_{11}	d_{12}	d_{13}	d_{14}
d_{20}	d_{21}	d_{22}	d_{23}	d_{24}
c_{00}	c_{01}	c_{02}	c_{03}	c_{04}
c_{10}	c_{11}	c_{12}	c_{13}	c_{14}

(b). Vertical Code.

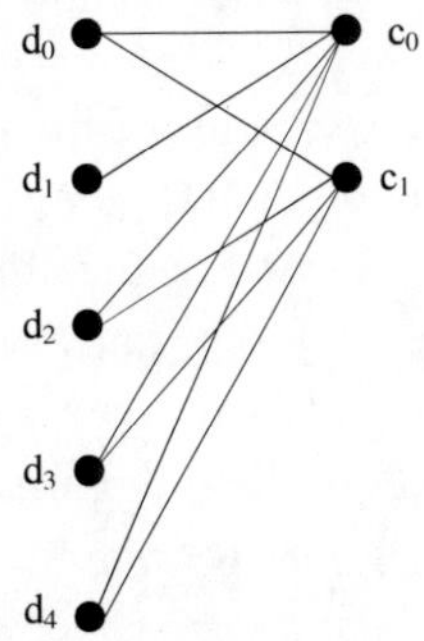

(c) Graph definition.

Figure 3. Examples that illustrate the definition of symbols.

In Figure 3(a), if the parity symbol p_2 is calculated from five data symbols d_{00}, d_{21}, d_{12}, d_{33} and d_{44}, the set of $\{d_{00}, d_{21}, d_{12}, d_{33}, d_{44}, p_2\}$ is called chain p_2. And $p_2 = d_{00} \oplus d_{21} \oplus d_{12}$

$\oplus\ d_{33} \oplus d_{33}$. If any five of the seven symbols can reconstruct the original five data symbols, the code is "optimal". The storage efficiency is 5/(5+2) = 5/7. For each chain, if the symbols can be put into distinctive devices as shown in Figure 3(a), it is a "horizontal code". Otherwise, as illustrated in Figure 3(b), it is a "vertical code" where parity symbols are labeled as c_{ij}. In Figure 3(c), there are five connections with parity symbol c_0, and the in-degree is 5.

The basic rationale of the XOR based code is described here. Suppose there are n data symbols, $d_0, d_1, \ldots, d_{n-1}$. The parity symbols are $c_i = d_0 \oplus d_1 \oplus \ldots \oplus d_{n-1}$.

Figure 3(c) shows how the two parity symbols are generated graphically. Since $c_0 = d_0 \oplus d_1 \oplus d_2 \oplus d_3 \oplus d_4$ and $c_1 = d_0 \oplus d_2 \oplus d_3 \oplus d_4$, they form two chains:

chain c_0: $\{ d_0, d_1, d_2, d_3, d_4, c_0\}$; and

chain c_1: $\{d_0, d_2, d_3, d_4, c_1\}$.

If there is only one symbol missing in a chain, it can be recovered by XORing other symbols in the same chain. For example, assume $\{d_1\ d_2\ d_3\ d_4\}$ = 10111, then $c_0 = 0$. If d_2 is missing, it can be recovered: $d_2 = c_0 \oplus d_0 \oplus d_1 \oplus d_3 \oplus d_4 = 1$. This is also the principle of RAID-5. Obviously, if any two or more symbols are missing in one chain, then neither of them can be reconstructed since two or more unknown variables cannot be solved by only one equation. For example, if d_1 and d_2 are missing in the chain c_0, they cannot be recovered from the remaining data. However, if we know one of the two missing symbols in advance, then the other one can be recovered. For example, if d_1 is an imaginary symbol of 0, then d_2 can be recovered by XORing other data in chain c_0: $d_2 = c_0 \oplus 0 \oplus d_0 \oplus d_1 \oplus d_3 \oplus d_4$. This idea has been used to construct the code schemes and find the starting chain. Alternatively, if one of the two missing symbols can be recovered by other chains, then the other one also can be recovered. For example, if d_2 can be recovered from chain c_2, then d_1 can be recovered from chain c_1 successively. The latter approach is used in recovering the remaining missing data. XOR based codes can be described as follows:

1. *Coding process*: Select data symbols in different devices according to an algorithm, XOR them and generate the parity symbols.
2. *Decoding process*: For the chain-based codes: 1. After the first step, find the starting chain where only one data symbol is missing and recover it. 2. Find the successive set in which only one data symbol is unknown, recover it. 3. Iteratively repeat the second step until all the missing data are recovered. Matrix based codes can be decoded using matrices.

The main difference among the XOR based codes lies in the construction of parity symbols and the performance of each individual procedure. There are trade-offs. Some codes perform better based on certain perspectives, while for others properties may dictate better performance.

Another view of the XOR based codes is predicated on the use of matrix theory. Suppose the n data devices and r parity devices can be mapped into $n + r$ rows and n columns. If any n of $n + r$ rows are invertible, the code has r fault tolerance.

For example, suppose there are only 3 data devices and one parity device. The code algorithm is $c = d_0 \oplus d_1 \oplus d_2$. The algorithm can be rewritten using the matrix:

Although graphs, arrays, matrices and equations can be used to illustrate the codes, none of them can clearly and concisely explain all the XOR based codes. For example, an array form generally is used to illustrate RAID systems (Plank 2005); a graph explains the Tornado code (Luby et al. 1997) more clearly and matrix expressions can explain the Feng's codes (Feng et al. 2005a; Feng et al. 2005b) more concisely.

$$\begin{pmatrix} 1 & 0 & 0 \\ 0 & 1 & 0 \\ 0 & 0 & 1 \\ 1 & 1 & 1 \end{pmatrix} \times \begin{pmatrix} d_0 \\ d_1 \\ d_2 \end{pmatrix} = \begin{pmatrix} d_0 \\ d_1 \\ d_2 \\ c \end{pmatrix}$$

If d_1 device is erased, we can delete the second row and recover d_1 as follows:

$$\begin{pmatrix} d_0 \\ d_1 \\ d_2 \end{pmatrix} = \begin{pmatrix} 1 & 0 & 0 \\ 0 & 0 & 1 \\ 1 & 1 & 1 \end{pmatrix}^{-1} \times \begin{pmatrix} d_0 \\ d_2 \\ c \end{pmatrix}$$

Actually, the code above is a RAID-5 scheme. We can also map the matrix into an array form illustrated as:

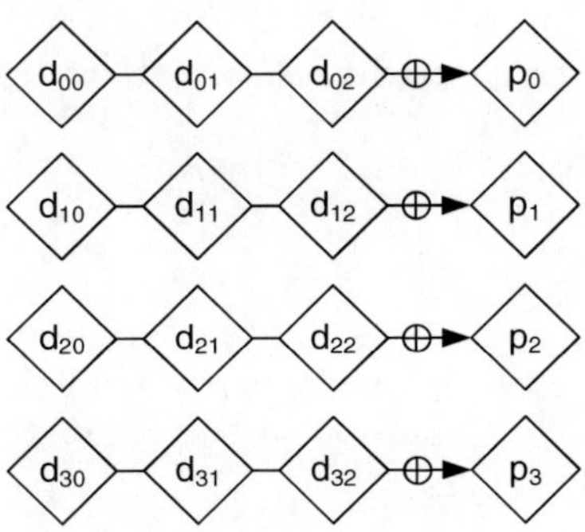

In the following sections, we will explain the different code schemes using different forms. Our intent is to make them more understandable to our readers. We will also insert the original equations in each code scheme if it is necessary.

Since the symbol index operates on modular arithmetic, we define <*a*>prime = *b* if and only if $b \equiv a$ mod prime and $0 \le b \le$ prime – 1, where "prime" is the minimum prime number which is larger than *n*. For example, if *n* is 9, the "prime" is 11. For convenience, we will use <*a*> instead of <*a*>prime or a directly so there is no confusion. For example, if we say "3×(*n*-1)", we mean "<3× (*n*-1)>prime".

3. SPECIALIZED TOLERANCE CODES

Some STCs, such as EVENODD (Blaum et al. 1995), Row-Diagonal Parity (RDP) codes (Corbett et al. 2004), STAR (Huang and Xu 2008), and X-code (Xu and Bruck 1999), are MDS codes, but they have a limited fault tolerance that is not higher than 3. Some other codes can provide higher fault tolerance, but they suffer from other disadvantages. For instance, the WEAVER code (Hafner 2005a) has a higher fault tolerance up to 12. However, it is a non-MDS code and has low storage efficiency, usually less than 50%.

In this section, we will introduce the EVENODD code first, and then the X code and the RDP code since the latter two codes are evolved from the EVENODD code. After that, the S code and the STAR code are discussed. The S code is an extension of the X code and the rationale behind the STAR code was derived from the EVENODD and X code. We will also address the Liberation code whose principle comes from matrix and isomorphism functions. The P code, which has a flexible association that falls between the parity and data locations will be reviewed. Finally "the "Weaver code derived from graph theory, and the HoVer code which provides a framework for any of the codes introduced herein will be investigated. The HoVer code is unique in that any algorithm that meets code specifications can be applied as the framework code.

3.1. EVENODD Code

In EVENODD code (Blaum et al. 1995), there are n data and 2 parity devices. Each device consists of n-1 symbols. The whole device can be analog to an (n–1) row × (n+2) column array. In order to make the encoding and decoding processes clearer, one imaginary row is added below the array. The data in the imaginary row are all set to 0. Now the array is an n×(n+2) array. The encoding procedure is expressed in the equations below. Let

$$S = \bigoplus_{t=1}^{n-1} d_{n-1-t,t}, \quad Q_i = \bigoplus_{t=0}^{n-2} d_{i-t,t} \tag{1}$$

Q_i means the parity of data symbols in the chain. For each row x, the parity symbols are obtained according to the following:

$$p_i = \bigoplus_{t=0}^{n-1} d_{i,t}, \tag{2}$$

and

$$q_i = S \oplus Q_i, \tag{3}$$

where i = 0, … , n-2.

Suppose n = 5. The coding process can be illustrated in Figure 4. The symbols, colored yellow are in the imaginary row at the bottom, and all are set to 0. All their symbols in one chain are connected by a line. We say the slope of one straight line is a means that there two symbols (i, j) and $(i\text{-}a, j+1)$ are on the line. The lines in Figure 4(a) have a slope of 0, and the lines in Figure 4(b) have a slope of 1.

If we only look at the data and P device, Eq. (2) and Figure 4 present a RAID-5 system. Actually for all the XOR based MDS codes, the first parity device is always a *P* device. Thus, in the following sections, we will skip the *P* device generation.

Regarding the *Q* device, there are five chains with a slope of 1 that are used to derive the *Q* parity results as shown in Figure 4(b). Since there are only four symbols in the *Q* device, one of the five parity results, *S*, has to be hidden within the others by an XOR operation. For example, first the parity result *S* is obtained from d_{31}, d_{22}, d_{13}, d_{04}, and then the parity q_0 can be calculated by XORing the d_{00}, d_{32}, d_{23}, d_{14} with *S*, as illustrated in Eq. (3).

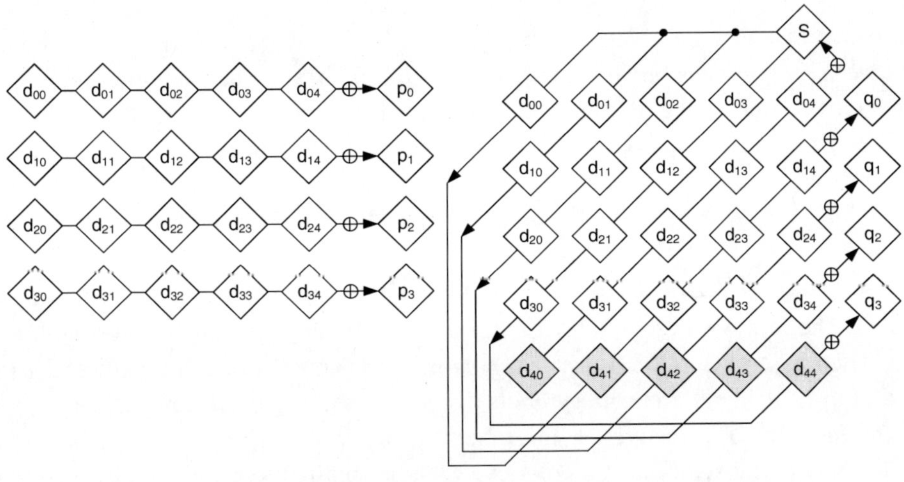

(a). Chains with Slope 0. (b). Chains with Slope 1.

Figure 4. Encoding scheme of EVENODD, *n*=5 (Plank 2008).

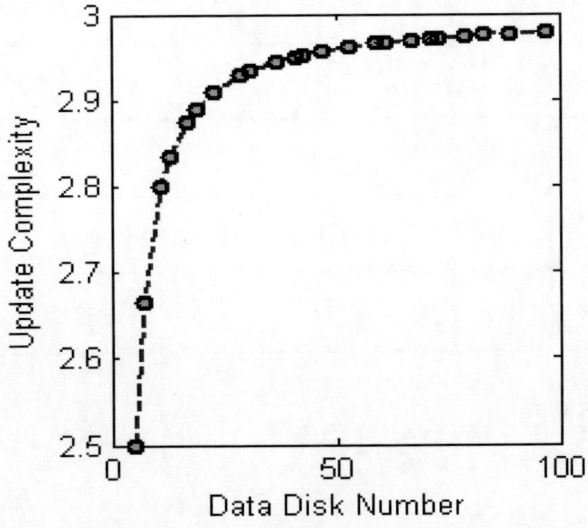

Figure 5. EVENODD Update Complexity.

In Figure 4, when any symbol besides d_{31}, d_{22}, d_{13} or d_{04} is modified, one P symbol and one Q symbol have to be re-generated since they participate directly in the parity calculation. The update complexity is 2. When any symbol in d_{31}, d_{22}, d_{13}, or d_{04} is modified, we have to modify one P symbol and all the Q symbols, because they are involved in all Q parity calculations. The update complexity is n. Then the total update complexity is $(((n\times(n-1)-(n-1))\times 2 + (n-1)\times n)/(n\times(n-1))$. It is close to three when the number of disks increases, as shown in Figure 5.

EVENODD can recover any two failed devices. The detailed decoding process can be found in reference (Blaum et al. 1995). Here we only illustrate the rationale behind the decoding process through an example and focus on a typical case where both of the failed devices are data devices. Other cases such as when the parity device is erased can be handled easily.

Example #1

Suppose n is 5. The storage system is like a 4x7 array. Each column stands for one device. Suppose columns 0 and 2 have been erased. It means each chain has two erased data symbols. As discussed in Section 2, a chain with one unknown symbol is needed as a starting chain. And prior to selecting a starting chain, all parity symbols with a slope of 1 must be recovered. Since all the symbols in the Q device are mixed with S, it is mandatory to recover S first.

As observed from Figure 4, the result of XORing all the symbols in the P device is $p_0 \oplus p_1 \oplus p_2 \oplus p_3$. The four S's in the Q device can be cancelled by XORing all the symbols in the Q device. Then by XORing the parity symbols in the P and Q devices together, the result is equal to that of XORing all the data symbols except the set of $\{d_{40}, d_{31}, d_{22}, d_{13}, d_{04}\}$, which is S. After the value of S is recovered, the other parities with a slope of 1 can be derived by XORing the S with the symbols in the Q device. Suppose the recovered Qi and S symbols are shown in Figure 6. After all the parity symbols are recovered, we should first find the starting chain to initiate the decoding process.

?	0	?	1	0	1	0
?	1	?	0	0	0	0
?	1	?	0	0	0	1
?	1	?	1	1	1	0
0	0	0	0	0	S=1	

Figure 6. A decode example (n = 5).

?	0	?	1	0	1	0
1	1	?	0	0	0	0
?	1	0	0	0	0	1
?	1	?	1	1	1	0
0	0	0	0	0	S=1	

(a)

?	0	?	1	0	1	0
1	1	0	0	0	0	0
1	1	0	0	0	0	1
?	1	?	1	1	1	0
0	0	0	0	0	S=1	

(b)

?	0	0	1	0	1	0
1	1	0	0	0	0	0
1	1	0	0	0	0	1
0	1	?	1	1	1	0
0	0	0	0	0	S=1	

(c)

0	0	0	1	0	1	0
1	1	0	0	0	0	0
1	1	0	0	0	0	1
0	1	0	1	1	1	0
0	0	0	0	0	S=1	

(d)

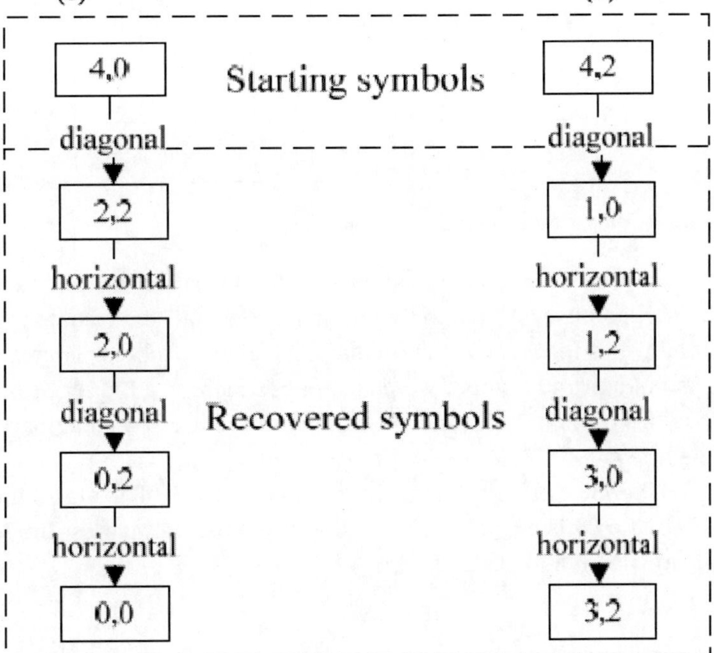

(e). Recovery sequence.

Figure 7. Illustration of the recovery procedure of the EVENODD code.

Through an iterative looping process, two deleted imaginary symbols d_{40} and d_{42} can be found, which lead to two starting chains namely diagonal chain S $\{d_{31}, d_{22}, d_{13}, d_{04}, S\}$ and diagonal chain Q_1 $\{d_{33}, d_{24}, d_{10}, d_{01}, d_{10}, Q_1\}$. The first two points d_{22} and d_{10} can be located and recovered by adjusting factor S and the parity symbol. According to Eq. (1) and Eq. (3), $d_{22} = 0$ and $d_{10} = 1$ as shown in Figure 7(a). Symbols in the same chain are color coded for convenience. The second iterative loop derives two recovered symbols with horizontal parity as shown in Figure 7(b).

Therefore, d_{20} = 1 and d_{12} = 0. After that, let's consider the chains that contain the symbols d_{20}, d_{12}, which are chain $\{d_{20}, d_{11}, d_{02}, d_{34}, d_{43}, Q_2\}$ and chain $\{d_{12}, d_{03}, d_{30}, d_{21}, d_{44}, Q_3\}$. The only missing symbols in these chains are d_{02} and d_{32} as shown in Figure 7(c). Thus it is straightforward to get d_{02} = 0 and d_{30} = 0. Again, using horizontal parity, it concludes that d_{00} = 0 and d_{32} = 0 in Figure 7(d). Now all the data symbols are recovered. Figure 7(e) gives the symbol recovering sequence.

This example illustrates that one iterative process is composed of one slope and one horizontal line search. Suppose device X and Y failed. For each iterative process, the difference in rows between each consecutive horizontal line is equal to (*x*-*y*), which is the difference in the failed column indexes. For example, the yellow line in Figure 7(b) is 2 and the next yellow line in Figure 7(d) is 0. The difference is 2.

				P	Q
?	0	?	1	1	0
d10	1	d12	0	0	0
?	1	?	0	0	1
0	0	0	0	S=1	

Figure 8. Illustration of the recovery procedure, n=4.

However, if the two recovered symbols are on the same row, it means there are no unknown symbols in a consecutive horizontal chain. Since all the remaining chains have two unknown symbols, we cannot recover the remaining failed symbols. Therefore, if the number of data devices is not a prime number we can suppose there is a bigger prime number p close to n. Each device has p-1 symbols, and we assume there are p-n imaginary devices to form the $p \times (p-1)$ data array.

The EVENODD code can only tolerate two failed devices. The algorithm in the extended version of the EVENODD is capable of tolerating up to eight erasures. Interested readers can find more details in Blaum and Bruck (1996).

3.2. X-Code

Because of the factor S, the EVENODD code cannot achieve the optimal update complexity of 2. Instead, it is approaching 3 as the number of devices increases. Xu and Bruck (1999) has studied an array code based on the concept behind the EVENODD code, and proposed the X-code that can achieve an optimal update complexity.

Different from the construction of EVENODD, the parity of the X-code is calculated by adding two parity rows rather than two parity columns. Therefore it is a vertical code. The X-code has two advantages over the EVENODD code. As a vertical code, it maintains an optimal updating complexity by combining information and parity symbols within columns.

The second advantage is that the X-code makes the computational load evenly distributed among all the columns.

The performance bottleneck that results from repeated writing operations has been eliminated. However, it cannot be shortened easily like horizontal codes can, which limits its application. In X-code, there are n devices, and n is a prime number. Each device has n symbols. Data stored in the first n-2 symbols in each disk can be viewed as an array of $(n-2)\times n$. The last two rows in each device are used to store parity symbols. Similar to the EVENODD scheme, parity symbols are constructed from data symbols along diagonals of certain slopes. Before we develop an encoding process, let's provide an algebraic description of the X-code. Assume d_{ij} is the symbol at the i-th row, j-th column. The parity symbols of the X-code are constructed according to the following encoding rules:

$$c_{0,j} = \sum_{k=0}^{n-3} d_{k,j+k+2}, \tag{4}$$

$$c_{1,j} = \sum_{k=0}^{n-3} d_{k,j-k-2}, \tag{5}$$

where $j = 0, 1, \ldots\ldots, n-1$.

Figure 9 illustrates how the X-code generates the parity symbols. Just like what we did previously in EVENODD, we add one imaginary row at the bottom (marked in yellow). Geometrically, we can think of the chain slopes as 1 and -1. The construction shows that the two parity rows are derived independently. Each data symbol can only affect exactly two parity symbols. Therefore the X-code has an update complexity of 2.

Suppose two devices failed. We can start an iterative recovery process by finding the chain with only one unknown symbol. Actually, for the X-code, we can find four starting chains. Each of them can start one iterative process. We illustrate the process in example #2.

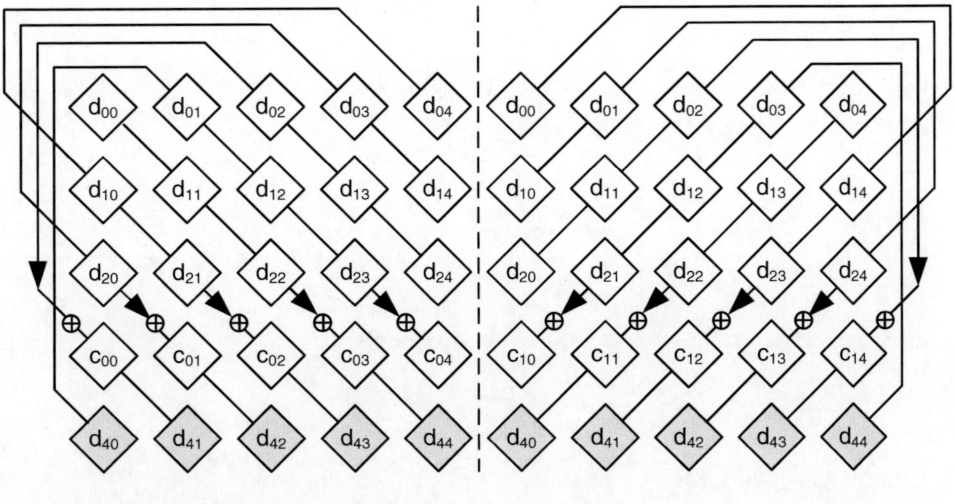

Figure 9. Illustration of X-code coding procedure, n=5.

Example #2

Suppose device 1 and 3 failed. We can recover four symbols at first, d_{23}, d_{11} and d_{21}, d_{13} in Figure 10 (a) and 10(b). Then iteratively, we can recover the next four symbols as shown by Figure 10(c) and 10(d). In the last loop, the remaining last two erased symbols are recovered, as shown in Figure 10(e) and 10(f). Figure 10(g) shows the recovery sequence.

Similar to EVENODD, the distance between two recovered symbols in the same column d is $2\times(x\text{-}y)$. Since the $2\times(x\text{-}y)$ should be a prime number with any number of rows n, then n should be a prime number.

If the number of disks is not a prime number, we have to choose a bigger prime number p close to n and add $p\text{-}n$ devices. Each data device has p symbols. In the added $p\text{-}n$ devices, the data symbols colored yellow can be set to a fixed value as shown in Figure 11. Figure 11 also shows that the X code cannot be shortened. Even though the number of data devices is not a prime number the number of parity symbols in one row should be a prime number.

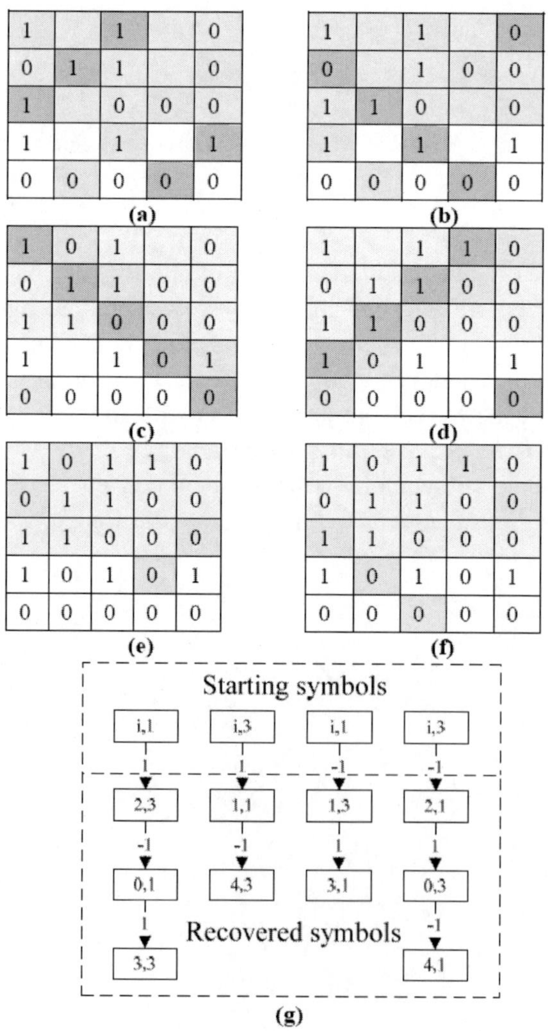

Figure 10. Illustration of the recovery procedure.

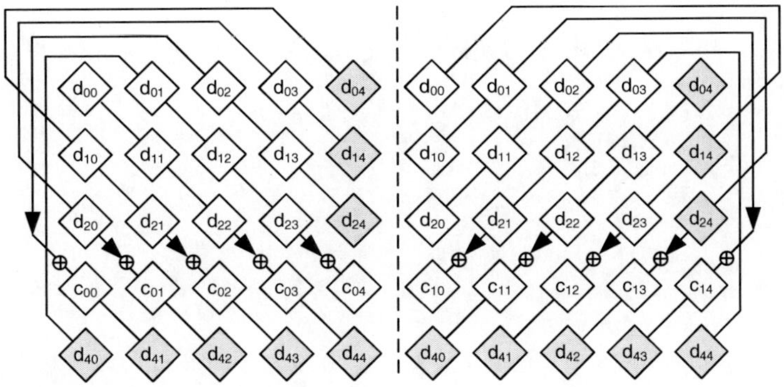

Figure 11. Illustration of X-code (n is not prime).

3.3. Row-Diagonal Parity Code (RDP)

RDP code (Corbett et al. 2004) is very similar to the EVENODD scheme. It uses the same slopes 0 and 1 to construct the *P* and *Q* devices. The key difference lies in how to construct the *Q* device. In the EVENODD code, the parity symbols in the *P* devices are independent from the symbols in the Q device. But in RDP code, the symbols in *P* device are involved in *Q* generation. The parity symbols are constructed according to the following encoding rules as shown in Figure 12.

$$p_i = \bigoplus_{t=0}^{n-2} d_{i.t} \quad , \tag{6}$$

$$q_i = (\bigoplus_{t=0}^{n-1} d_{i-t,t}) \oplus p_{i+1} \quad . \tag{7}$$

where i = 0, 1, ……, n-2.

For RDP, the P device should be generated first, and then the symbols in *Q* device will be generated from those in the data devices and *P* device.

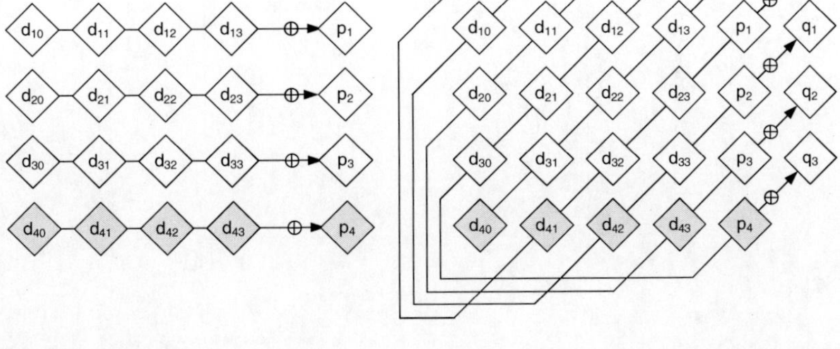

Figure 12. RDP encoding procedure (n=4).

Example #3

Let's define $q_i = Q_i + p_i + 1$ first. Q_i means the parity of the data symbols in the chain. As shown by Figure 12,

$q_0 = Q_0 + p_1,$
$q_1 = Q_1 + p_2,$
$q_2 = Q_2 + p_3,$
$q_3 = Q_3 + p_4,$
$q_4 = Q_4 + p_0.$

Suppose all imaginary data symbols are zero. Then p_4 is zero. And $(Q_0 + Q_1 + Q_2 + Q_3 + Q_4)$ is equal to $(p_0 + p_1 + p_2 + p_3 + p_4)$ since both of them are the XOR result of all data symbols. Then:

$$\begin{aligned} & q_0 + q_1 + q_2 + q_3 \\ & = Q_0 + p_1 + Q_1 + p_2 + Q_2 + p_3 + Q_3 + p_4 \\ & = (Q_0 + Q_1 + Q_2 + Q_3) + (p_1 + p_2 + p_3 + p_4) \\ & = (Q_0 + Q_1 + Q_2 + Q_3 + Q_4) + (p_1 + p_2 + p_3 + p_4 + p_0) + Q_4 + p_0 \\ & = Q_4 + p_0 \\ & = q_4. \end{aligned}$$

This means that the missing chain parity q_4 is hidden in the Q device.

In Figure 12, when any symbol other than d_{31}, d_{22}, d_{13} is modified, we have to modify one P symbol and two Q symbols. These symbols participate directly in the parity calculation. The update complexity is 3. When any symbol in d_{31}, d_{22}, d_{13}, or d_{04} is modified, we only need to modify one P symbol and one Q symbol, because the three symbols lie on the "missing chain". The update complexity is 2. Then the total update complexity is $(((n\times(n\text{-}1)\text{-}(n\text{-}1)) \times 3 + (n\text{-}1) \times 2)/(n\times n))$. It is close to 3 when the number of disks increases, as shown in Figure 13.

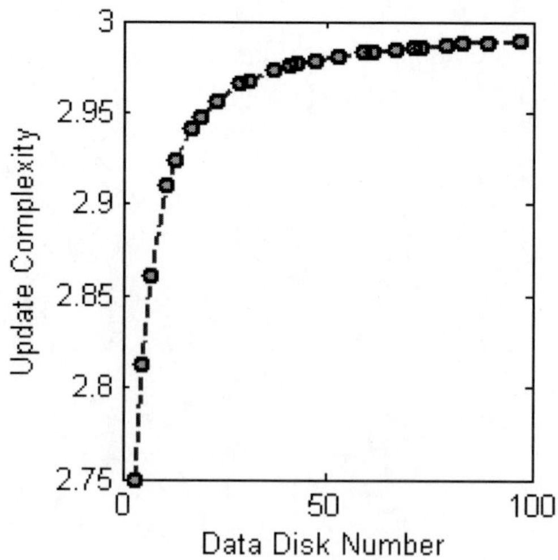

Figure 13. RDP Update Complexity.

The decoding procedure is very similar to that in EVENODD because it adopts the horizontal and diagonal slopes. However, for RDP, there is a missing chain. The iterative process has to stop if there is no parity symbol for the chain. Thus in each iteration, we have to decide whether the next one lies on the missing chain. We illustrate such process by the example below.

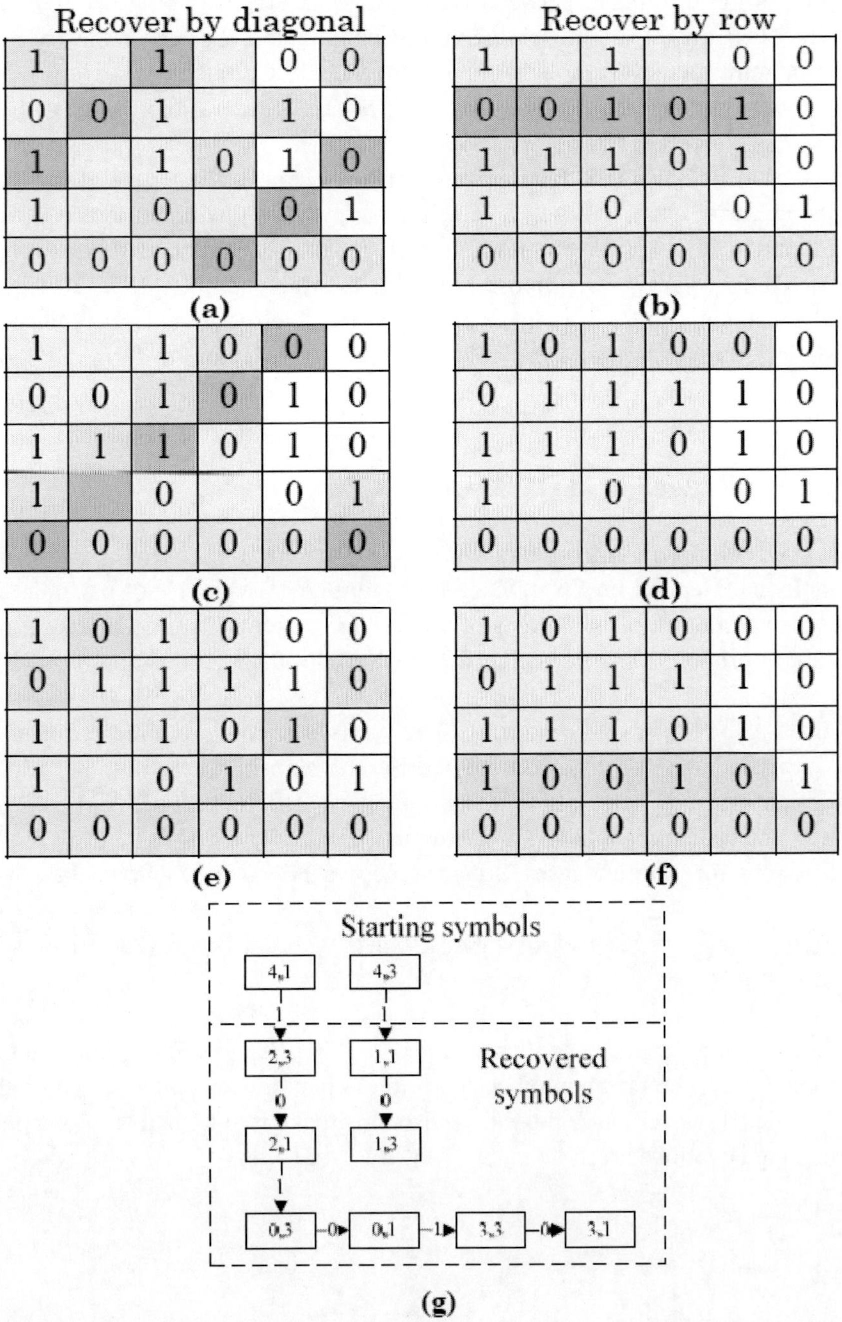

Figure 14. Illustration of the RDP decoding procedure.

Example #4

Assume that data devices 1 and 3 are erased. First, we need to find a starting point. The idea is the same as discussed in EVENODD. In this example, we can begin with two starting chains. Two missing symbols d_{23} and d_{11} are reconstructed immediately as shown in Figure 14(a). Then, the row parity can be used to recover two more missing symbols d_{21} and d_{13} in the two rows, as shown in Figure 14(b).

Now, here is the difference between RDP and EVENODD. Using the RDP procedure, since the symbol d_{13} lies in the missing chain, it cannot start the next recovery process and stops. The remaining failed symbols have to be retrieved by another iterative process one by one as shown in Figures 14(c), 14(d), 14(e) and 14(f). Figure 14(g) presents the recovery sequence.

Figure 14(g) also shows that there are two starting chains at first. One of them eventually hits the missing parity chain symbol. There is actually only one chain that can traverse the erased columns. In contrast to the EVENODD code, the RDP code needs to run more iteration. For RDP, usually we suppose "n+1" is a prime number. If n+1 is not a prime number, we can suppose there is a bigger prime number p close to n+1, and then each device has p-1 symbols. And we assume there are p-n-1 imaginary devices to form the $(p\text{-}1) \times (p+1)$ array.

3.4. S-Code: an Extension of the X-Code

The X-code requires an array equal to size $n \times n$, where n is a prime number. Its' slopes are 1 and -1. In an extended EVENODD code, Blaum and Bruck (1996) has pointed out that other slopes also can be selected to construct a storage system. Inspired by these ideas, Katti and Ruan (2005) proposed an alternative for the X-code. It is another family of MDS array codes.

The S-code also requires an $n \times n$ array. The parity symbols are allocated in the last two rows. The parity symbols come from two different slopes. Therefore the S code is an extended version of the X code wherein two lines with different slopes look skewed. The S code does not require that n should be a prime number. It only requires that "n" should satisfy different conditions for different cases. Suppose the two slopes are a and b, and:

$$\gcd(a, n) = \gcd(b, n) = \gcd(b-a, n) = \gcd(a-b, n) = \gcd(a, b) = \gcd(2a-b, n) = \\ =\gcd(2b-a, n) = 1,$$

here "gcd" stands for "greatest common divisor". Also the smallest prime factor of n is neither two nor three. The prime number condition is just one special case where the S code reduces to X code. However, such flexible restrictions make it not feasible in practice.

The encoding equations are:

$$c_{0,k} = \sum_{i+aj=(n-2)+ak} d_{ij} \tag{8}$$

and

$$c_{1,k} = \sum_{i+bj=(n-2)+bk} d_{ij} \quad , \tag{9}$$

where $0 \le i \le n\text{-}3$, $0 \le j \le n\text{-}1$.

Figure 15 presents a geometrical view of the S code, where the upper blue line corresponds to the line of slope a.

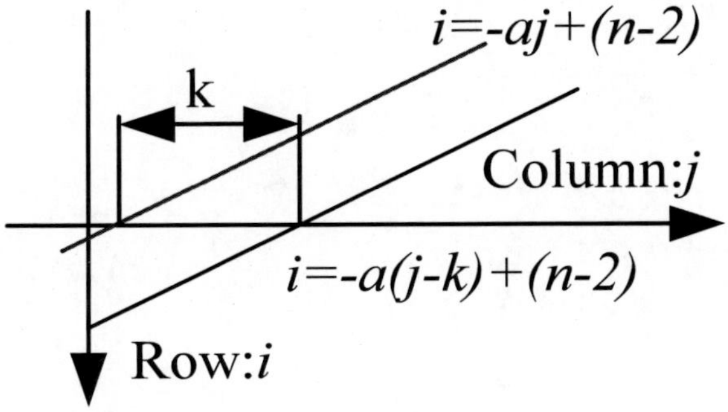

Figure 15. Geometrical view of the S Code.

Suppose the blue line is $j = -a \times j + (n-2)$. Then the black line in Figure 15 is $j = -a \times (j-k) + (n-2)$. It can be rewritten as $i + a \times j = (n-2) + a \times k$, which is exactly the chain expression in Eq. (8). In practice, we can get the (n-2) + a×k at first, and then we have to find i and j to satisfy the equation.

k	x	(i,j):
0	3	(0,4),(1,1),(2,3)
1	0	(0,0),(1,2),(2,4)
2	2	(0,1),(1,3),(2,0)
3	4	(0,2),(1,4),(2,1)
4	1	(0,3),(1,0),(2,2)

(a) $c_{0,k}$ row , Slope = 2.

k	x	(i,j)
0	3	(0,1),(1,4),(2,2)
1	1	(0,2),(1,0),(2,3)
2	4	(0,3),(1,1),(2,4)
3	2	(0,4),(1,2),(2,0)
4	0	(0,0),(1,3),(2,1)

(b) $c_{1,k}$ row, Slope = 3.

Figure 16. x=(n-2)+a×k; i+a×j=x.

Example #5

Suppose n = 5, the slopes are 2 and 3. The symbols involved in each chain are shown in Figure 16. For example, when k = 0 and x = 3, we can get the data symbols d_{04}, d_{11}, d_{23}. For each symbol, $i + a \times j$ is 3. That is, 0+2×4 = 1 + 2×1 = 2 + 3×2. The chain c_{00} is shown in Figure 17(a), the imaginary row colored yellow is the last row like X-code. From another perspective, if we only consider the column index of each symbol in the chain, it can be viewed as a modular operation, starting from index 2 with step 2.

With the help of imaginary rows, the chain will traverse all five symbols one by one as shown in Figure 17(b). In the interest of time we will not review the remaining coding chains in Figure 17 in detail. Since the S code can be treated as an extension of the X code, it has a similar decoding method. The only difference is that for each iterative process, the S code determines whether it can continue or stop in each step. In example #6 we illustrate the S decoding procedure.

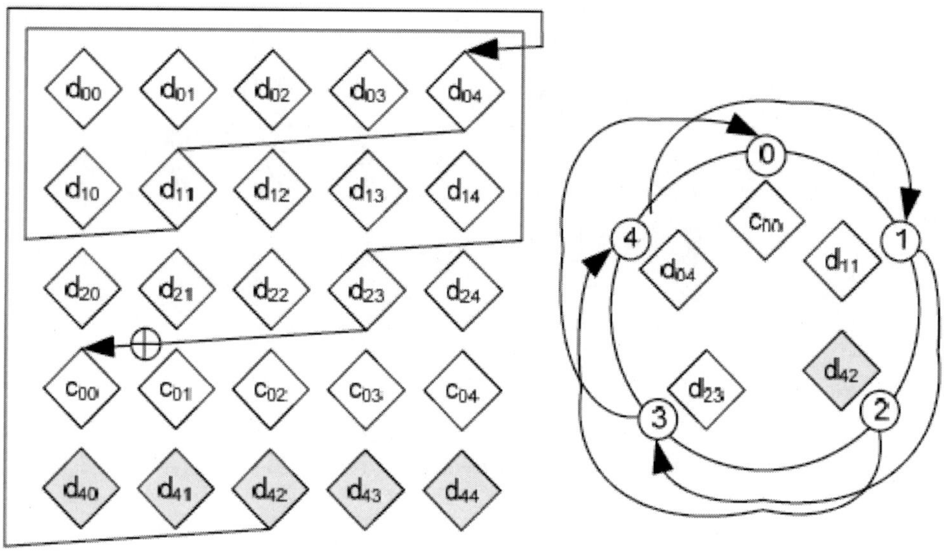

(a) (b)

Figure 17. Example of chains in the S code.

Example #6

Suppose data devices 1 and 3 are failed. We can find four starting chains beginning with: chain c_{01}, c_{04}, c_{10}, and c_{13}. Figure 18(a) and Figure 18(b) illustrate how the initial chains can recover d_{03}, d_{31}, d_{43}, d_{01} independently. At this point, we can initiate the recovery of the next four recovery chains from the four recovered symbols. However, the recovery of two consecutive chains from d_{32} and d_{43} has to stop since there are no unknown symbols in the chains. Then in the next iteration, only two symbols d_{11} and d_{13} can be recovered, as shown in Figure 18(c) and (d). Such operations continue until all the lost symbols are retrieved. The results are shown in Figure 18(e), (f), (g), and (h). Figure 18(i) gives the recovery sequence.

The S code allows for more flexibility in the number of disks. Since the S code procedures require different configurations and restrictions, its performance is no better than that of the X code. It is more cumbersome in operation than the X code.

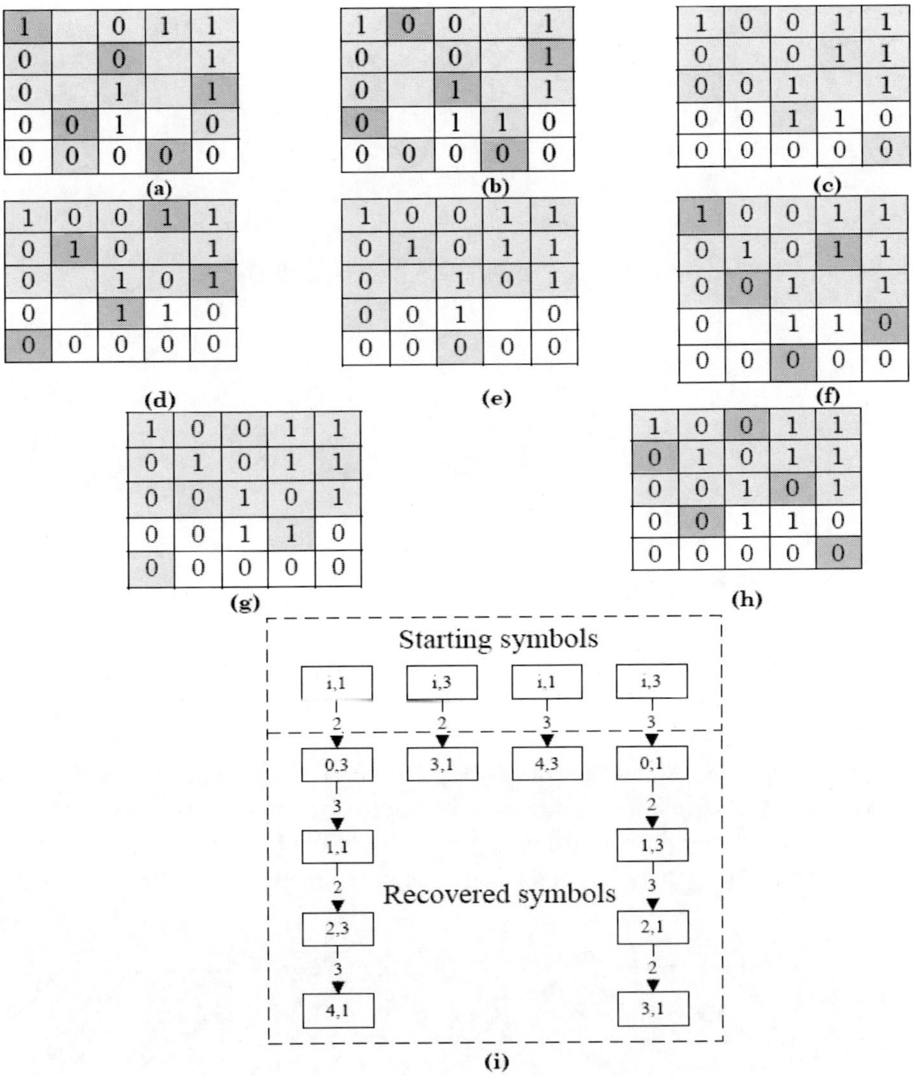

Figure 18. Example of the S-code decoding procedure.

3.5. The STAR Code

While there are a number of coding schemes for the two-erasure-tolerant case, it is not easy to construct an MDS three-erasure-tolerant code. The general EVENODD (Blaum and Bruck 1996) illustrates one. However, it is somewhat complicated. Based on the idea of the general EVENODD and X codes, Huang (Huang and Xu 2008) constructs a new code that can tolerate 3 erasures by adding one more parity device. Different from the general EVENODD code in which the slope is 2, the new added parity column diagonal of the slope is -1. Such a small difference (slope = -1 instead of 2) leads to a much more efficient decoding algorithm for recovering 3 erasures. Suppose parity symbols are stored in the *P*, *Q*, and *R* devices. The following encoding rules are shown in Figure 18.

$$S_1 = \bigoplus_{t=1}^{n-1} d_{n-1-t,t}, \tag{10}$$

$$S_2 = \bigoplus_{t=1}^{n-1} a_{t-1,t} \tag{11}$$

For each row i, the redundant symbols are obtained as follows:

$$p_i = \bigoplus_{t=0}^{n-1} d_{i,t}, \tag{12}$$

and

$$q_i = S_1 \oplus (\bigoplus_{t=0}^{n-1} d_{i-t,t}), \tag{13}$$

$$r_i = S_2 \oplus (\bigoplus_{t=0}^{n-1} d_{I+t,t}) \tag{14}$$

The *P* and *Q* devices are the same as those in the EVENODD code. Actually, if we only consider the *P* and *Q* devices, the STAR code is exactly the same as the EVENODD code, in which the parity results are derived from slopes 0 and 1. Figure 19 shows the construction of the *Q* and *R* devices. The generation of the *R* device is symmetrical with that of the *P* device.

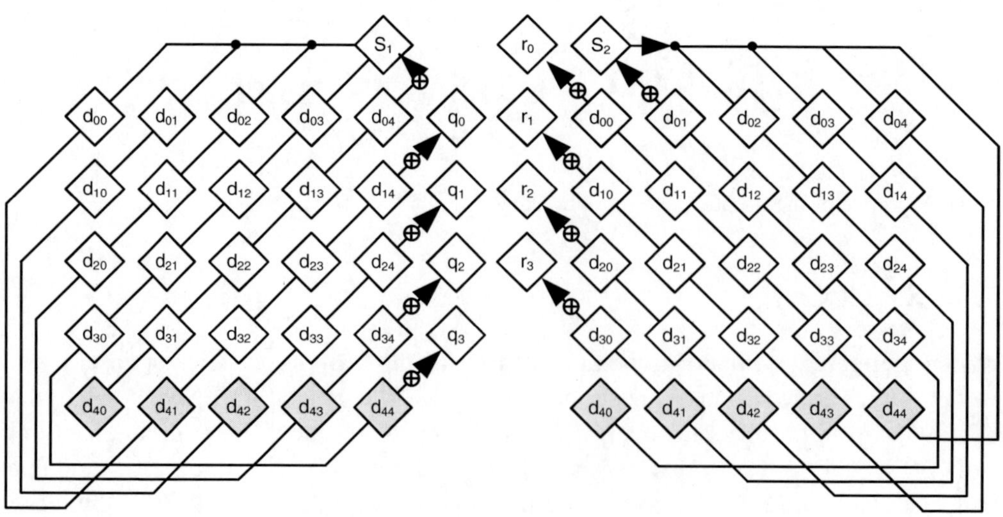

Figure 19. The STAR code encoding procedure: n=5.

In Figure 19, when any symbol which is not on chain "*S*" is modified, we have to modify one *P*, *Q* and *R* symbol. The update complexity is 3. When any symbol on an *S* chain, for example, d_{31}, d_{22}, d_{13}, d_{04} or d_{00}, d_{11}, d_{22}, d_{33}, d_{44}, is modified, we have to modify one *P* symbol and all the *Q* or *R* symbols because they are involved in all *Q* or *R* parity calculations. The update complexity is *n*. The total update complexity is:

$$\frac{((n\times(n-1)-2\times(n-1))\times 3+2\times(n-1)\times n}{n\times(n-1)}.$$

It approximates 5 when the number of disks increases, as shown in Figure 20. The ideal complexity is three. The reason for the increase in update complexity is that both the *P* and *Q* devices need the S factor to generate parity symbols.

Since the STAR code is based on the same concept of the EVENODD and X codes, its encoding procedure is easier to understand. However, in the STAR code, we cannot find any starting chain with only one unknown symbol directly. The decoding procedure is more complicated than the two-erasure schemes.

The idea of the STAR code is to reduce a triple-erasure to a two-erasure mode, and then the EVENODD algorithm can be implemented. In this section, we will use an example to illustrate how the decoding procedure works. We only consider data devices since the procedure is typically applied to them.

When three data devices are erased, similar to the EVENODD code, we can recover the adjustors S_1 and S_2 by XORing the *P* and *Q* devices or the *P* and *R* devices. Then we can recover the parity symbol for each diagonal chain for future processing.

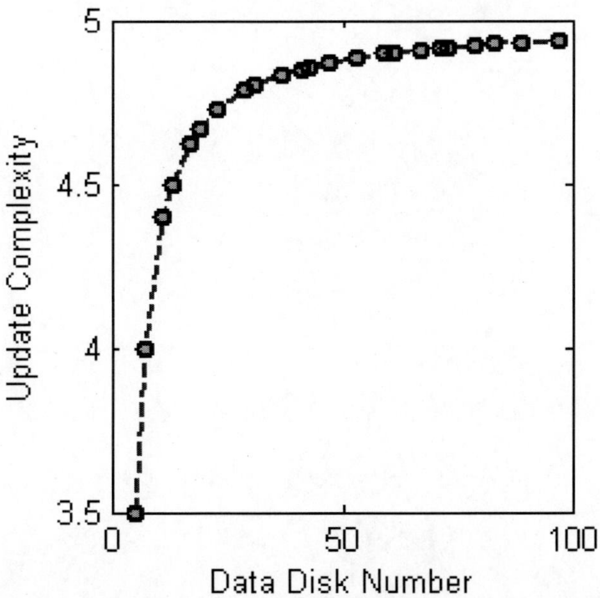

Figure 20. STAR Update Complexity.

Example #7

Figure 21 presents a decoding procedure that consists of four steps. We assume $n = 5$, and three data devices *x*, *y*, *z* have been erased, where $x<y<z$ and $y-x = z-y$. This is called the

symmetrical case. The STAR code needs three steps to recover the middle failed device, and then it becomes the EVENODD decoding case.

Step #1. Calculate Syndromes

Since all the parity symbols in the Q and R devices are combined with the S factor, we should recover them first. By XORing the remaining data symbols with the parity symbols, the "syndrome" is obtained. Actually it is the parity of the three failed symbols. Syndrome sp_i, sq_i or sr_i denotes the XOR sum of the parity symbols p_i, q_i or r_i and their corresponding non-erasure data symbols. For example, in Figure 21, $sp_0 = p_0 \oplus (d_{03} \oplus d_{04})$, $sq_0 = q_0 \oplus (d_{23} \oplus d_{14})$ and $sr_0 = r_0 \oplus (d_{04} \oplus d_{43})$, etc. Figure 21(a) shows the recovered adjustor. Figure 21(b) gives the syndrome results according to the chain in Figure 19.

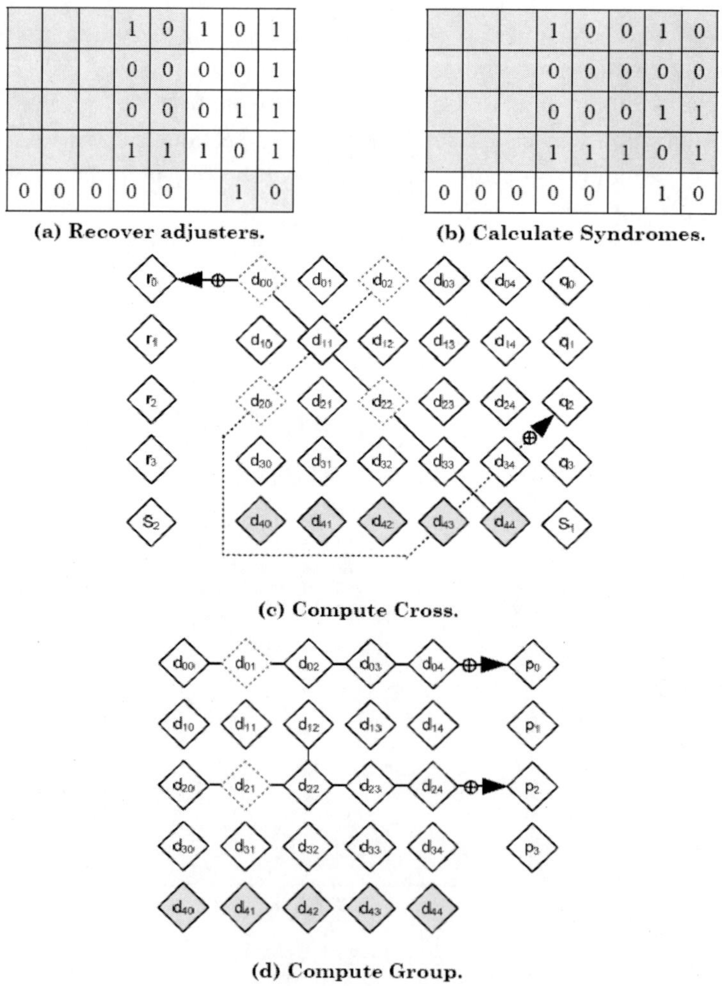

Figure 21. Example of the STAR code decoding procedure for Symmetrical Case.

Step #2. Recover the temporary Group in the Middle Erasure Column

After all the syndromes are derived, some preliminary work for the middle erasure device recovery can be done.

Since the three failed devices are symmetrical to the middle device, we can find two diagonal chains where the cross point lies on the middle device. For example, in Figure 21(c), we find two chains where syndromes sq_2 and sr_0. sq_2 equal the sum of XORing three unknown symbols d_{20}, d_{11} and d_{02}. Similarly, sr_0 is equal to the sum of d_{00}, d_{11}, and d_{22}. The two chains cross on d_{11}.

Obviously, if we XOR sq_2 and sr_0, then d_{11} would be cancelled. We denoted the result as a "Cross". In this example, the cross is equal to the sum of XORing d_{00}, d_{22}, d_{20}, d_{02} marked in red. Since d_{00}, d_{02} lie on row 0, and d_{22}, d_{20} lie on row 2, we can XOR the cross with the corresponding horizontal parity sp_0, sp_2.

The result is the sum of the two failed symbols in the middle device, namely: $d_{01} \oplus d_{21}$ as shown in Figure 21(d) in red. We call the result "Group" since it is the group of two data symbols in the middle erased device.

We follow the same approach and shift the cross downward row by row in order to get other XOR groups. In this example, we can get: $d_{11} \oplus d_{31}$, $d_{21} \oplus d_{41}$, and $d_{31} \oplus d_{01}$.

Step #3. Recover the Middle Erasure device

In Step #2 we derived the groups. Now we rewrite them as follows:

$\{d_{41} \oplus d_{21}; d_{21} \oplus d_{01}; d_{01} \oplus d_{31}; d_{31} \oplus d_{11}\}$.

Since symbol d_{41} is an imaginary symbol with zero value, symbol d_{21} can be recovered directly. Next, symbol d_{01} can be recovered since the sum of XORing the pair d_{01} and d_{21} is available. Consequently, symbols d_{31} and d_{11} are recovered. In the end, all the symbols in the middle erased column have been recovered. The recovered sequence is shown as follows:

$\{d_{41} \rightarrow d_{21} \rightarrow d_{01} \rightarrow d_{31} \rightarrow d_{11}\}$.

When three erased data devices are not symmetrical, there are two methods that can be employed. One can be found in reference (Huang and Xu 2008).

Another approach is to look at the asymmetrical case as a symmetrical case because of the circular property of the code. Actually, we can put the data device on one circle. If three devices fail, we can always find one failed device which can be looked at as if it were in the middle. For example, suppose $n = 5$, data devices 0, 1, 3 are erased, and then device 3 can be in the middle. If device 0, 1, and 4 are erased, device 0 can be in the middle as shown in Figure 22(a) and (b). Then the same method described above can be used. The illustrations for case 1 are shown in Figure 22(c) and (d).

After the middle device has been recovered, the problem is reduced to an EVENODD decoding of the two erasures case. More details about STAR decoding can be found in reference (Huang and Xu 2008).

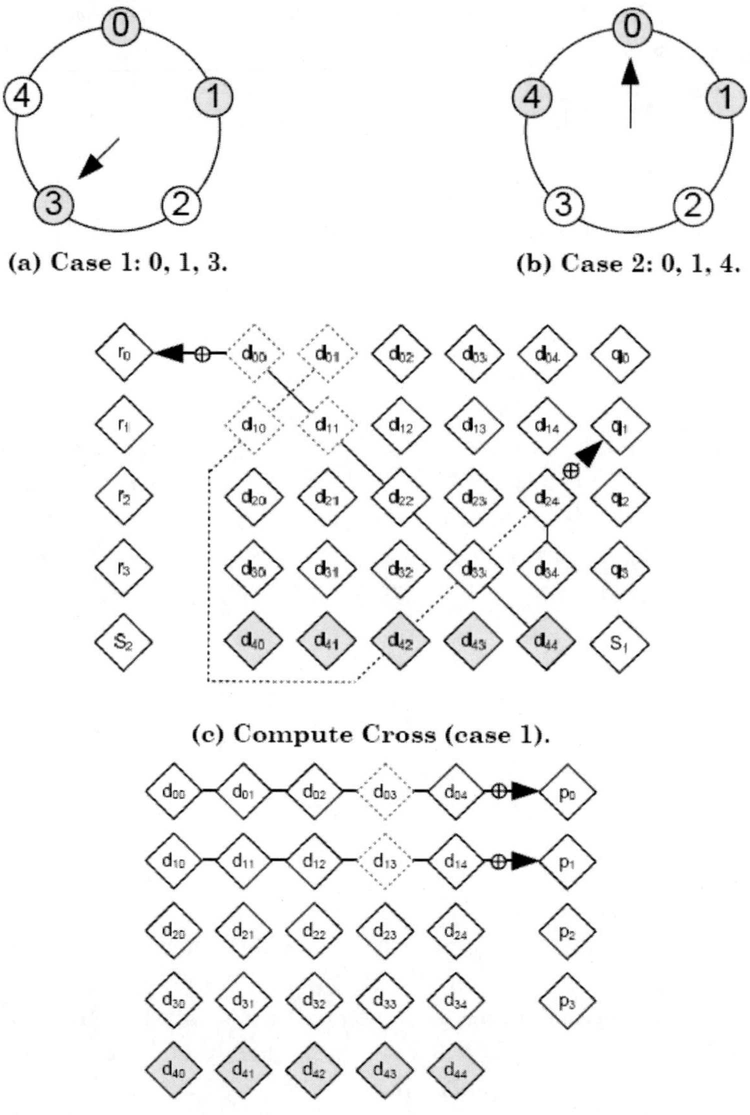

(a) Case 1: 0, 1, 3.

(b) Case 2: 0, 1, 4.

(c) Compute Cross (case 1).

(d) Compute Group (case 1).

Figure 22. Example of STAR code decoding procedure for Asymmetrical Case.

3.6. The Liberation Code

Because the EVENODD and RDP have patents and patents have a profound impact on those who implement storage systems, and are the main reason why RAID-6 systems are still being implemented with Reed-Solomon coding (Plank 2008), sometimes people want to find alternative coding techniques that exhibit near-optimal performance. Plank (2008) has put forward a Liberation code that can be applied to RAID-6 systems.

Here the reader needs some basic knowledge of matrix multiplication.

$$\begin{pmatrix} a_{00} & a_{01} \\ b_{00} & b_{01} \end{pmatrix} \cdot \begin{pmatrix} d_{00} \\ d_{01} \end{pmatrix} = \begin{pmatrix} (a_{00} \times d_{00}) \oplus (a_{01} \times d_{01}) \\ (b_{00} \times d_{00}) \oplus (b_{01} \times d_{01}) \end{pmatrix}$$

The equation above is one example of how to derive results by matrix multiplication. We use XOR "$\oplus$" instead of addition "+" because they are the same if the operation is on the GF(2) field, and a, b are 0 or 1, and d is a data symbol. Suppose there is only one data device and one parity device, each one has two symbols. Figure 23(a) shows the relationship. Then its matrix form is illustrated in Figure 23(b). We also denote the relationship as G×D=C, where G is the generation matrix, D is the data matrix, and C is the parity matrix.

Actually, every code above can be described using this method. The example in Figure 4 can be described as the matrix in Figure 24. For reference, we redraw the EVENODD example here.

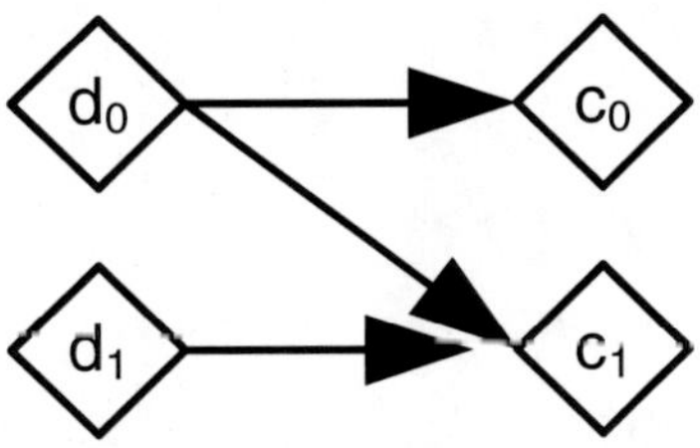

$$\begin{pmatrix} 1 & 0 \\ 1 & 1 \end{pmatrix} \cdot \begin{pmatrix} d_0 \\ d_1 \end{pmatrix} = \begin{pmatrix} d_0 \\ d_0 \oplus d_1 \end{pmatrix}$$

(a) (b)

Figure 23. Example of matrix.

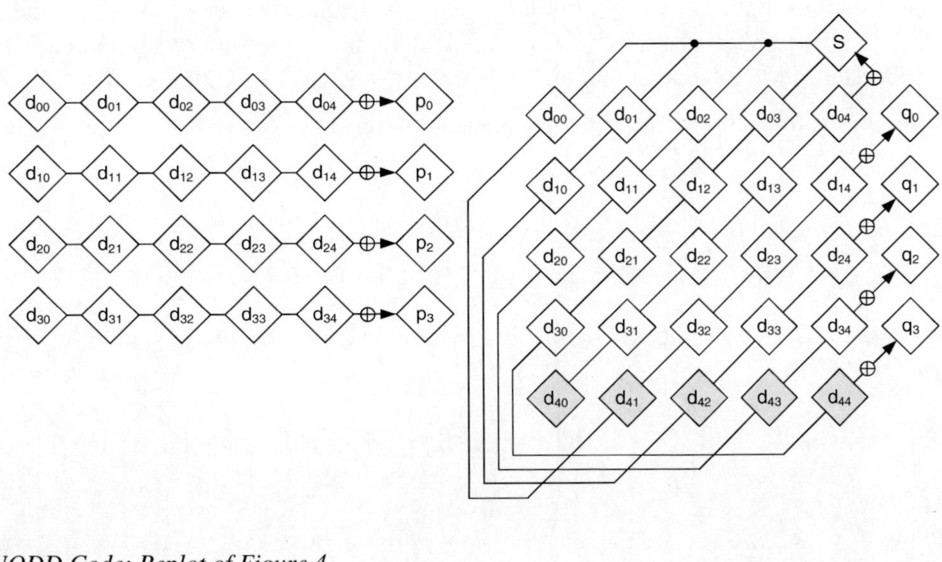

EVENODD Code: Replot of Figure 4.

In Figure 24, if we use $G \times D = C$, we can get $q_1 = d_{10} \oplus d_{01} \oplus d_{31} \oplus d_{22} \oplus d_{13} \oplus d_{33} \oplus d_{04} \oplus d_{24}$. It follows the "1" sequence in the yellow row.

$$(d_{00}\ d_{10}\ d_{20}\ d_{30}\ d_{01}\ d_{11}\ d_{21}\ d_{31}\ d_{02}\ d_{12}\ d_{22}\ d_{32}\ d_{03}\ d_{13}\ d_{23}\ d_{33})^T$$

(a) Transposition Matrix: *D*.

1000 1000 1000 1000 1000
0100 0100 0100 0100 0100
0010 0010 0010 0010 0010
0001 0001 0001 0001 0001

(b) Generated Matrix for *P* device.

1000 0001 0011 0110 1100
0100 1001 0010 0101 1010
0010 0101 1010 0100 1001
0001 0011 0110 1100 1000

(c) Generated Matrix for *Q* device.

Figure 24. Matrix for EVENODD (n=5).

Not considering optimization, the number of "1"s in one row determines the encoding performance. The number of "1"s in one column determines the update complexity. Plank gives another algorithm to generate the matrix for the *Q* device (Plank 2008).

Define I_j^w as the w×w identity matrix whose columns have been rotated to the right by j columns. For example, $I^3 = \begin{pmatrix} 1 & 0 & 0 \\ 0 & 1 & 0 \\ 0 & 0 & 1 \end{pmatrix}$, then $I_1^3 = \begin{pmatrix} 0 & 1 & 0 \\ 0 & 0 & 1 \\ 1 & 0 & 0 \end{pmatrix}$.

Define $O_{i,j}^w$ as a w×w matrix where every symbol is zero, except the symbol in position (i, j), for example, $O_{1,2}^3 = \begin{pmatrix} 0 & 0 & 0 \\ 0 & 0 & 1 \\ 0 & 0 & 0 \end{pmatrix}$, where i, j are equal to i mod w and j mod w respectively.

Assume there are *n* data devices and two parity devices; each of them has *w* symbols. We can illustrate the generation matrix in Figure 25. Actually every code above can be illustrated in the same way.

Since the Liberation code is applied to the RAID-6 system, P device must be the same as that in the EVENODD code. Then the first $w \times w$ row should be the identity matrix, as Figure 25 shows. Then the challenge lies in the definition of the X_i matrices such that the number of "1"s is minimal, yet maintains the properties of MDS, since a small number of "1"s are important for fast encoding and updating.

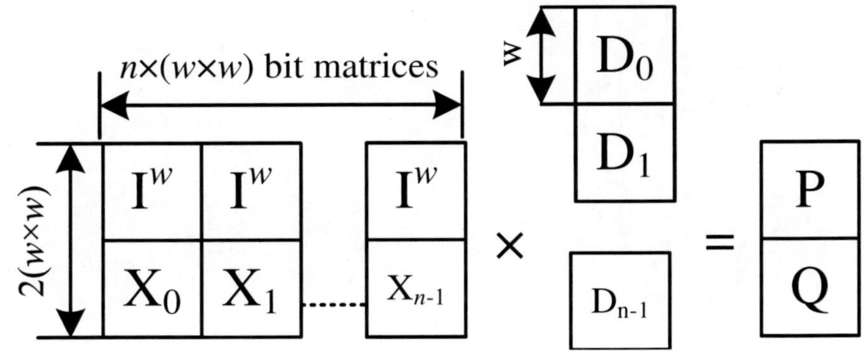

Figure 25. Illustration of the Liberation code.

The encoding algorithm is then defined as follows:

(1) $X_0 = I^w$

(2) For $0 < j < n$, $X_j = I_j^w + O_{y,y+j-1}^w$, where $y = j\,(w-1)/2$.

Figure 26 shows the X_j result when $n = 5$ and $w = 5$

10000 01000 00100 00010 00001
01000 00100 00010 00011 10000
00100 00110 00001 10000 01000
00010 00001 10000 01000 01100
00001 10000 11000 00100 00010

Figure 26. Matrix for Q with n=5 and w=5.

The matrix for the Q device has $(n \times w + n - 1)$ "1"s, plus the $n \times w$ "1"s for the P device. There are $2 \times n \times w + n - 1$ "1"s in total, and the average number of "1"s on each column is $(2 \times n \times w + n - 1)/(n \times w)$, which is the update complexity as shown in Figure 27.

An intuitive decoding process can be derived from the Plank algorithm (Plank 1999). By deleting the corresponding rows and calculating an inverse matrix, a symbol can be recovered. Plank has also proposed an optimal solution called bit matrix scheduling that reduces the amount of computation significantly. More details can be found in references (Plank 2008; Luo et al. 2009).

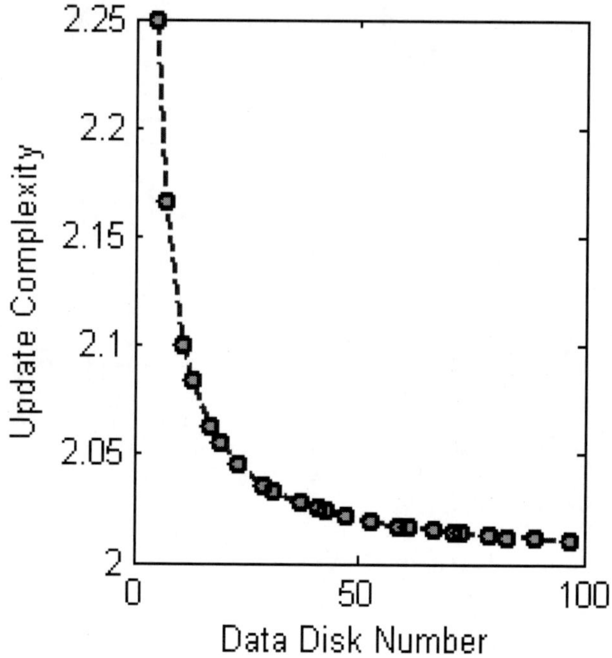

Figure 27. Update Complexity of the Liberation Code.

3.7. The P-Code

The X and S codes require an array equal in size to $n \times n$, where n is normally a prime number. Jin et al. (2009) proposed a new vertical construction named the P code, which reduces the number of disks from n to n-1, assuming n is a prime number. In this section we will further investigate the use of the P code. For the P code, there are n-1 devices, and n is a prime number. Each device has (n-1)/2 symbols. The first row is assigned to the parity symbol. Each symbol is labeled with an index from 1 to n-1. The remaining (n-1)/2 -1 rows are filled with data symbols. We label data symbols with a two-integer-tuple, (x, y), in which (x+y) mod n is equal to a parity index in the same device. The symbol can be used to generate p_x and p_y parity symbols. Figure 28(a) shows the construction, where n = 7. Then the entire matrix is 6-rows-by-3-columns.

Parity symbols lie in the first row. Each parity symbol is labeled from 1 to 6. Then the two data symbols in device 1 are labeled as "(2, 6)" and "(3, 5)". The tuple is just marked for a parity chain. The symbol labeled with (2, 6) can be used to generate P_2 and P_6. And since each data symbol only takes part in a two parity symbol generation, the update complexity is 2. The encoding algorithm is described as follows: Select data symbols whose labels include integer j; and construct the j-th parity symbol by XORing all the selected data symbols together;

P_1	P_2	P_3	P_4	P_5	P_6
(2,6)	(3,6)	(1,2)	(1,3)	(1,4)	(1,5)
(3,5)	(4,5)	(4,6)	(5,6)	(2,3)	(2,4)

(a). Construction.

1	1	0	1	1	0
0	1	0	1	1	1
1	1	1	0	1	0

(b)Encoding Example.

Figure 28. Encoding of the P-code.

Figure 28(b) shows an example. We can derive the parity chains as follows. Each chain can generate the corresponding parity symbol.

chain (1) = {(1,2), (1,3), (1,4), (1,5)},
chain (2) = {(2,6), (1,2), (2,3), (2,4)},
chain (3) = {(3,5), (3,6), (1,3), (2,3)},
chain (4) = {(4,5), (4,6), (1,4), (2,4)},
chain (5) = {(3,5), (4,5), (5,6), (1,5)},
chain (6) = {(2,6), (3,6), (4,6), (5,6)}.

Now, suppose devices 3 and 4 in Figure 28(b) are erased, Example #8 shows the recovery procedure.

Example #8

Suppose columns 3 and 4 in Figure 28 are erased. There are four symbols missing. The set is {(1, 2), (4, 6), (1, 3), (5, 6)}. Similar to other XOR codes, the key idea in the decoding procedure is to find the starting chain with only one unknown symbol. In this example, chain (5) and chain (2) have only one unknown symbol: (5, 6) and (1, 2). We can recover them directly. Based on (1, 2), we can find chain (1) and recover (1, 3). Based on (5, 6), we can find chain (6) and recover (4, 6). The missing parity symbol can be recovered by the encoding procedure. Now the two failed devices are recovered. The recovery sequence is shown in Figure 29. The starting chain index can be derived from the following rule. If the missing device index i is even, the recovery chain index is $i/2$, else it is $(i+n)/2$.

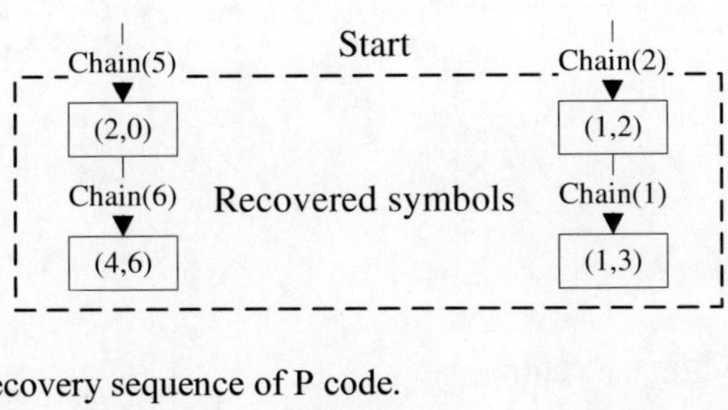

Figure 29. The decoding recovery sequence of P code.

Compared to the X and S codes with an $n \times n$ array, the P code only needs $(n-1)\times(n-1)/2$ symbols. However, the X and S codes can start four recovery chains, while the P code can only initiate two. In addition, the P code requires a more complicated algorithm to derive the symbols in the chains.

3.8. The Weaver Code

Both the STAR and the general EVENODD code have shown that when more than three disks have been erased, it is difficult to construct an MDS erasure code, particularly for the decoding scheme. Applying graph theory on the vertical array code, and with the help of a brute force computer search, Hafner (2005a) proposed the "Weaver code". This code is a non-MDS code and can tolerate up to 12 disk failures. Unfortunately, there is not much theoretical analysis for the implementation of the Weaver Code recorded since there exists no clearly specified coding or decoding algorithm. Compared with other codes discussed above, the Weaver code can be considered as a framework. The Weaver code is non storage-efficient due to extra overhead. However, such extra overhead can help to achieve better performance and offers some other advantages. Hafner describes three design principles related to the construction of the Weaver code. The first is the "Fixed in-degree" principle.

Every parity symbol is derived from a fixed number of data symbols independent of the number of data devices. "Fix in-degree" is a major difference compared with the other codes above. This property limits the computational costs of many operations such as short writing and rebuilding. For example, EVENODD always needs to read n-2 data symbols to recover the failed symbols. In contrast, when $n \geq 6$, the Weaver code only needs to read four data symbols. This reduces the work load and leads to better performance. The disadvantage is the increase in the number of parity symbols.

The second principle is "Rotation". If some data symbols are selected to derive one parity symbol, it can be achieved by just shifting the column index of data symbols to derive the next parity in the same row. For example, in Figure 30, if c_0 comes from d_0, d_2, and d_6, then c_3 must come from d_3, d_5 and d_1. Generally speaking, code constructions are specified by the simple rotation of a base configuration on a module operation that is rotationally symmetric.

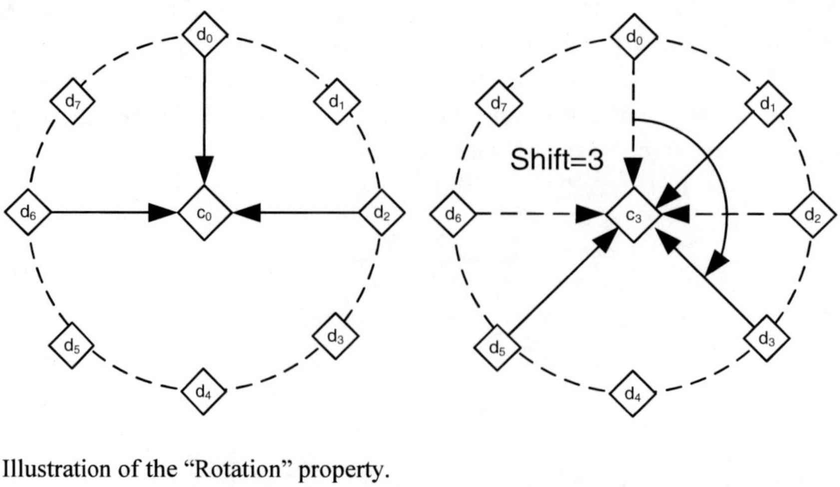

Figure 30. Illustration of the "Rotation" property.

The third principle is the “Vertical Code”, as illustrated in Figure 31.

S_0	S_1		S_j		S_{n-1}
$d_{0,0}$	$d_{0,1}$	...	$d_{0,j}$	...	$d_{0,n-1}$
$d_{1,0}$	$d_{1,1}$	...	$d_{1,j}$	...	$d_{1,n-1}$
...	...	...	...	...	...
$d_{r-1,0}$	$d_{r-1,1}$	...	$d_{r-1,j}$	...	$d_{r-1,n-1}$
$c_{0,0}$	$c_{0,1}$	...	$c_{0,j}$	...	$c_{0,n-1}$
$c_{1,0}$	$c_{1,1}$	...	$c_{1,j}$	...	$c_{1,n-1}$
...	...	...	...	...	...
$c_{q-1,0}$	$c_{r-1,1}$	...	$c_{r-1,j}$	...	$c_{q-1,n-1}$

Figure 31. Array structure of the Weaver Code.

Here we illustrate a coding procedure for the Weaver code by using an example.

Example #9

Figure 31 gives a geometrical layout that presents the vertical nature of the Weaver code. Each device has r data symbols and q parity symbols. $d_{i,j}$ is a data symbol; $c_{i,j}$ is a parity symbol. We denote the set of data symbol indices to derive $c_{i,j}$ as $\beta(i, j)$. In Figure 32, $c_{0,0}$ comes from $d_{0,0}$, $d_{1,0}$, $d_{2,0}$, $d_{0,1}$, $d_{1,1}$, then $\beta(0, 0) = \{d_{0,0}, d_{1,0}, d_{2,0}, d_{0,1}, d_{1,1}\}$, that is:

$$c_{i,j} = \oplus d_{u,v}\text{, where } d_{u,v} \in \beta(i, j) \tag{15}$$

We define the Weaver Code as W(n,k,t). Here, n is the number of devices/disks in a code, k is the maximum parity in-degree; and t is the fault tolerance, $k \le t$.

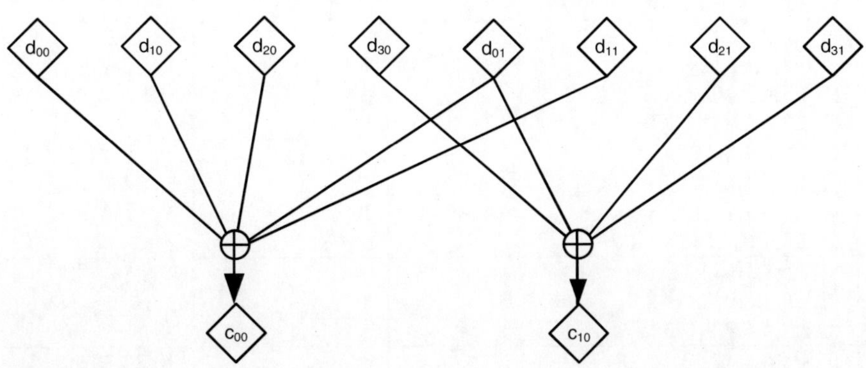

Figure 32. Graphic view of the Weaver code.

Next, let’s consider how to construct the rotation property of the Weaver code. We define any set with column offset s as $k+s$, and we denote it as the following operation:

$$k + s = \{(u, (v + s) \bmod n) : (u, v) \in k\}$$

Suppose $n = 5$ in Figure 32, and $k(0,2) = \{ d_{02}, d_{12}, d_{22}, d_{03}, d_{13}\}$. The next parity symbol is derived by shifting the column index. Then, the rotation property can be described as:

$$k(i, j) = k(i, 0) + j,$$

where $0 \leq j \leq n\text{-}1$, $0 \leq i \leq q\text{-}1$, $q = t/\gcd(k,t)$.

Now we can describe the construction of $W(n,k,t)$ as follows:

(1) Find a set $k(i,0)$ to generate c_{i0}
(2) Derive the parity symbols c_{ij} by using the rotation property, where $0 \leq j \leq q\text{-}1$.
(3) Verify that such scheme can satisfy specifications by applying computer brute force.

When $k=t$, Table 1 is obtained through brute force computation, in which "n0+" means n ≥ n0. Other kinds of Weaver code forms can be found in reference (Hafner 2005a). Table 1 gives a partial listing of parity defining sets ($k1(0)$ and offset s) and valid stripe sizes n for fault tolerance $1 \leq t \leq 10$.

Table 1. Parity defining sets for *W(n,t,t)* (Hafner 2005a)

t	K(0)	Off sets	Stripe Size n
1	{1}*	0	2+
2	{1,2}*	0	4+
3	{1,2,3}*	1	6,8+
	{1,2,4}	2	7+
4	{1,3,5,6}	1	10+
	{1,2,3,6}*	0,2,3	11+
5	{1,3,4,5,7}	2	12,15+
	{1,5,6,8,9}	3	13+
	{1,2,3,4,7}	1	14+
	{1,2,3,6,9}*	2	15+
6	{1,5,8,9,10,12}	2	17,19,21+
	{1,6,8,9,11,12}	7	17,20+
	{1,2,3,6,9,10}	0	18+
	{1,2,3,4,6,9}*	5	19+
7	{1,4,5,6,7,8,11}	4	20,23-24,26,28+
	{1,2,4,7,10,12,13}	3	20,24+
	{1,3,5,6,7,11,12}	1	22+
	{1,2,3,4,6,9,14}*	6	23+
8	{1,2,4,8,10,11,12,13}	0	26,28+
	{1,2,6,7,8,9,12,14}	0	27+
	{1,2,3,4,6,7,9,14}*	0	28+
9	{1,4,5,6,7,12,13,15,18}	2	30,32,34+
	{1,4,5,8,11,12,13,14,15}	6	31+
	{1,2,3,4,6,7,9,14,15}*	5	32+
10	{1,2,5,6,7,10,13,15,19,20}	3	35,40+
	{1,2,4,7,8,9,13,14,17,18}	0	37+
	{1,2,3,4,6,7,9,14,15,19}*	3	40+

Below are some interesting observations based on Table 1.

- The first entry is a simple RAID-1 mirror code using a non-standard layout with a load-balancing property. Every device is equally burdened by data and a parity mirror.
- In the first three rows, when $n = 2t$, these codes actually are MDS codes. For a larger t, such MDS construction codes can be derived.

Since there is no general coding algorithm, there is no general decoding algorithm. Each decoding procedure is completely done by brute force.

3.9. HoVer Codes

The codes discussed so far can be classified into two categories, Horizontal and Vertical. However, there are other options. For example a hybrid solution may insert parity codes horizontally and vertically. Actually, such ideas have led to a new family of XOR based erasure codes called the HoVer Codes (Hafner 2005b).

Similar to the Weaver code, the HoVer code also has no specific encoding and decoding algorithm. It is a frame that is constructed following certain principles. It defines a layout of codes, and any code scheme that has such a layout can be considered as a HoVer code. Horizontal and vertical codes can be viewed as special cases when the parameter is set to certain special values.

X(0,0)	**...**	**X(0,n-1)**	**H(0,0)**	**...**	**H(0,h-1)**
..					
X(r-1,0)	**...**	**X(r-1,n-1)**	**H(r-1,0)**		**H(r-1,h-1)**
V(0,0)	**...**	**V(0,n-1)**			
..					
V(v-1,0)	**...**	**V(v-1,n-1)**			

Figure 33. The layout of the HoVer Codes.

Figure 33 shows a layout of the HoVer codes. Normally, a HoVer code can be presented as HoVer(t, v, h, r, n), where:

- t: the fault tolerance;
- v: the number of vertical parity rows;
- h: the number of independent parity columns;
- r: the number of data symbols in each device; and
- n: the number of data devices.

For example, the EVENODD code is a HoVer (2, 0, 2, *n*-1, *n*) code, and the X-code is a HoVer (2, 2, 0, *n*-2, *p*) code.

There is not a general algorithm for HoVer codes. Actually, it is impossible to give such an algorithm. Similar to the Weaver code, Hafner (2005b) has specified several conditions and conjectures:

- Condition 1: Each data symbol is connected to exactly t parity symbols;
- Condition 2: For any HoVer(t, v, h, r, n) code, $r \leq v \times \dfrac{n-t}{t-h}$;
- Conjecture 1: If n is a prime number and $n>t$, there exists a HoVer(t, v, h, r, n) code for every pair (r, n) that satisfies the condition 2. In general, for any n sufficiently large, there is a number $r_0=r_0(n, t, h, v)$ such that there exists a HoVer code provided $r \leq r_0$. Alternatively, for any r, there is a number $n_0=n_0(t, t, h, v)$ such that there exists a HoVer code provided $n \geq n_0$.
- Theorem 1: Suppose there exists a code HoVer(t, v, 1, r_0, n). For a certain fixed choice of $v \geq 1$, $t \geq 2$, n and r_0, there exists a HoVer(t, v, 1, r, n) code for every $1 \leq r \leq r_0$.

Unfortunately, to date, there is neither any specification on how to compute parity symbols nor a general algorithm for the HoVer code. Beyond the above conjecture and theorems, Hafner has given several sample coding schemes based on experimental studies (Hafner 2005b). Some of them can be demonstrated through theoretical analysis; others can only be verified by exhaustively checking all possible failure scenarios for a presumed fault tolerance t.

Table 2 shows one example, HoVer code (2, 1, 1, 4, 7). The numbers in the table illustrate the data symbols in the same chain when computing the Vertical parity symbols.

Table 2. HoVer Code example: *H*(*2,1,1,4,7*)

3	4	5	6	0	1	2	H(0)
4	5	6	0	1	2	3	H(1)
5	6	0	1	2	3	4	H(2)
6	0	1	2	3	4	5	H(3)
0	1	2	3	4	5	6	H(4)
V(0)	V(1)	V(2)	V(3)	V(4)	V(5)	V(6)	

The encoding procedure in Table 2 can be described as:

$$H(i) = \bigoplus_{k=0}^{n-1} d_{i,k} \tag{16}$$

$$V(j) = \bigoplus_{k=0}^{r-1} X(k,(j+r-k)) \tag{17}$$

Since this coding scheme is similar to the EVENODD code and X-code, it is easy to find a similar decoding procedure.

4. General Tolerance Codes

When the fault tolerance exceeds three, decoding algorithms become very complex. To date, there are two methods to develop general codes. One is based on a matrix and finite field. The other is based on a graph.

Section 4.1 will discuss code schemes based on a matrix. Sections 4.2 will introduce codes based on a graph. In this section, readers need to have an understanding of some basic matrix theories.

4.1. The Matrix Based Code

4.1.1. Basic Knowledge of a Finite Field

We introduce some concepts about the finite field here. We do not intend to give a precise definition. Our intent is to help readers understand the following sections.

First, in this paper, the coefficient operation of a polynomial is based on XOR, for example, $1\times x+1\times x=(1\oplus 1)\times x=0\times x=0$, *and* $x+x+x=x$.

Second, readers can suppose that the generation polynomial of a finite field GF() is equal to 0. For example, suppose the generation polynomial is x^2+x+1 for GF(22). We can suppose $x^2+x+1 = 0$. If we add $x+1$ on both sides, we get $x^2=x+1$. For x^3, we get $x\times x^2 = x\times (x+1) = x^2+x = 1$. For x^4, it is $x^4 = x^3\times x = 1\times x = x$, etc. Then, for x^2, we suppose it is $x+1$ in a polynomial form, or we assume it is 11 if we put all coefficients together, or we assume it is "3" in the decimal form. Then x^2 can be described in different forms. The results are shown in Table 3.

Table 3. Isomorphism of the finite field

Generated Element	Polynomial Element	Binary Element	Decimal Element
x^0	1	01	1
x^1	x	10	2
x^2	x+1	11	3
x^3	1	01	1

The third, "inverse" means that the product of x and y is equal to one. For example, since $x \times (x^2)$ is equal to "1" in Table 3, x is the inverse of x^2. However, it is difficult to find the inverse easily in practice. Normally it can be derived in three different ways: Brute-force search, the Extended Euclidean algorithm, or a look-up table.

4.1.2. Information Dispersal Algorithm

Before explaining encoding and decoding procedures for general codes based on Reed-Solomon, we will introduce a famous traditional algorithm called Information Dispersal Algorithm (IDA) (Rabin 1989) because it can derive other XOR based algorithms (Huang et al. 2007a) although IDA is not based on any XOR operations. Instead of giving a thorough discussion, here we only introduce the idea of the IDA since it will be enough to help our readers understand the general code schemes based on Matrix.

Rabin (1989) introduced the IDA, which breaks the data into n pieces and adds r pieces in such a way that any n pieces can reconstruct the original data set. Later in 1999, an easier expression of this method was proposed (Plank 1999; Plank 2005). It is described as G•D=C as shown below:

$$\begin{pmatrix} 0^0 & 0^1 & 0^2 & \cdots & 0^{n-1} \\ 1^0 & 1^1 & 1^2 & \cdots & 1^{n-1} \\ \vdots & \vdots & \ddots & & \\ & & & \ddots & \\ (n-r+1)^0 & \cdots & & & (n-r+1)^{n-1} \end{pmatrix} \times \begin{pmatrix} d_0 \\ d_1 \\ \vdots \\ \vdots \\ d_{n-1} \end{pmatrix} = \begin{pmatrix} c_0 \\ c_1 \\ \vdots \\ \vdots \\ c_{r-1} \end{pmatrix}$$

There is another form based on a Cauchy matrix shown below. The key property of the matrix is that any n rows in the matrix can form a new non-singular $n \times n$ matrix, and the inverse matrix can be derived.

$$\begin{pmatrix} \frac{1}{x_0 - y_0} & \frac{1}{x_1 - y_0} & \cdots & \cdots & \frac{1}{x_{n-1} - y_0} \\ \frac{1}{x_0 - y_1} & \frac{1}{x_1 - y_1} & & & \frac{1}{x_{n-1} - y_1} \\ \vdots & & \ddots & & \vdots \\ \vdots & & & \ddots & \vdots \\ \frac{1}{x_0 - y_{r-1}} & \frac{1}{x_1 - y_{r-1}} & \cdots & \cdots & \frac{1}{x_{n-1} - y_{r-1}} \end{pmatrix} \times \begin{pmatrix} d_0 \\ d_1 \\ \vdots \\ \vdots \\ \vdots \\ d_{n-1} \end{pmatrix} = \begin{pmatrix} c_0 \\ c_1 \\ c_2 \\ \vdots \\ c_{r-1} \end{pmatrix}$$

When any number of disks less than r fails, (i.e. $m \leq r$), the corresponding row in the matrix above can be cut off. The new matrix then has $n+(r-m)$ rows and n columns. Any n rows can be drawn from the remaining matrix to form a new non-singular matrix G. Its inverse matrix G-1 is computed, for example, by Gaussian Elimination, and then the original data D can be recovered by G-1•C. Hence all lost symbols can be recovered.

Obviously the largest tolerant number is r. If $m>r$ disks fails, the remaining matrix rank has only $n-(m-r)$, and the failed disks cannot be recovered. This observation implies that more rows based on the Vandermonde/Cauchy are required if we want to tolerate more failures.

Compared to the XOR methods, however, such a matrix based calculation is computationally very expensive if applied as it is (Plank 2008). The calculation of checking blocks requires an n-way dot product. Thus, creating r check blocks of size B has a complexity of $O(B \times n \times r)$, which grows rather quickly with n and r. Meanwhile, the decoding procedure involves an $O(n^3)$ matrix inversion and another n-way dot product for each symbol block to be decoded. Furthermore, GF multiplication is more expensive than integer multiplication. For these reasons, Reed-Solomon coding is usually deemed appropriate only for limited values of n and r.

Therefore, although the matrix leads to ideal general erasure codes, it is not a preferable option in practice since with a large n and r its computation time increases sharply when the size n grows. Then it leads to three questions:

- Is the symbol in GF(2) derivable from GF(q) or GF(2m)? Can the symbols in GF(q) or GF(2m) be converted into GF(2)?
- Could the performance be improved by some optimization?
- Is the expensive calculation of the inverter matrix mandatory in the reconstructive procedure?

Because the key rationale behind the IDA is to construct a matrix such that any n rows can get the $n\times n$ non-singular matrix, answers to the above questions led to some general XOR based codes.

4.1.3. Isomorphism matrix

Regarding the first question, probably the most intuitive approach is to replace the symbol in GF(q) or GF(2m) with GF(2) by isomorphism (Blomer et al. 1995; Huang et al. 2007a). For example, consider three data and two parity devices in GF(22) as shown in Table 4. Each device has two symbols.

Table 4. Layout of Example

d_1	d_3	d_5	p_1	q_1
d_2	d_4	d_6	p_2	q_2

The original matrix expression is described as follows, where D means the data device and P and Q are the parity devices:

$$\begin{pmatrix} P \\ Q \end{pmatrix}^T = \begin{pmatrix} D_1 \\ D_2 \\ D_3 \end{pmatrix}^T \times \begin{pmatrix} 1 & 1 \\ 1 & x \\ 1 & x+1 \end{pmatrix} = \begin{pmatrix} D_1 \\ D_2 \\ D_3 \end{pmatrix}^T \times \begin{pmatrix} 1 & 1 \\ 1 & 2 \\ 1 & 3 \end{pmatrix}$$

Considering that any symbol in GF(22) is multiplied by (x+1) (Huang 2007a). For general cases, the symbol in the GF(22) field can be rewritten as $ax+b$. For example, $1 = 0\times x + 1\times 1$, where $a = 0$ and $b = 1$. Then:

$$(ax+b) \times (x+1) = a\times (x^2+x) + b\times (x+1) = a + b\times (x+1).$$

Therefore ($ax+b$) $\times$ (x+1) can be represented as the following matrix:

$$(ax+b)(x+1) = \begin{pmatrix} a & b \end{pmatrix}\begin{pmatrix} 0 & 1 \\ 1 & 1 \end{pmatrix}$$

Hence, the symbol x+1 can be converted into $\begin{pmatrix} 0 & 1 \\ 1 & 1 \end{pmatrix}$ in GF(2). Following the same procedure, the symbols 0, 1, 2, 3 in GF(22) can be represented as the 2×2 matrix of $\begin{pmatrix} 0 & 0 \\ 0 & 0 \end{pmatrix}$, $\begin{pmatrix} 1 & 0 \\ 0 & 1 \end{pmatrix}$, $\begin{pmatrix} 1 & 1 \\ 1 & 0 \end{pmatrix}$, and $\begin{pmatrix} 0 & 1 \\ 1 & 1 \end{pmatrix}$.

Then the expression above can be rewritten as:

$$\begin{pmatrix} p_1 \\ p_2 \\ q_1 \\ q_2 \end{pmatrix}^T = \begin{pmatrix} d_1 \\ d_2 \\ d_3 \\ d_4 \\ d_5 \\ d_6 \end{pmatrix}^T \times \begin{bmatrix} 1 & 0 & 1 & 0 \\ 0 & 1 & 0 & 1 \\ 1 & 0 & 1 & 1 \\ 0 & 1 & 1 & 0 \\ 1 & 0 & 0 & 1 \\ 0 & 1 & 1 & 1 \end{bmatrix} \tag{18}$$

For example,

$q_1 = d_1 \oplus d_3 \oplus d_4 \oplus d_6$, and

$q_2 = d_2 \oplus d_3 \oplus d_5 \oplus d_6$.

With such method, the Vandermonde/Cauchy matrix can be converted into XOR operations (Huang 2007a).

4.1.4. Companion Matrix

An alternative method is to use a companion matrix. Each multiplication of a symbol is implemented by a companion matrix A. We take the same GF(22) example. Suppose the primitive polynomial is x^2+x+1. We derive the companion matrix A directly. $A = \begin{bmatrix} 0 & 1 \\ 1 & 1 \end{bmatrix}$, then $A^2 = \begin{bmatrix} 1 & 1 \\ 1 & 0 \end{bmatrix}$, $A^3 = \begin{bmatrix} 1 & 0 \\ 0 & 1 \end{bmatrix}$, and $0 = \begin{bmatrix} 0 & 0 \\ 0 & 0 \end{bmatrix}$. Then the expression in Eq. (18) can be rewritten as:

$$\begin{pmatrix} p_1 \\ p_2 \\ q_1 \\ q_2 \end{pmatrix}^T = \begin{pmatrix} d_1 \\ d_2 \\ d_3 \\ d_4 \\ d_5 \\ d_6 \end{pmatrix}^T \times \begin{bmatrix} 0 & 1 & 0 & 1 \\ 1 & 1 & 1 & 1 \\ 0 & 1 & 1 & 1 \\ 1 & 1 & 1 & 0 \\ 0 & 1 & 1 & 0 \\ 1 & 1 & 0 & 1 \end{bmatrix} \tag{19}$$

For example,

$q_1 = d_2 \oplus d_3 \oplus d_4 \oplus d_5$, and
$q_2 = d_1 \oplus d_2 \oplus d_3 \oplus d_6$.

Blaum has proposed another example for GF(28) to convert the Reed-Solomon code in GF(28) to GF(2) by using a companion matrix (Blaum et al. 1995). More details about companion matrices can be found in reference (Roman 1992).

4.1.5. Feng's Matrix

The isomorphism and companion methods can convert a normal finite field operation to a GF(2) XOR operation. The idea behind those methods is to replace an original symbol in this field with a matrix in another field. Since there are lots of "1"s in the result, the converted matrix requires a lot of XOR operations.

Based on the concept of LDPC (Low Density Parity Code) and a permutation matrix, Feng gives two other construction methods to convert the Vandermonde/ Cauchy matrix to a sparse matrix (Feng 2005a; Feng 2005b).

Let $n = m+1$ be a prime number, I_m be an $m \times m$ identity matrix, and O_m be an $m \times m$ zero matrix.

Define the cyclic permutation matrix

E_n as $E_n = \begin{bmatrix} \vec{0} & 1 \\ I_m & \vec{0}^T \end{bmatrix}$, where $\vec{0}$ is a $1 \times m$ vector of "0"s.

According to the matrix theory, $E_n^n = I_n$.

For example, if n=5, $E_5 = \begin{bmatrix} 0 & 0 & 0 & 0 & 1 \\ 1 & 0 & 0 & 0 & 0 \\ 0 & 1 & 0 & 0 & 0 \\ 0 & 0 & 1 & 0 & 0 \\ 0 & 0 & 0 & 1 & 0 \end{bmatrix}$, and $E_5^5 = I_5$ on the GF(2).

Define $J_n = \begin{bmatrix} I_{n-1} \\ 0 \end{bmatrix}$ as a n rows by n-1 columns matrix, the Reed Solomon Vandermonde matrix can be extended to GF(2) by the following procedure (Feng 2005a):

- Replace '1' in RS matrix with I_m;
- Replace '0' in RS matrix with O_m;
- Replace the other number with $J^T \times E_m \times J$.

Then the Reed-Solomon matrix is converted to a Reed-Solomon like matrix as follows:

$$H=\begin{pmatrix} I_m & I_m & I_m & \cdots & I_m \\ I_m & J^T EJ & J^T E^2 J & \cdots & J^T E^n J \\ \vdots & \vdots & \vdots & \ddots & \vdots \\ I_m & J^T E^{(r-1)} J & J^T E^{(r-1)2} J & \cdots & J^T E^{(r-1)n} J \end{pmatrix} \tag{20}$$

For example, considering $n = 5$, and $r = 3$, we can get:

$$E_5=\begin{pmatrix} 0 & 0 & 0 & 0 & 1 \\ 1 & 0 & 0 & 0 & 0 \\ 0 & 1 & 0 & 0 & 0 \\ 0 & 0 & 1 & 0 & 0 \\ 0 & 0 & 0 & 1 & 0 \end{pmatrix}, \quad E_5^4=\begin{pmatrix} 0 & 1 & 0 & 0 & 0 \\ 0 & 0 & 1 & 0 & 0 \\ 0 & 0 & 0 & 1 & 0 \\ 0 & 0 & 0 & 0 & 1 \\ 1 & 0 & 0 & 0 & 0 \end{pmatrix}, \text{ and } \quad J_5=\begin{pmatrix} 1 & 0 & 0 & 0 \\ 0 & 1 & 0 & 0 \\ 0 & 0 & 1 & 0 \\ 0 & 0 & 0 & 1 \\ 0 & 0 & 0 & 0 \end{pmatrix}, \text{ then}$$

$$\text{H}=\left(\begin{array}{c|c|c|c|c} 1000 & 1000 & 1000 & 1000 & 1000 \\ 0100 & 0100 & 0100 & 0100 & 0100 \\ 0010 & 0010 & 0010 & 0010 & 0010 \\ 0001 & 0001 & 0001 & 0001 & 0001 \\ \hline 1000 & 0000 & 0001 & 0010 & 0100 \\ 0100 & 1000 & 0000 & 0001 & 0010 \\ 0010 & 0100 & 1000 & 0000 & 0001 \\ 0001 & 0010 & 0100 & 1000 & 0000 \\ \hline 1000 & 0001 & 0100 & 0000 & 0010 \\ 0100 & 0000 & 0010 & 1000 & 0001 \\ 0010 & 1000 & 0001 & 0100 & 0000 \\ 0001 & 0100 & 0000 & 0010 & 1000 \end{array}\right)$$

For a Cauchy-like matrix, another solution is also given as follows (Feng 2005b)]:

- Replace '1' with I_m;
- Replace '0' with O_m;
- Replace $1/(x_i$-$y_j)$ with $(E_j$+E_i+$j)$-1, where $0 \le I < n$, $0 \le j < r$.
- Add I on the left column and $\vec{1}$ on the bottom rows.

Then the Cauchy matrix can be converted to a Rabin-like matrix:

$$\begin{pmatrix} I & 0 & \cdots & 0 & (I+E^r)^{-1} & (I+E^{r+1})^{-1} & \cdots & (I+E^{r+n-1})^{-1} \\ 0 & I & \cdots & 0 & (E+E^r)^{-1} & (E+E^{r+1})^{-1} & \cdots & (E+E^{r+n-1})^{-1} \\ \vdots & \vdots & \ddots & \vdots & \vdots & \vdots & \ddots & \vdots \\ 0 & 0 & \cdots & I & (E^{r-1}+E^r)^{-1} & (E^{r-1}+E^{r+1})^{-1} & \cdots & (E^{r-1}+E^{r+n-1})^{-1} \\ \vec{0} & \vec{0} & \cdots & \vec{0} & \vec{1} & \vec{0} & \cdots & \vec{0} \\ \vec{0} & \vec{0} & \cdots & \vec{0} & \vec{0} & \vec{1} & \cdots & \vec{0} \\ \vdots & \vdots & \ddots & \vdots & \vdots & \vdots & \ddots & \vdots \\ \vec{0} & \vec{0} & \cdots & \vec{0} & \vec{0} & \vec{0} & \cdots & \vec{1} \end{pmatrix}$$

4.1.6. Decoding Method

Since the encoding methods in subsection 4.1 are based on matrix theory, it is intuitive to use a Gauss-Jordan elimination method to get an inverse matrix and recalculate the failed symbol. Feng proposed a decoding method (Feng 2005a; Feng 2005b). It first computes syndromes and then uses a recursive algorithm to solve the linear equations. Hafner (2005c) introduced another method to reconstruct failed disks from matrices.

After an inverse matrix is derived, failed disks can be recalculated. Some optimal methods also exist. Huang (2007b) gave a heuristic method based on matching. Plank (2008) presented another method called "bit matrix scheduling" based on the fact that one plus one is zero in GF(2).

The methods above are all based on matrix operations. To date, we cannot find any concise and simple decoding algorithms like that for EVENODD. It is too complicated to be described in a clear, direct way. Therefore we do not intend to introduce them in detail in this paper. Readers can find a detail description in references (Plank 2008; Feng 2005a; Feng 2005b; Huang et al. 2007a).

4.2. Graph Based Codes

4.2.1. The Tornado Code

Luby et al. (1997; 2001) presented a practical loss-resilient code framework, which achieves an optimal linear encoding/decoding time. The Tornado code is based on the idea that lost data symbols can be recovered by computing the sum of parity symbols and existing data symbols as illustrated in Section 2. The method is straightforward and easy to understand.

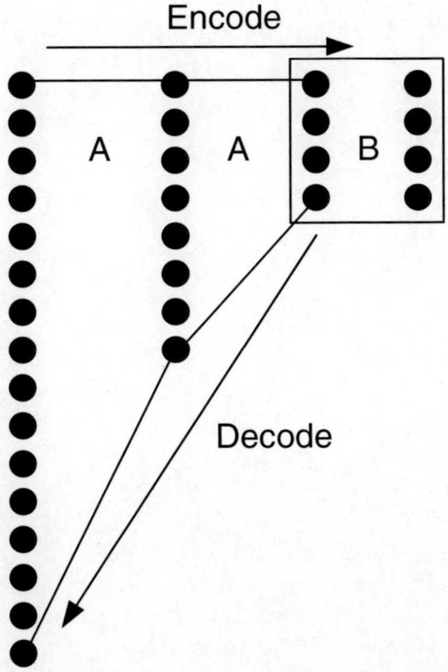

Figure 34. Tornado code architecture (Luby et al. 1997)].

Suppose there are two coding methods A and B. The input message has n symbols, and method A output $\beta \times n$ parity symbols. Method B is a conventional loss-resilient code with (1-β). The Tornado code cascades A until the number of output symbols is roughly $\sqrt{n}$. Then method B is used on the last stage. The total redundant symbols are $n \times \beta/(1-\beta)$. Now the Tornado code is a code whose rate is 1-β. Figure 34 illustrates the architecture of the code.

The Tornado code is actually a framework. The only requirement is that the code should be processed in linear time n (Luby et al. 1997). For example, method A uses an LDPC error correction code and method B uses the Reed-Solomon correction code.

The decoding procedure begins with the last stage. After input symbols are recovered by method B, the recovery process is executed iteratively until the original data symbols are recovered. The time complexity of both the encoding and decoding processes is linear in $O(n)$.

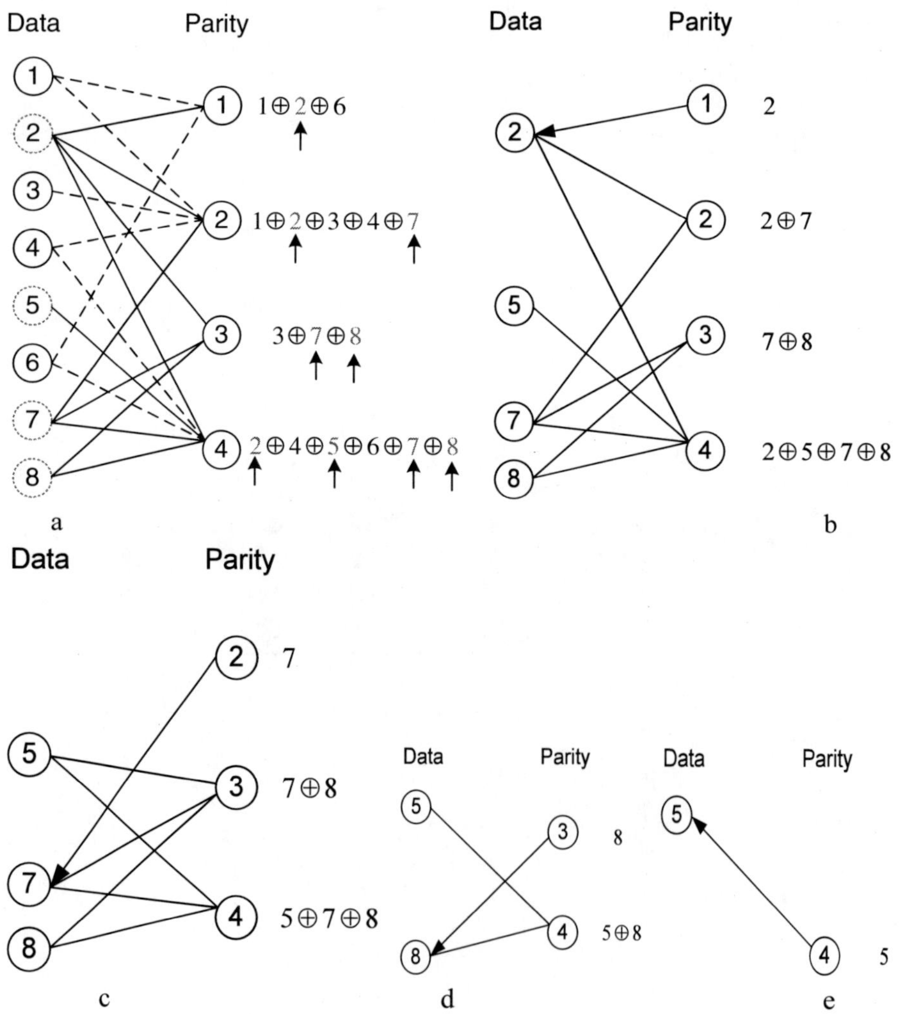

Figure 35. Example of the Tornado code recovery process.

Example #10

Figure 35 shows an example of a recovery process for one stage. Suppose β is 0.5. Then eight input symbols can derive four parity symbols in Figure 35(a). Suppose four data symbols 2, 5, 7 and 8 are failed.

For parity symbol 1, there is only one unknown variable. To illustrate the recovery procedure clearer, Figure 35(a) is redrawn as Figure 35(b), which only retains the unknown input symbols and the connected parity symbols. The input symbol 2 can be recovered directly by the XOR input symbol 1, 6 and parity symbol 1.

Then, by removing the recovered symbol, its neighbors, and all edges adjacent to its neighbors, Figure 35(c) is obtained. The input symbol 7 is recovered. By repeating a similar process iteratively, the input symbols 8 and 5 are recovered as shown by the Figure 35(d) and Figure 35(e).

The Tornado code is a probability code, and the lost symbols can be recovered with a certain degree of high probability. The Tornado code can reconstruct lost data symbols if the number of retrievable nodes in the encoded symbol set is at least equal to the number of symbol nodes.

In addition to the Tornado code, other codes such as the LT code, Online Code, Raptor code and other LDPC codes have been proposed.

We will not introduce them here since they are mainly used in wide-area content-distribution frameworks. They are beyond the scope of this paper and interested readers can find the survey of LDPC codes in references (MacKay 2005; Thul 2003; Liva 2006; Brack 2007; Dimakis et al. 2011).

4.2.2. Pyramid Codes

In practice, sometimes many product services have been successfully built on top of append-only storage systems. Write operations are only appended to the end of existing symbols and symbols are rarely modified once written (Kreyszig 2005).

Huang constructed Pyramid codes for such applications to trade space for access efficiency (Huang et al 2007b). It has been pointed out that such codes can be optimal under certain conditions. There are two classes of these types of codes: Basic and General.

The main idea behind the basic code is divide- and -conquer. We illustrate this idea using the example shown in Figure 36.

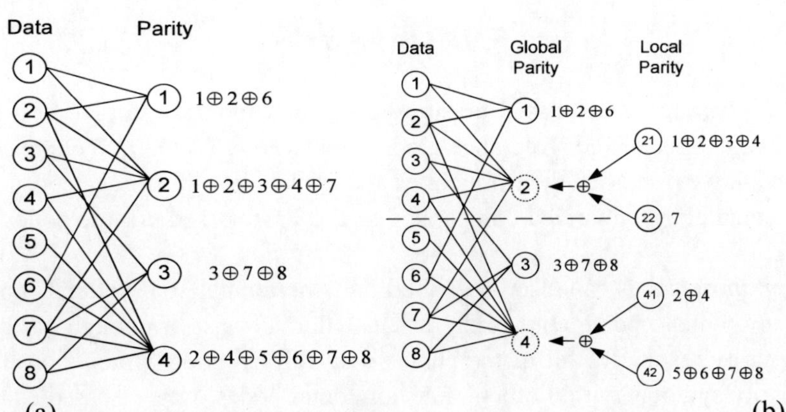

Figure 36. Example of Basic Pyramid code.

Example #11

Suppose there are 8 data symbols. One coding scheme can generate four parity symbols, as shown in Figure 36(a). Now we pick up parity symbols 2 and 4. We divide the data symbols into two subgroups. One is S_1:{1,2,3,4} and the other is S_2:{4,5,6,7}. We generate the symbols 21 in S_1 by the same algorithm. The process uses the symbols in S_2.

Now, six parity symbols are obtained instead of four. The symbols c_{21}, c_{22}, c_{41}, c_{42} are called "Local Parity" and c_1, c_2 are called "Global Parity".

The advantage of the application of the Basic Pyramid code illustrated above is that it can improve read performance. For example, if d_7 is failed, the original method needs to read at least data symbols d_3, d_8 and parity symbol c_3 to recover it. But the basic pyramid code only needs to read one parity symbol c_{22} to recover d_7. That is why c_{22} is called "local parity".

Therefore, the basic pyramid encoding procedure can be described as follows:

- Find one erasure code scheme.
- Divide the parity symbols into two groups: Global and Local.
- Generate the global parity symbols as usual.
- Divide the data symbols into subgroups.
- Generate the local symbols in each subgroup.

The decoding procedure depends on specific coding algorithms, which are not addressed in this paper.

The general pyramid code is not a simple extension of the basic version. For general codes, the group can be nested. For example, the groups in Figure 36 can be divided into {1,2,3,4}, {3, 4, 5, 6}, and {5 ,6 ,7 ,8}. The encoding procedure for the general code requires knowledge in multiple areas including graphing, vector spacing, and matricing.

The encoding process is very complex and there is no existing graphical illustration of the method. Therefore, only the essential rationale behind the algorithm is described here. First develop a construction matrix, which requires multiple software applications and high powered hardware. The matrix should have such a property: that if we cross out the rows which correspond to erased data, the remaining matrix should still be invertible. A detailed analysis of this algorithm can be found in reference (Huang et al. 2007b).

5. DISCUSSIONS

This survey groups XOR based erasure codes for canonical storage systems into two categories: STCs (Specialized Tolerance Codes) and GTCs (General Tolerance Codes). In Sections 3 and 4, we discussed their principles and applications in storage construction. Table 5 presents a complete summarized comparison of the XOR based erasure codes discussed in this paper.

Generally, those codes can also be divided into "horizontal" or "vertical" codes. Vertical codes have an optimal update complexity because they consist of enough parity symbols to embody the parity results for all of the chains. For vertical codes, thus, there is no need to hide some parity symbols within others. For horizontal codes, we presented examples of the EVENODD, X, and RDP codes. Due to the hidden parity symbol, the EVENODD and RDP codes' update complexity are close to 3 in contrast to the X code's optimal complexity which is 2. Vertical codes also have a better coding and decoding performance than that of horizontal codes. For example, the encoding performance of the EVENODD code is n-1/2-1/(2×(n-1)) XOR operations per coding word. It is roughly n-(1/2) (Plank 2008). The encoding performance of the X code is n-2 XOR operations.

Horizontal codes, however, are more flexible than vertical codes since the number of devices can be shortened. An example was shown in the discussion of the X code.

The STCs normally have a clear structure and straight forward decoding algorithms. Most codes are two-erasure codes since they are simple and intuitive when used to recover failed devices. Among all of the two-erasure codes, the only difference lies in the construction of the parity chain. Therefore, two-erasure codes have similar coding and decoding performance (Plank et al. 2009). However, tradeoffs have been made among different codes to cope with various situations in practice. For example, the X, P, and S codes have the optimal update complexity. The EVENODD, RDP, and Liberation codes can be implemented on RAID-6 systems. When recovering more failed devices, algorithms become more complex. To date, the STAR code possesses the best performance among the three-erasure-tolerant codes. The generalized EVENODD code can tolerate 8 failed devices. The Weaver code tolerates up to 12 failed devices but with a cost of low storage efficiency.

When the number of failed disks increases further, the capacity of these STCs is not sufficient. GTCs offer solutions. Generally Matrix or Graph methods are used in GTCs but there are no concise, clear algorithms. The examples can be found in Feng's method or the Tornado code.

GTCs can tolerate more failures, but their algorithms are much more complex and difficult to be implemented in canonical storage systems. XOR operations normally depend on the number of "1"s in a matrix. When using matrix methods for encoding in GTCs, the general code coming from the Reed-Solomon matrix normally has an MDS property. But it has more "1"s in the matrix. They can be optimized by reducing the "1"s through some algorithms. It has been a hot topic in recent years to seek an optimal method for developing algorithms. Even if a sparse matrix can be constructed for an encoding procedure, it is still challenging to find an optimal decoding procedure. Finding optimal methods for the coding and decoding matrices is another topic. We do not intend to introduce them in this paper. Some optimal methods for coding and decoding a matrix can be found in reference(Luo et al. 2009; Huang et al. 2007a; Hafner et al. 2005c).

Based on large irregular graphs, the Tornado code is designed to function on a larger scale with hundreds of thousands of blocks in high performance environments. It is very different from Reed-Solomon codes. LDPC codes can be implemented using the Tornado architecture. However, LDPC codes are probability codes, and they cannot guarantee that the lost symbol can be recovered in some cases (Luby et al. 1997). LDPC codes are more suitable for channel transmission.

From the perspective of encoding and decoding procedures, most STCs give a clear encoding and decoding algorithm except for the Liberation, Weaver, and HoVer codes. Actually, the Weaver and HoVer codes can be considered as frameworks based on computer simulation. Regarding GTCs, the encoding and decoding algorithms are more complicated except for the Tornado code. Some of the literature discussed architectural design rules and included examples for the Tornado, Weaver, HoVer, and Pyramid codes. Any coding scheme that satisfies those rules may be included in code designs. Some of the literature presented the algorithms very clearly, such as those related to the EVENODD, X, and RDP codes.

Table 5. A complete comparison of the XOR based erasure codes for storage construction

	Fault Tolerant	Encode complexity	Update Complexity	MDS	Storage Efficiency	H/V *Note1	Frame work	Year	Patent
EVENODD	2	$O(n^2)$	~3,	Yes	n/(n+2)	H	No	1994	Yes
Tornado	General	$O(n\ln(1/e))$	N/A	No	$1-\beta$	N/A	Yes	1997	Yes
X code	2	$O(n^2)$	2	Yes	n/(n+2)	V	No	1999	No
RDP	2	$O(n^2)$	~3	Yes	n/(n+2)	H	No	2004	Yes
S code	2	$O(n^2)$	2	Yes	n/(n+2)	V	No	2005	No
WEAVER	≦12	N/A	N/A	No	<50%	N/A	Yes	2005	No
HoVer	*Note2	N/A	N/A	No	N/A	N/A	Yes	2005	No
RS-Like	General	$O(n^2)$	>r	Yes	n/(n+r)	H	No	2005	No
Rabin-like	General	$O(n^2)$	>r	Yes	n/(n+r)	H	No	2005	No
Pyramid	General	N/A	N/A	No	N/A	N/A	Yes	2007	No
STAR code	3	$O(n^2)$	~5	Yes	n/(n+3)	H	No	2008	No
Liberation	General	$O(n^2)$	>r	Yes	n/(n+r)	H	No	2008	No
P code	2	$O(n^2)$	2	Yes	n/(n+2)	V	No	2009	No

Among the codes discussed in this survey, the Tornado is the fastest since it is linear to the number n. Besides, Plank (2004)] has looked into the optimal and nearly-optimal LDPC erasure codes for small values of n, and provided some simulation results without proof.

To date, there is no specialty code that can take full advantages of parallel implementations. Although some of the codes can be executed in parallel, they need complicated control mechanisms, such as mutex, spin lock controls in multi-core systems, or RAM read/write collision control in VLSI designs. We believe a fast, concise and efficient parallel design for decoding will be a hot topic in the future.

In addition to the codes discussed in this survey, there is another family of codes called network codes (Dimakis et al. 2010). Recently Dimakis et al. (2011) conducted a survey of the network codes for distributed storage on April 2010. The network codes are not designed for a canonical storage system. Some of them focus on security in an un-trusted environment (Oliveira et al. 2010). Therefore we do not cover them in our survey.

Conclusion

It is common to use redundancy to improve the system availability and fault tolerance for storage systems. Due to their high performance, XOR based erasure codes have been intensively studied in recent years. However no one code has become the standard in the storage industry. To date, such a "perfect" code has not been developed (Hafner 2005a). In this paper, we provided a survey of the XOR code schemes used to construct storage systems. Design principles and algorithm construction were analyzed and detailed examples were provided. We hope this survey will inspire more active research in this area.

References

Bindel, D., Chen, Y., Eaton, P., Geels, D., Gummadi, R., Rhea, S., Weatherspoon, H., Weimer, Wells, W., C., Zhao, B., and Kubiatowicz, J. 2000. Oceanstore: An extremely wide-area storage system. in *Proceedings of the 9th International Conference on Architectural Support for Programming Languages and Operating Systems,* 190-201.

Blaum, M., and Bruck, J. 1996. MDS array codes with independent parity symbols. *IEEE Transactions on Information* Theory, 529-542.

Blaum, M., and Roth, R. 1993. New array codes for multiple phased burst correction. *IEEE Transactions on Information Theory*, 66-77.

Blaum, M., Brady, J., Bruck, J., and Menon, J. 1995. EVENODD: An Efficient Scheme for Tolerating Double Disk Failures in RAID Architectures. *IEEE Transactions on Computers*, 192-202.

Blomer, J., Kalfane, M., Karp, R., Karpinski, M., Luby, M., and Zuckerman, D. 1995. *An XOR-Based Erasure-Resilient Coding Scheme*. Technical report, The International Computer Science Institute (ICSI).

Brack, T., Alles, M., Lehnigk-Emden, Kienle, T. F., Wehn, N., Berens, F., Ru EGG, A. 2007. A Survey on LDPC Codes and Decoders for OFDM-based UWB Systems. In Proc. *IEEE Vehicular Technology Conference* (VTC), 1549 - 1553 , Dublin, Ireland, April.

Cassuto, Y., and Bruck, J. 2006. Cyclic low density MDS array codes. in proceedings of the 2006 *IEEE International Symposium on Information Theory*, 2794-2798, Seattle, WA, U.S.A., July.

Corbett, P., English, B., Goel, A., Grcanac, T., Kleiman, S., Leong, J., and Sankar, S. 2004. Row-Diagonal Parity for Double Disk Failure Correction., *Proc. of USENIX FAST 2004*, Mar. 31 to Apr. 2, San Francisco, CA, USA

Dimakis, G,. Godfrey, P. B., Wu, Y., Wainwright M., and Ramchandran, K. 2010. Network Coding for Distributed Storage Systems. *IEEE Transactions on Information Theory*, 56, 9, 4539-4551.

Dimakis, G., Ramchandran, K., Wu, Y., Suh, C. 2011. A Survey on Network Codes for Distributed Storage., *The Proceedings of the IEEE*, (submitted) http://arxiv.org/abs /1004.4438.

Feng, G. L., Deng, R. H., Bao, F., and Shen, J.C. 2005. New Efficient MDS Array Codes for RAID Part I: Reed-Solomon-Like Codes for Tolerating Three Disk Failures. *IEEE Transactions on Computers*, 1071-1080.

Feng, G. L., Deng, R. H., Bao, F., and Shen, J. C. 2005. New Efficient MDS Array Codes for RAID Part II: Rabin-Like Codes for Tolerating Multiple (greater than or equal to 4) Disk Failures. *IEEE Transactions on Computers*, 1473-1483.

Feng, J., Chen, Y., and Summerville, D. 2010. EEO: an Efficient MDS-Like RAID-6 Code for Parallel Implementation., *IEEE Sarnoff 2010,* Princeton, New Jersey, April 11-13.

Hafner, J. L. 2005a. WEAVER Codes: Highly Fault Tolerant Erasure Codes for Storage Systems. the Fourth USENIX Conference on File and Storage Technologies (FAST '05), 211-224.

Hafner, J. L. 2005b. HOVER Erasure codes for disk arrays. *IBM Research Report*, RJ10352 (A0507-015) July 12.

Hafner, J. L., Deenadhayalan, V., Rao, K. K., and Tomlin., J. A. 2005c. Matrix Methods for Lost Data Reconstruction in Erasure Codes.. *Proc. of 4th USENIX Conference on File and Storage Technologies* (FAST '05), March.

Hellerstein, L., Gibson, G. A., Karp, R. M., Katz, R. H., and Patterson, D. A. 1994. Coding techniques for handling failures in large disk arrays. *Algorithmica,* 182-208.

Huang, C., and Xu, L.H. 2008. STAR: An Efficient Coding Scheme for Correcting Triple Storage Node Failures. *IEEE Transactions on Computers*, 889-901

Huang, C., Chen, M. H., and Li, J. 2007b. Pyramid codes: Flexible schemes to trade space for access efficiency in reliable data storage systems. in *NCA-07: 6th IEEE International Symposium on Network Computing Applications*, Cambridge, MA, July.

Huang, C., Li, J., and Chen, M. 2007a. On Optimizing XOR-Based Codes for Fault-Tolerant Storage Applications. the 2007 *IEEE Information Theory Workshop*, ITW '07, 218 – 223.

IBM. 2009. About RAID levels and ClearCase. 2009. http://www-01.ibm.com /support /docview.wss?uid=swg21149421#Level%206%20--%20Independent%20Data%20Disks.

Jin, C., Jiang, H., Feng D., and Tian, L. 2005. P-Code: A New RAID-6 code with optimal properties. the *23rd International Conference on Supercomputing*, New York, June.

Katti, R., and Ruan, X.Y. S-code: new distance-3 MDS array codes. the IEEE International Symposium on Circuits and Systems, 2005. (ISCAS 2005.), 23-26.

Kempf, J. 2007. *Erasure Coding for the Software Engineer: A Case Study*. Technical report, Rensselaer Polytechnic Institute.

Kreyszig, E., 2005. *Advanced Engineering Mathematics*, 9th edition, Wiley.

Lin, S., and Costello, D. J. 2004. Error Control Coding, Fundamentals and Applications., Prentice Hall Press.

Liva, G., Song, S., Lan, L., Zhang, Y., Lin ,S., and Ryan, W. E. 2006. Design of LDPC codes: A survey and new results. *J. Commun. Softw. Syst*., vol. 2, no. 3, 191-211, Sep.

Luby, M. G., Mitzenmacher, M., Shokrollahi, M. A., and Spielman, D. 2001. A. Efficient erasure correcting codes. *IEEE Transactions on Information Theory*, 569-584.

Luby., M. G., Mitzenmacher, M., Shokrollahi, M. A., Spielman, D. A., and Stemann, V. 1997. Practical loss-resilient codes. in *Proceedings of the 29th annual ACM symposium on Theory of computing*, (STOC '97), New York, NY, USA, 150-159.

Luo, J., Xu, L., and Plank, J. S. 2009. An Efficient XOR-Scheduling Algorithm for Erasure Codes Encoding. the IEEE International Conference on Dependable Systems and Networks (DSN-2009), Lisbon, Portugal.

Mackay, D.J.C. 2005. Fountain codes. *IEEE Proc.Comm*., vol.152, 1062–1068.

Oliveira, P., Lima, L., Vinhoza, T., Barros, J., Medard, M. 2010. Trusted Storage over Untrusted Networks., *IEEE Globecom Communication Theory Workshop* December.

PATTERSON, D.A., GIBSON, G.A., AND KATZ,R. 1988. A Case for Redundant Arrays of Inexpensive Disks., *SIGMOD International Conference on Data Management*, 109-116, Chicago.

Pinheiro, E., Weber, W. D., and Barroso, L. A. 2007. Failure Trends in a Large Disk Drive Population. in *Proceedings of the 6th USENIX Conference on File and Storage Technologies* (FAST'07), San Jose, CA, February.

Plank, J. S. 1999. A Tutorial on Reed-Solomon Coding for Fault-Tolerance in RAID-like Systems. *Software: Practice and Experience*, vol. 27, no.9, 995–1012, Sept.

Plank, J. S. 2004. Enumeration of Small, Optimal and Near-Optimal Parity-Check Erasure Codes. Technical report, Department of Computer Science, University of Tennessee, (UT-CS-04-535).

Plank, J. S. 2005b. Note: Correction to the 1997 Tutorial on Reed-Solomon Coding. *Software, Practice and Experience,* Vol. 35, Issue 2, February, 189-194.

Plank, J. S. 2008. The RAID-6 liberation codes., in *Proceedings of the 6th USENIX Conference on File and Storage Technologies* (FAST'08), 97-110

Plank, J. S., Luo, J., Schuman, C. D., Xu, L., and Wilcox-O'hearn, Z. 2009. A Performance Evaluation and Examination of Open-Source Erasure Coding Library for Storage. *7th USENIX Conference on File and Storage Technologies* (FAST '09), San Francisco, CA, February.

Plank., J.S. 2005a. Erasure Codes for Storage Applications. Tutorial, *4th Usenix Conference on File and Storage Technologies (FAST-2005),* San Francisco, CA, December.

Rabin, M. 1989. Efficient Dispersal of Information for Security, Load Balancing and Fault Tolerance. J. ACM, 32(4), 335-348, Apr.

Reed, S., and Solomon, G. Polynomial codes over certain finite fields. 1960. *Journal of the Society for Industrial and Applied Mathematics,* 300-304.

Roman, S. 1992. *Coding and Information Theory*. Springer-Verlag, 293-294.

Schroeder, B., and Gibson, G. A. 2007. Disk Failures in the Real World: What Does an MTTF of 1,000,000 Hours Mean to You?. in *Proceedings of the 6th USENIX Conference on File and Storage Technologies (FAST'07),* San Jose, CA, February.

Thul, M., Kienle, F., Wehn, N. 2003. A Survey on LDPC- and Turbo-Decoder Implementations. In P*roc. International Conference on Software, Telecommunications and Computer Networks (SoftCOM),* pages 122-126, Venice, Italy, October.

WIKIPEDIA. 2011. http://en.wikipedia.org/wiki/Redundant_array_of_independent_disks.

Xu, L., and Bruck, J. 1999. X-Code: MDS Array Codes with Optimal Encoding. *IEEE Transactions on Information Theory* Vol.45, 272-276.

In: Biology and Communication in the Information Age
Editor: Rafiqul Islam
ISBN: 978-1-62808-252-4

Chapter 13

SENSITIVITY OF THE MURINE LIVER LIPID TO THE LOW-INTENSITY X-RAY RADIATION AT LOW DOSES

M. A. Klimovich, M. V. Kozlov and L. N.Shishkina*
Emanuel Institute of Biochemical Physics of Russian Academy of Sciences, Moscow, Russia

ABSTRACT

The parameters of the physicochemical regulatory system of lipid peroxidation (LPO) in liver of female outbreed mice at the autumn and spring-summer seasons after action of X-radiation at doses less than 1.5 mGy under changing dose rate. The initial values result in significant effect on the change in the relative weight of the liver, the relative content of phosphatidylcholine and lysoforms of phospholipids in the liver of mice after X-rays exposure. Significant decrease and substantial influence of the dose rate dynamics during irradiation are revealed for the molar ratio of [sterols]/[phospholipids], the ratio of phosphatidylcholine / phosphatidylethanolamine, and the ratio of the sums of the more easily oxidizable to the more poorly oxidizable fraction of phospholipids. Experimental data allow us to conclude that the biological effects of X-rays at low doses under changing dose rate have a complex, nonlinear character.

Keywords: X-rays, low-dose, changing dose rate, the liver of mice, lipid peroxidation, phospholipids, cholesterol

1. INTRODUCTION

The biological effect of the radiation at low doses on the body is still one of the most controversial branch of radiobiology. The lack of clear radiobiological consequences under

* E-mail: shishkina@sky.chph.ras.ru.

exposure of animals at low doses causes the search of parameters, changes of which would assess the state of metabolism in a complex system as a whole.

Previously, the high sensitivity of the physicochemical parameters of the regulatory system of lipid peroxidation (LPO) to damaging factors including the radiation action at low doses has shown [1 - 3]. It was revealed that the radiobiological response under X-ray exposure at low doses is due to the functional state of the membrane composition and physicochemical properties of lipids which significantly depend on many factors: age, season, diet et al [4, 5]. It is shown that parameters of the LPO regulatory system, such as the lipid composition and their antioxidant activity, in particular, have an influence on the ratio of damage in DNA and membranes, and the repair of the membranes after X-ray exposure at low doses [3, 6]. The liver hepatocytes, performing a key role in the maintenance the homeostasis of the tissue mammals, are highly sensitive to environmental pollution [1, 7]. However, there are no published data about the impact of X-ray at low doses under changing dose rate on lipid metabolism in the liver of laboratory animals.

The aim of this work is to identify changes in the intensity and the state of the lipid peroxidation parameters in the murine liver and the relative weight of liver after exposure of the low-intensity X-rays at low doses under changing dose rate depending on the initial parameter values in of the liver of the intact mice.

2. Experimental Part

The objects of study were mice (females) of mature outbreed mice (3 - 3.5 months age). The total number of animals was 85 individuals, before exposure was 13 weeks. The antioxidant status and the LPO intensity in tissues of mice were modified by carrying out experiments in different seasons. Experiments were repeated twice: in September-October (the experiment № 1) and in May-June (the experiment № 2). Mice were divided per 10 animals in each group which had the adequate weight. The animals were irradiated in special containers previously formed groups per 10 animals; each mouse was located in a single cell, while maintaining the movement freedom. The animals were irradiated with a single dose from 12-00 to 13-30 to exclude the influence of daily variations in the liver lipid antioxidant activity [5]. X-ray source was a microwave electron cyclotron resonance accelerator, the device and principle of operation is described in detail in [8, 9]. X-ray emission of the electrons detected by the detector model of the firm BYCRON 2,5 M2, 5/3LP-X. The absence of a microwave - the components in the irradiation zone was monitored with a dosimeter. X-ray dose rate was modified due to changes in the dynamic pressure of the vacuum in the source chamber. The intensity of the X-ray source after the first few minutes of operation rises to a maximum dose rate, which then decreases within irradiation. Typical curves of the dose rate dynamics within irradiation are shown in Figure 1.

Control animals were mice from the same party with them were the same manipulations as with experimental animals, excluding the X-ray action. Decapitation of the experimental and control mice were produced within 30-31 and 37-38 days after irradiation, from 10 to 11 in the morning to avoid daily variation of the studied parameters.

The murine organs were placed in ace cooled weighting bottles immediately after decapitation. Lipids were isolated from liver by the Blay and Dyer method in the Kates modification [10].

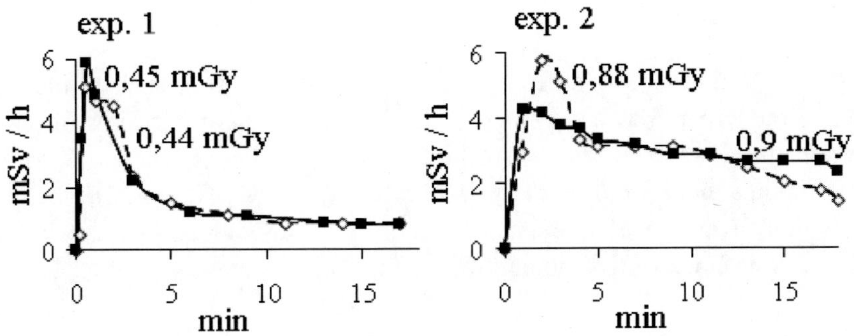

Figure 1. Dynamics of the dose rate during irradiation.

Qualitative and quantitative composition of phospholipids (PL) was analyzed by thin layer chromatography, as it was described in [11]. We used the type of silica gel G (Sigma, USA), glass plates 9 × 12 cm and mixture of solvents chloroform : methanol : glacial acetic acid : water (25 : 15 : 4 : 2) as a mobile phase. Spectrophotometrical measurements were carried out by KFK-3 (Russia) at the wavelength λ = 800 nm. This technique is described in detail in [12].

In addition to the quantitative analysis of different fractions of PL the following generalized parameters of the lipid composition were also evaluated: the PL proportion in the total lipid composition (% PL); the phosphatidylcholine to phosphatidylethanolamine (PC/PE) ratio in PL content; the ratio of the sums of the more easily oxidizable to the more poorly oxidizable PL (ΣEOPL/ΣPOPL).

The ΣEOPL/ΣPOPL value was calculated by the formula [13]: ΣEOPL/ΣPOPL = (PI+PS+PE+PG+CL+PA)/(LPPC+SM+PC), where PS is phosphatidylserine, PI is phosphatidylinositol, , PG is phosphatidylglycerol, CL is cardiolipin, and PA is phosphatidic acid, LPPC is lysoforms of PL, SM is sphingomielin. The liver PL of individuals sometimes had the additional fractions of PL, which are probably oxidized species of PC (PC^1) and PE (PE^1) at very low quantities.

The sterol content in lipids was determined spectrophotometrically at 625 nm wavelength [14]. Serva Company (FRG) cholesterol was used for plotting the calibration curve. The amount of the LPO products was estimated by their reaction with 2-tiobarbituric acid (TBA-reactive substances, TBA-RS) [15].

The protein content was measured by the modified microbiuretic method at λ = 540 nm [16]. The relative weight (index) of organ was calculated as the ratio of liver weight in mg to body weight in g and expressed as per mille (‰).

The antiperoxide activity (APA) of lipids, i.e., the ability of lipids to decompose peroxides, was determined by the difference in the concentrations of peroxides in the oxidized methyloleate and in the lipid solutions in this methyloleate [17].

All measurements were performed for each animal individually. The experimental data were processed by the commonly used variation statistic method [18] and by means a computer software package KINS [19]. The experimental data are presented in the tables and figures as the arithmetic means with the indication of the mean square errors of the arithmetic means (M ± m).

3. Results and Discussion

The radiation action at low doses under changing dose rate increased the values of the liver indices in the experience as compare with than in the control groups of animals within one month after the X-ray under dose of the 0.45 mGy in the experiment 2, within 38 days after the irradiation at doses of 0.98 mGy, 0.99 mGy, 1.14 mGy (Figure 2). The growth of the liver index within a month after exposure occurs for these groups of the irradiated mice, which had the lower index of liver in control.

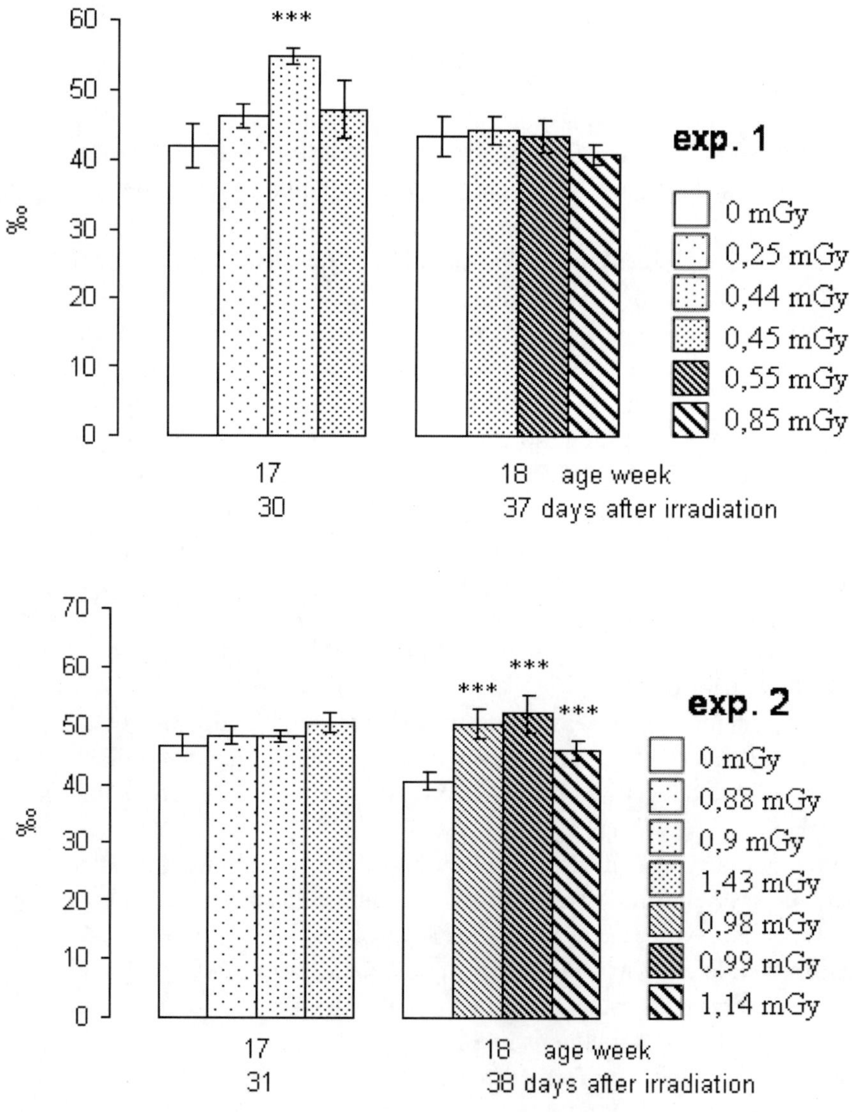

Figure 2. The values of the relative weight (index) of liver in groups of the control and irradiated mice within after action in the different experiments.

As you might expect, the different parameters of the LPO regulatory system have different sensitivity and the ability to normalization after irradiation at low doses under changing dose rate. Data about LPO intensity and relative content of lysoforms and SM in PL are presented in Figure 3 and 4 correspondingly. As known, the liver is characterizing the relative low intensity of LPO [9, 20]. The lowest content of TBA-reactive substances in liver homogenates and the lower share of lysoforms in PL in groups of control animals were revealed in experiment 2, which indicates about the lower intensity of LPO in spring-summer season.

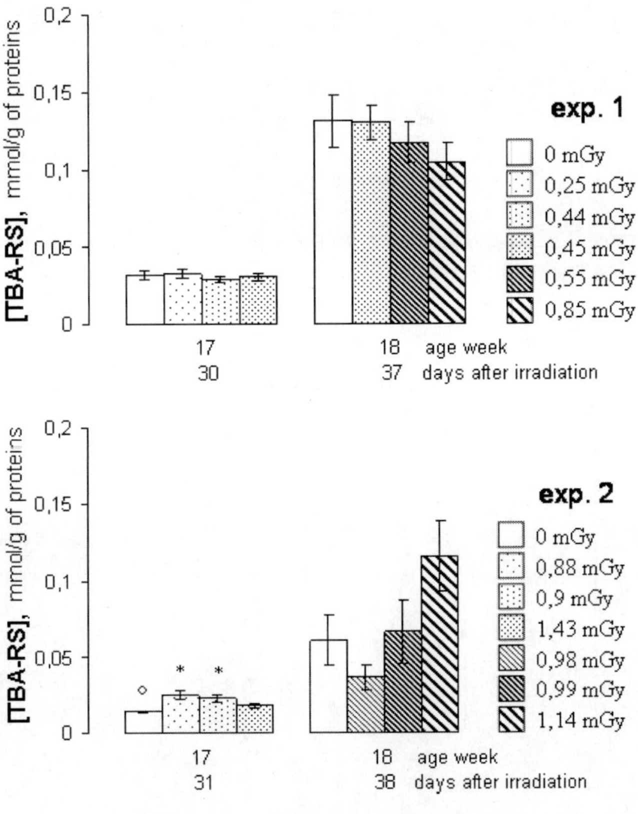

Figure3. The content of TBA - reactive substances in the liver homogenate in groups of the control and irradiated mice within after action in the different experiments.

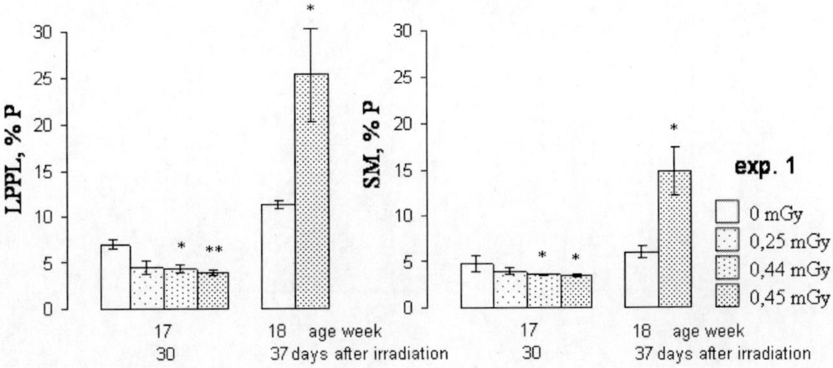

Figure 4. (Continued).

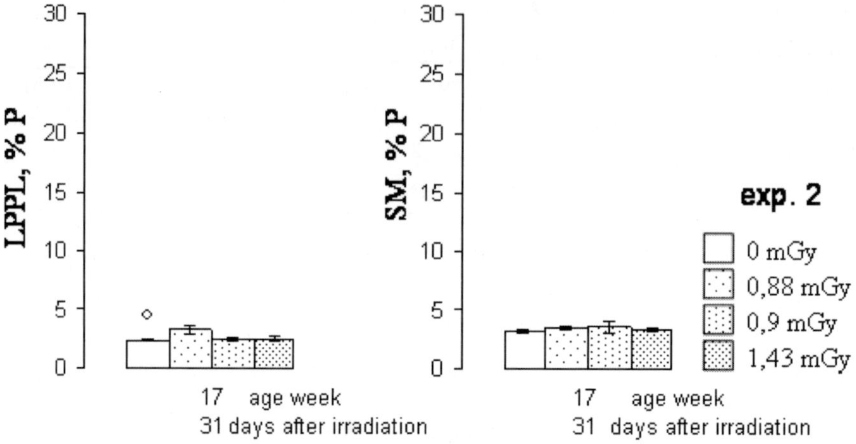

Figure 4. The lysoforms of PL and SM content in the murine liver phospholipids in groups of the control and irradiated mice within after action in the different experiments.

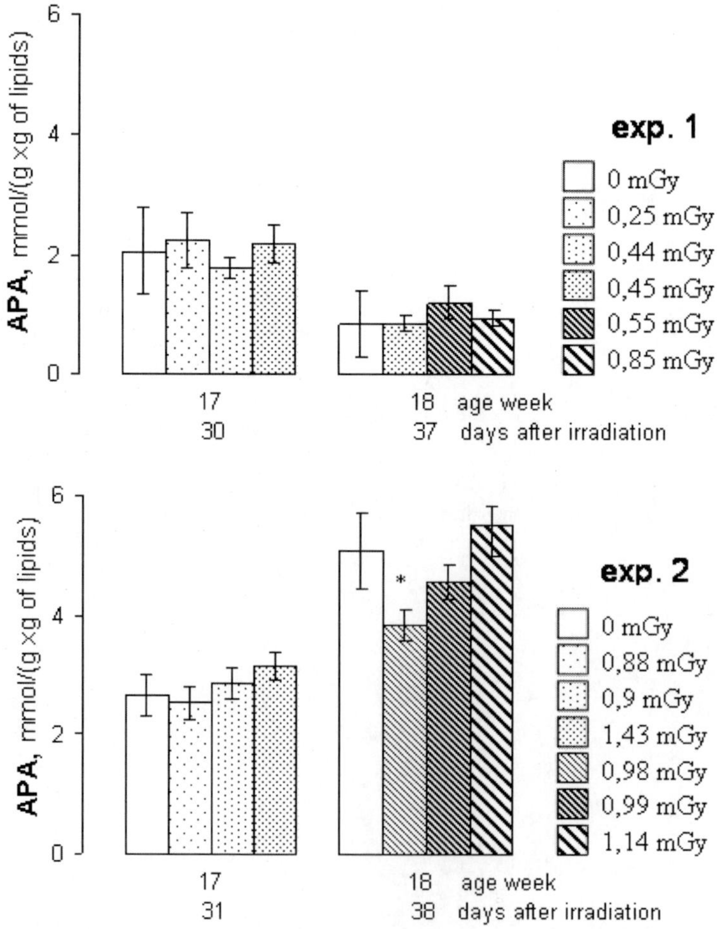

Figure 5. The antiperoxide activities of the liver lipids in groups of the control and irradiated mice within after action in the different experiments.

Besides, the significant growth of the TBA - reactive substances content 1.6-1.8 times was also found in experiment 2 within 31 days after the irradiation of mice at doses of 0.88 - 0.9 mGy. However, the LPO intensity has only the tendency to the growth when the X-ray dose is increased from 0.98 to 1.14 mGy.

It is need to note that the APA of the murine liver lipids in all experiments was not significantly different between corresponding age control groups (Figure 5). Besides, the APA variability in the liver lipids within the group of irradiated mice is lower than that in age control groups. Perhaps, this is due to the reduction of resistance the organism and its adaptive capacity after irradiation [21, 22].

It was earlier revealed that the APA lipids or the presence of peroxides in their affect the processes of biosynthesis and degradation of the PL [23, 24]. Since, most of the liver lipids of mice in these experiments had the ability to decompose peroxides, so the mean group values for the liver lipids parameters are shown on the figures and tables for mice the lipids of whose had APA.The ratios of the PL minor fractions, as well as the content of PL and sterols in the composition of total lipids are presented in Tables 1 and 2.

Table 1. The composition of the minor fraction (%P) of phospholipids and the sterols and phospholipids proportions in the composition of total lipids in the murine liver lipids (experiment 1)

mGy	0	0,88	0,9	1,426
n	5	10	10	10
Age, week	17			
	The minor fraction of phospholipids (% P) of the murine liver			
PC[1]	*0,6±0,1*	0,8±0,15	0,6±0,1	0,6±0,1
PI	*7,7±0,3*	6,7±0,6	7,1±0,4	7,8±0,3
PS	*1,7±0,1*	3,0±0,8	1,6±0,1	1,5±0,1
PE[1]	*0,37±0,07*	0,4±0,1	0,4±0,1	0,35±0,05
PG	*2,2±0,2*	2±0,2	1,8±0,2	2,0±0,1
CL	*3,5±0,3*	3,4±0,2	3±0,2	3,4±0,3
PA	*1,4±0,4°*	1,3±0,3	0,64±0,08*	0,85±0,06*
	Share phospholipids_of total lipids murine liver			
%	*41,4±3,7*	38,1±2,1	52,8±3,6	51,1±2,3*
	Share sterols of total lipids murine liver			
%	*10,2±1,7*	6,4±0,6***	6,6±0,6***	7,5±0,7**

Differences between mean group values of the control mice. significant at:
○ p <0,05; ○○p<0,01; ●p<0,001
Differences between mean group values of irradiated and control mice significant at:
* p <0,05; ** p <0,01; *** p <0,001
(%) of phospholipids, and % PL and sterols in the mass fraction (%) of total lipids.
The individual phospholipids fractions in the tables presented in the mass fraction

As follows from the data presented in Tables 1 and 2, in experiment 1 the content of such minor fractions PL as PI (2 times), PG (tendency to diminution or reduction 2 times at different doses), PA (2-3 times) decreased as compared with that in the control age animals within 1 month after irradiation. While the PE share in PL liver of irradiated mice has in certain growth (Figure 7). However, within 37 days after irradiation of mice at the dose 0,45mGy content of minor fractions of PL are not significantly changed, but we could speak of a tendency to their increase. Besides, the reduction of the PE content (Figure 7) is obviously due to increase the lysoforms of PL (Figure 4) ($p < 0.05$).

Table 2. The composition of the minor fraction (%P) of phospholipids and the sterols and phospholipids proportions in the composition of total lipids in the murine liver lipids (experiment 2)

mGy	0	0,25	0,44	0,45	0	0,45
n	5	10	10	10	5	10
Age, week	17				18	
	The minor fraction of phospholipids (%P) of the murine liver					
PC[1]	*1,7±0,4*	1±0,1	1,1±0,1	1±0,2	*3,9±1,0*	-
PI	*6,7±0,8*	3,9±0,3**	3,4±0,4**	3,2±0,3***	*5,0±0,7*	6,1±0,8
PS	*1,6±0,3*	3±1	2,8±0,7	2,2±0,5	*6,7±2,4*	8,9±2,8
PE[1]	*0,8±0,5*	-	0,7±0,1	0,7±0,1	-	-
PG	*2,5±0,8*	2±0,3	1,7±0,2	1,3±0,2*	*1,8±0,6*	2,4±0,6
CL	*3,8±2*	3,6±0,4	3,6±0,4	3,1±0,4	*2,2±0,8*	4,6±1,3
PA	*6±2*	2±0,6*	2,1±0,6*	3,2±0,6	*3,0±0,9*	4,4±1,1
	Share phospholipids of total lipids murine liver					
%	*23±2*	24±3	27±3	26,5±2	*35±3,1*	13,9±1,9***
	Share sterols of total lipids murine liver					
%	*6±1,2*	2,4±0,6***	3±0,4***	2,1±0,3***	*8±1,7*	3,3±0,4**

The most substation changes brought about under the action of X-rays revealed on the composition of PL and sterol content, as well as the molar ratio of [sterols]/[PL] (Figure 6).

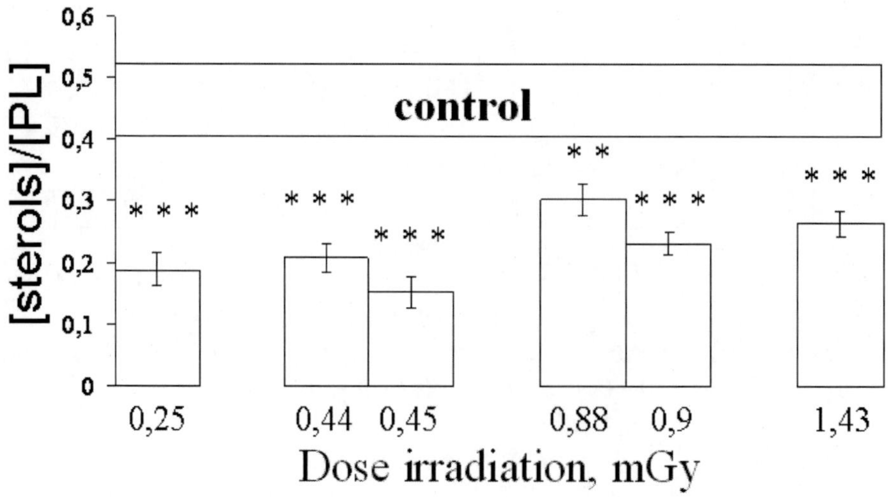

Figure 6. The [sterols]/[phospholipids] molar ratio in groups of the control and irradiated mice within one month after X-ray radiation at low doses under changing dose rate.

In the experiment 2 the liver PL composition was more resistant to the action of X-rays at low doses, which could be due to the low intensity of LPO. It is to note twofold diminution relative content of PA ($p<0.05$) within one month after irradiation at doses of 0.9 mGy and 1.43 mGy. All of these changes in the composition of the murine liver PL within one month after irradiation at doses less than 1.5 mGy should lead to an increase in the rigidity of the membrane and the diminution of the ability of the membrane lipid to oxidation.

As known, that cholesterol is an essential component of cell membranes of animals, which modifies the viscosity of the bilayer and the packing density of the PL in it [13, 25-27]. It was found that X-ray caused a significant the decrease content of the sterol content in the total liver lipid composition in both experiments (Tables 1 and 2).

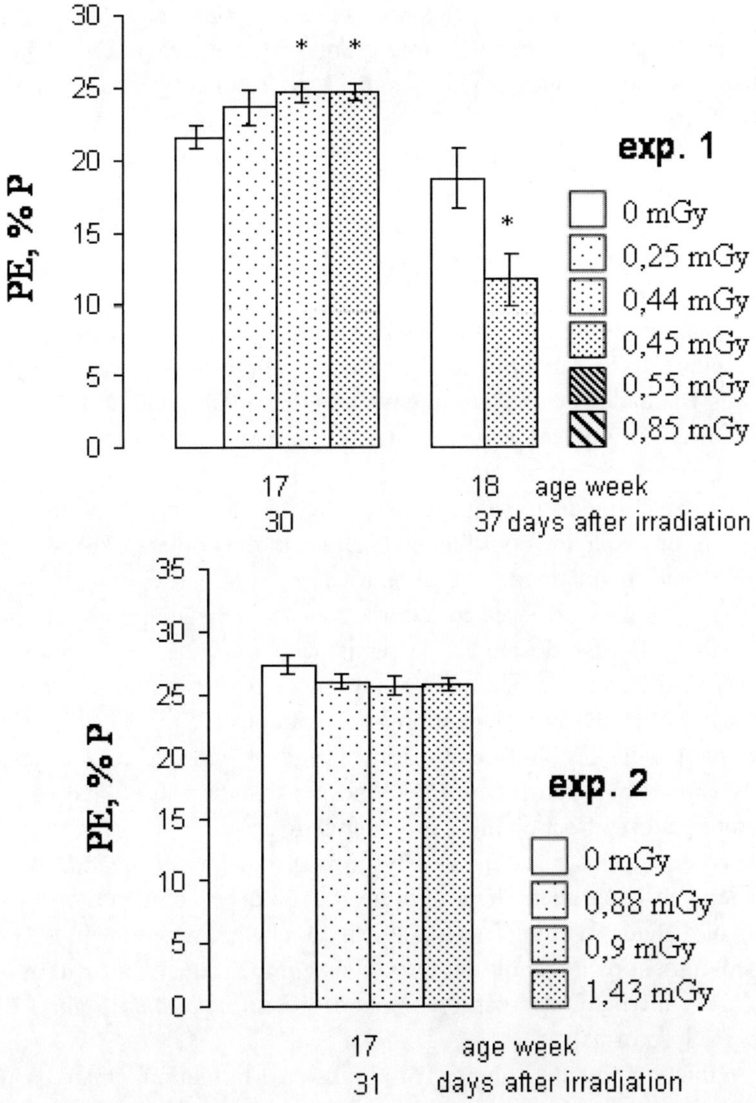

Figure 7. The phosphatidylethanolamine share in phospholipids of the liver in groups of the control and irradiated mice after the radiation action in the different experiments.

Since about 80% of cholesterol is synthesized in the endoplasmic reticulum of hepatocytes, so the reduction of its content in the total lipids content shows a synthetic dysfunction of liver. This assumption is confirmed by the significant change in the liver index of irradiated mice with respect to the corresponding age control group. Perhaps, reducing the ability of lipids to oxidation and increase the rigidity of the membrane system [28], the liver can be regarded as an adaptation of the organism to the impact of a weak radiation factor under changing dose rate. As shown, the molar ratio of [sterols]/[PL] is one of the few parameters values of whose are due to the structural state of membranes.

It is interestingly that this parameter is constant in the liver lipids of mice in the control age groups (Figure 6). It is found both the significant decrease in this ratio in the liver lipids of irradiated mice in all experimental groups and the significant influence the dynamics of the dose rate during irradiation of mice at doses of 0.44 - 0.45 mGy and 0.88 – 0.9 mGy (Figure 6). Changes of the PL share in the total lipid composition after the radiation action in all experiments are revealed under their more high content in the control groups (Tables 1 and 2).

As known, PC and PE are the major components in the PL liver of mice. Although changes both of PC and PE doses have at the same direction after irradiation at all investigated doses (Figures 7 and 8), the ratio of these fractions which is also characterized the structural state membrane system of liver [13]. And the ratio [PC]/[PE] reliable changes under that action (Figures 9).

It is revealed changes in the content of lysoforms of PL and SM dependence on the X-ray doses which had where at the same direction. As know, cholesterol, SM and lysoforms of PL involved in processes related to the arrangement of the membrane structure [26, 29, 30]. Because of the high affinity of SM and cholesterol, the metabolism of these two compounds in intact cells is the integrated nature [26, 31].

Analysis of the data allow us to suggest that under the X-ray action at doses less than 1.5 mGy under changing dose rate the coordinated changes in the content of these fractions in the liver lipid of mice were maintained. The similar effect was earlier found for lipids of red blood cells of laboratory mice exposed to chronic low-level γ-radiation at doses of 1.44 and 28.8 cGy, under the different dose rate, [32], as well as under the combined action of chemical toxicants at different doses and the low-intensity γ-radiation at doses of 1.44-1.6 cGy, when mice received ecdysteroid-containing preparations [32].

Since these changes in the lipid composition found in our studies and described in the literature, usually caused by intensification of lipid peroxidation in animal tissues, it can be regarded as a compensatory effect on the X-ray at low doses.

The simultaneous changes in content of lysoforms and SM in the PK liver after X-ray irradiation especially in experiment № 1 (Figure 4) could be compensatory after action of damaging factors of different nature. The analogical conclusion was done in [33].

The increased content of PC could be viewed as compensatory, but within 37 days after the X-ray in the first experiment decreased content of PC is due to the PC and PE degradation (Figures. 7 and 8) to lysoforms.

Significant reduction in the ability of lipids to oxidation, that is the ratio of ΣEOPL/ΣPOPL [13], revealed in the liver of irradiated mice in all groups, regardless of the value of this parameter in the age control groups (Figure 10).

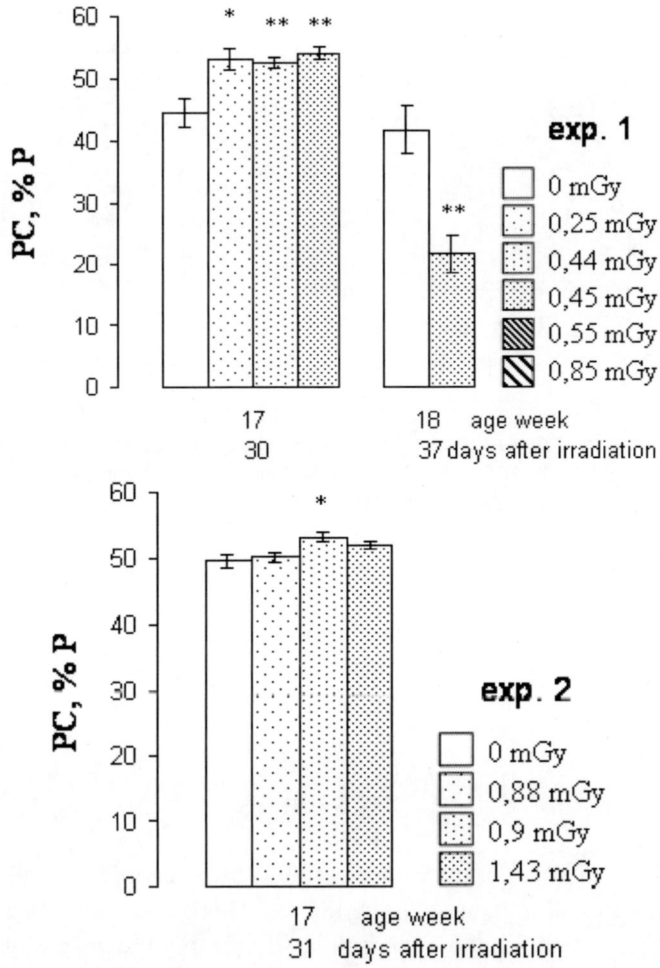

Figure 8. The phosphatidylcholine share phospholipids of the murine liver in groups of the control and irradiated mice within after action in the different experiments.

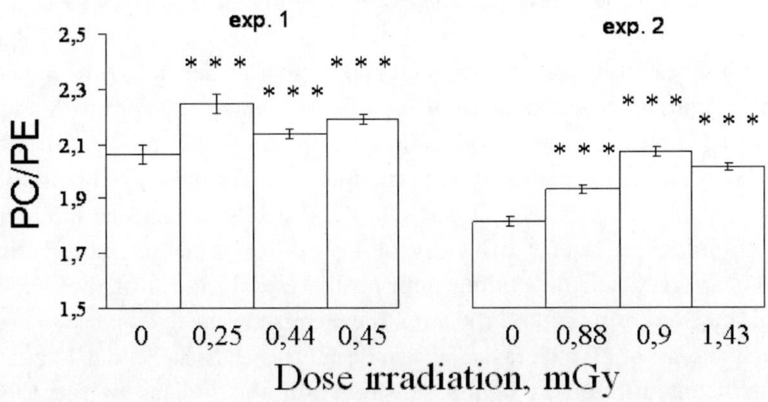

Figure 9. The phosphatidylcholine to phosphatidylethanolamine (PC/PE) ratio in the murine liver phospholipids composition within one month after X-ray radiation at low doses under changing dose rate.

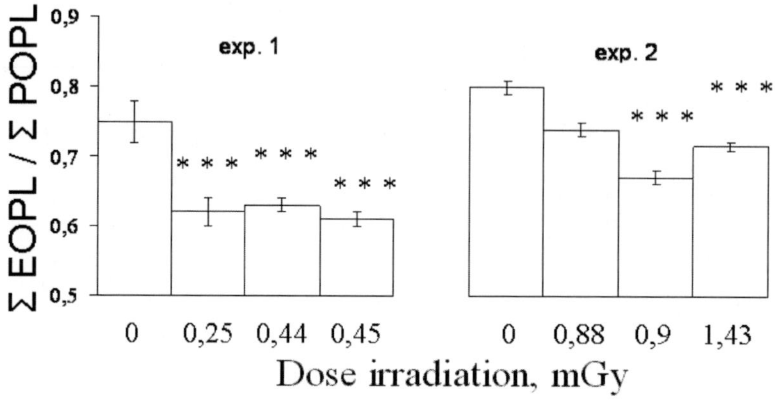

Figure 10. The ratio of the sums of the more easily oxidizable to the more poorly oxidizable fraction (ΣEOPL/ΣPOPL) of the murine liver phospholipids within one month after X-ray radiation at low doses under changing dose rate.

The reduction of the ΣEOPL/ΣPOPL ratio is accompanied by increasing ratio of the main PL fractions (PC/PE) in the liver of irradiated mice in all experimental groups (Figure 9 and 10). The reduction of ΣEOPL/ΣPOPL and an increase in the PC/PE ratio were earlier found in the PL liver of rodents inhabiting in the contaminated areas [1, 34]. Besides, it is need to note the significant influence of the dose rate dynamics within irradiation.

CONCLUSION

It was found that irradiation of mice caused a significant decrease content of sterols in the liver lipid composition in all variants of experiments. There was a significant decrease in the [sterols]/[PL] ratio in the liver lipids of the irradiated mice in all experimental groups, as well as a significant influence of the dose rate dynamics during irradiation of mice.

Significant reduction in the ability of lipids to oxidation (ΣEOPL/ΣPOPL) was found in the PL liver in all groups of irradiated mice, accompanied by the increase of ratio PC/PE. As the [sterols]/[PL] ratio value, these parameters is also depended on the dose rate dynamics during irradiation of mice.

Thus, data obtained allow us to conclusion about the high sensitivity of the ability of liver lipid to oxidation and the structural state of the liver membrane system to X-radiation action at doses less than 1.5 mGy under changing dose rate. Previously, biological effectiveness of this action has been shown by analyzing both the intensity of lipid peroxidation in plasma and organs [9], and the spleen index and the content of extracellular DNA in the blood plasma of mice [35]. The biological effects of X-ray at low doses under changing dose rate have complex non-linear character, depending both on the initial values of the parameters in the liver of control mice and time after exposure. The increase in the liver index, the relative content of PC and share lysoforms of PL after radiation significantly depends on the initial values for the most parameters. Significant differences in the sensitivity of the indices of the LPO regulatory system in the murine liver, depending on their level in groups of control mice and the dynamics of the dose rate during irradiation, and the lack of a linear dependence between their changes and the total dose of X-ray are evidence for absence of the possibility for the extrapolation of biological effects after irradiation of animals at sublethal and lethal doses to these under the low doses of radiation.

ACKNOWLEDGMENTS

The authors are grateful to the laboratory head of Prokhorov General Physics Institute, Russian Academy of Sciences Sergeychev K.F, and Karfidov D.M, Lukina N.A., for assistance in experiments on animals irradiation.

This work was supported by Basic Research Program of the Presidium of RAS "Fundamental sciences - medicine" (2006 - 2008).

REFERENCES

[1] Kudyasheva A.G, Shishkina L.N, Zagorskaya N.G, Taskaev A.I. Biochemical Mechanisms of Radiation Damages for Natural Populations of Rodent. - St. Petersburg.: *Science,* 1997. P. 156. (in Russian)

[2] Burlakova E.B., Goloshchapov A.N., Zhizhina, G.P., Konradov A.A. New aspects of the effects of low intensity radiation // *Radiation. Biology. Radioecology*. 1999. vol. 39, N. 1, pp. 26-34. (in Russian)

[3] Shishkina, L.N., Smotryaeva M.A. Relationship of Membrane and DNA Damage with Lipid Peroxidation under Weak Influence. // *Biophysic*s. 2000. vol. 45, N. 5, pp. 844-852. (in Russian)

[4] Burlakova E.B., Alesenko A.V., Molochkina E.M., Palmina N.P., Khrapova N.G. Bioantioxidants in radiation damage and malignant growth. Moscow: *Nauka*. In 1975. P. 294. (in Russian)

[5] Shishkina L.N., Burlakova E.B. The values of the antioxidant properties of lipids in radiation damage and repair of membranes / In.: Chemical and biological kinetics. New Horizons. M.: Chemistry. T. 2. *Biological Kinetics.* 2005. pp. 365-395.

[6] Smotryaeva M.A., Kruglyakova K.E., Shishkina L.N., Blyuhterova N.V., Goloshchapov A.N., Burlakova E.B. The structural and biochemical properties of mice blood components after low γ-irradiation of different intensity // *Radiation. Biology. Radioecology,* 1996. vol. 36, pp. 21-29. (in Russian)

[7] Blyuger A.F., Major A.J. Biochemistry of the liver in health and disease / In.: The basics of hepatology. Ed. Professor. A.F. *Blyugera. Riga. Zinatne*. 1975. pp. 44-83. (in Russian)

[8] Sergeichev K.F., Karfidov D.M., Lukina N.A. Electron Electron cyclotron resonance acceleration of relativistic energies by a microwave field in a mirror trap // *Plasma Physics*. 2007. vol. 33, N. 6, pp. 455-473. (in Russian)

[9] Klimovich M.A., Sergeichev K.F., Karfidov D.M., Lukina N.A., Shishkina L.N. Biological effectiveness of x-ray irradiation at low doses under changing dose rate // *Technologies of living systems*. 2010. N. 8. pp. 17-28. (in Russian)

[10] M. Kates: The Technique of lipidology. Mir, Moscow, 1975. P. 322 (Russian version)

[11] Biological membranes: the methods. Ed. S. Findlay, J.B. and B. X. Evans, transl. from English., *Springer-Verlag*, 1990. P. 423.

[12] Shishkina L.N., Kushnireva Ye.V., Smotryaeva M.A. «The combined effect of surfactant and acute irradiation at low dose on the lipid peroxidation process in tissues and DNA content in blood plasma of mice» //*Oxidation communications*. 2001. V. 24. N. 2. P. 276 - 286.

[13] Shishkina L.N., Kushnireva E.V., Smotryaeva M.A. A new approach to assessment of biological consequences of exposure to low-dose radiation // *Radiation. Biology. Radioecology*. 2004. vol. 44. N. 3. pp. 289-295. (in Russian)

[14] W. M. Sperry, M. Webb: A revision of the schoenheimer–sperry method for cholesterol determination, *J. Biol. Chem*., 1950. vol. 187, 97–106.

[15] T. Asakawa, S. Matsushita: Coloring conditions of thiobarbituric acid test for detecting lipid hydroperoxides. *Lipids*, 1980. vol. 15, 137–140.

[16] R. Itzhaki, D.M. Gill: A micro–biuretic method for estimating proteins. *anal. biochem*., 1964. vol. 9, 401–409.

[17] V. A. Men'shov, l. N. Shishkina, Z. N. Kishkovsky: Effect of biosorbents on the composition, content and antioxidative properties of lipids in the medium. *Appl. Biochem. Microbiolog*y, 1994. vol. 30, 359–369. (in Russian)

[18] G.F. Lakin: Biometry. 3rd Edition, Vysshaya shkola, Moscow, 1990. P. 293. (in Russian).

[19] E.F. Brin, S.O. Travin Modeling the mechanisms of chemical reactions // *Chem. Physics*. In 1991. vol. 10, N. 6, pp. 830-837. (in Russian)

[20] Shishkina L.N., Kushnireva E.V., Bespal'ko O.F., Polyakova N.V. The role of antioxidative status in development of the consequences of biological action under low dose and low intensity irradiation // *Radiation. Biol. Radioekol.* 2000. vol.40, N. 2, pp. 162-167. (in Russian)

[21] Romanovsky Yu.M., Stepanov N.V., Chernavskii D.S. Mathematical biophysics. Moscow: *Nauka*. 1984. P. 304. (in Russian)

[22] Klimovich M.A., Kozlov M.V., Hrustova N.V., Shishkina L.N. Influence of the lipid characteristics on the age changes of the interrelation in the metabolic regulatory system of the laboratory murine tissues // *Adv. gerontol.* 2010. vol. 23, N. 3, pp. 427–429. (in Russian)

[23] Urnysheva V.V Shishkina L.N. Effect of low doses of chemical toxicants on phospholipid composition in animal liver // *Biology Bulletin* 2004. vol. 31, N. 2, pp. 126-131. (in Russian)

[24] Kozlov M.V., Kushnireva E.V., Urnysheva V.V., Taran J.P., Shishkina L.N. The influence of characteristics of lipids on the regulation of biochemical processes in animal liver and blood // *Biophysics*. 2007. vol. 52. pp. 693-698. (in Russian)

[25] Klimov A.N., Nikulcheva N.G. Lipids, lipoproteins and atherosclerosis. St. Petersburg. Peter Press, 1995. 304 p. (in Russian)

[26] H. Ohvo – Rekila, B. Ramstedt, P. Leppimaki, J. P. Slotte: Cholesterol interactions with phospholipids in membranes. // *Prog. Lipid Res.,* 2002. 41, pp. 66–97.

[27] Filippov A., Oradd G., Lindblom G. The effect of cholesterol on the lateral diffusion of phospholipids in oriented bilayer // *Biophysical Journal*. 2003. vol. 84, N. 5, pp. 3079-3086.

[28] Diakowski W., Ozimek Ł., Bielska E., Bem S., Langner M., Sikorski A.F. Cholesterol affects spectrin-phospholipid interactions in a manner different from changes resulting

from alterations in membrane fluidity due to fatty acyl chain composition // *Biochem. Biophys. acta.* 2006. 1758. pp. 4 - 12.

[29] Gribanov. G.A. Structure and biological role of lysophospholipids // *Chemistry*. 1991. vol. 37, N. 4, pp. 2-10. (in Russian)

[30] Karpunin D.V., Akimov S.A., Frolov V.A. Pore formation in lysolipid- and cholesterol-containing planar lipid membranes, Biol. Membrany, 2005. vol. 22, pp. 429–432. // *Biol. membrane.* 2005. vol. 22. №. 5. P. 429-431. (in Russian)

[31] Ipatova O.M., Torkhovskaya T.I., Zakharova T.S., Khalilov E.M. Sphingolipids and cell signaling: Involvement in apoptosis and atherogenesis // *Biochemistry.* 2006. vol. 71, N. 7, pp. 713-722. (in Russian)

[32] Shishkina L.N., Shevchenko O.G. Lipids of the blood erythrocytes and their functional activity // *Biology bulletin reviews*. 2010. vol. 130, N.6, pp. 587-602. (in Russian)

[33] Shevchenko O.G., Shishkina L.N. Blood erythrocyte lipids as a model for estimating of biological consequences under the action of the physical and chemical factors // *Technologies of living systems*. 2009. V. 6, N. 8. pp. 21-32 (in Russian)

[34] Shevchenko O.G., Shishkina L.N., Kudyasheva A.G. Phospholipids composition in tissues of the root voles microtus oeconomus inhabiting territories with elevated radioactive background // *J. evolution. biochem. and physiology*. 2006. vol. 42, N. 3, pp. 218-224. (in Russian)

[35] Klimovich M.A., Smotryaeva M.A., Gaintseva V.D., Shishkina L.N. The creation of tests for the estimation of the biological consequences of x-ray radiation at low doses and changing dose rate on animals // Radiation biology. *Radioecology,* 2009. vol. 49, N. 4, pp. 473-477. (in Russian)

In: Biology and Communication in the Information Age
Editor: Rafiqul Islam
ISBN: 978-1-62808-252-4

Chapter 14

THE INFLUENCE OF THE DERIVATIVES OF 5-HYDROXYBENZIMIDAZOLE ON THE STRUCTURAL PROPERTIES OF ERYTHROCYTE MEMBRANE

O. G. Luneva*[1] and L. Ya. Gendel[2]

[1]Biological Faculty, M.V. Lomonosov, Moscow State University,
Vorob'evy Gory, Moscow, Russia

[2]N.M. Emanuel Institute of Biochemical Physics, Russian Academy of Sciences,
Moscow, Russia

ABSTRACT

The effect of synthetic antioxidants and bioregulators — 5-hydroxybenzimidazole derivatives on the erythrocyte membrane structure and the erythrocyte morphology was studied by the spin probe method and scanning electron microscopy. Analysis of the EPR spectra and electron micrographs showed that derivatives containing side substituents differing in hydrophobic properties can have a concentration-dependent varying-intensity modifying effect on the structural organization of the plasma membrane of erythrocytes and on their shape. Within the concentration range 10-7-10-3 M, all the substances studied have an echinocytogenic effect, which is most pronounced for the derivative containing benzyl and ethoxy substituents at positions 2 and 5, respectively. The data obtained suggest that these nonelectrolytes are distributed in the outer monolayer of the erythrocyte membrane.

Keywords: Membrane transport, nonelectrolytes, 5-hydroxybenzimidazole derivatives, erythrocyte morphology

* E-mail: oluneva@yandex.ru.

INTRODUCTION

Various organic nonelectrolytes are currently widely used in practice. Among them are drugs, antioxidants, and other biologically active compounds [1-7]. It is known that exogenous hydrophobic nonelectrolytes can interact with the plasma membrane of blood cells, including erythrocytes, and are thus involved in membrane transport.

It was shown [4, 7-12] that the membrane transport of these and other exogenous organic substances and their distribution within the membrane can be accompanied by modifications of the erythrocyte membrane structure and the erythrocyte morphology. There is a hypothesis [8] that the changes in the surface architectonics of erythrocytes characterize specific features of the distribution of an intercalating substance in the membrane, in particular, in its inner or outer monolayer. The distribution of organic nonelectrolytes in the membrane is significantly affected by their hydrophobic properties, which, in their turn, depend on the presence of some hydrophobic constituents or others in the structure of the substance [7].

Information on the intramembrane distribution of an exogenous biologically active substance is necessary to understand the physicochemical mechanisms and location of its membranotropic action. Although this information is very important, this is still meager.

The biological activity of a new group of organic nonelectrolytes —5-hydroxybenzimidazole derivatives—has been studied. It is revealed that they have activities of synthetic antioxidants and bioregulators [13]. The membranotropic properties of these compounds with respect to the erythrocyte membrane have not been assessed.

In this context, the purpose of this work was to investigate the physicochemical features of interaction of organic nonelectrolytes differing in hydrophobic properties such as 5-hydroxybenzimidazole derivatives with the plasma membrane of erythrocytes by the spin probe method and scanning electron microscopy.

EXPERIMENTAL

The object of investigation was erythrocytes from the blood of adult male mongrel albino rats (with heparin as an anticoagulant). Erythrocytes were isolated as in our previous work [11]. In the experiments, erythrocytes were suspended in buffer A (145 mM NaCl, 5 mM KC1, 4 mM Na_2HP0_4, 1 mM NaH_2P0_4, 1 mM $MgS0_4$, 1 mM $CaCl_2$, l0 mM glucose, pH 7.4). The cell concentration in the suspension was $(6\text{-}6.5)\cdot 10^6$ cells per microliter. Erythrocyte ghosts were obtained according to a published procedure [14].

5-Hydroxybenzimidazole derivatives containing different side substituents were used [15]. As a spin probe, 2,2,6,6-tetramethyl-4-palmitoylamidopiperidine-l-oxyl (probe 1) was used [16]. EPR spectra were recorded with a RE-1307 radiospectrometer at (18±2)°C under standard conditions. The 5-hydroxybenzimidazole derivatives and the spin probe were used as ethanol solutions; the ethanol concentration in samples did not exceed 2 vol. %.

In the experiments using the spin probe method, samples were prepared as follows. One of the 5-hydroxybenzimidazole derivatives was added to a ghost suspension, the mixture was incubated for 20 min, the spin probe was then added, and EPR spectra were recorded. Samples incubated under identical conditions in the absence of the derivative studied were used as reference samples. In studying the effect of the 5-hydroxybenzimidazole derivatives on the erythrocyte morphology, the cells were incubated for 10-60 min with one of the 5-

hydroxybenzimidazole derivatives. Samples incubated under identical conditions in the absence of the derivative studied served as reference samples. Fixation of samples in 2.5% solution of glutaraldehyde (Serva, Germany) in phosphate buffer (pH 7.4) and washing of cells in phosphate buffer were performed according to published recommendations [17]. The samples were coated with a platinum layer in an EICO IB-3 ion coater. Microscopic examination was carried out with a Hitachi-405A (Japan) and Amray (USA) scanning electron microscopes. The results obtained were statistically processed by the computer program Statistica 5 using the Student's t-test and the Wilcoxon test.

RESULTS AND DISCUSSION

The table presents the 5-hydroxybenzimidazole derivatives studied in this work.

Table 1. 5. Hydroxybenzirnidazole derivatives and their structural formulas

Derivative no.	Derivative	Structural formula
I	5-Hydroxybenzimidazole	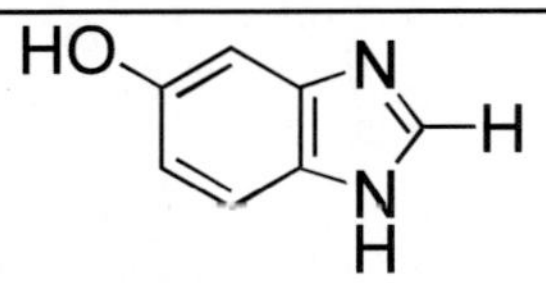
II	5-Ethoxybenzimidazole	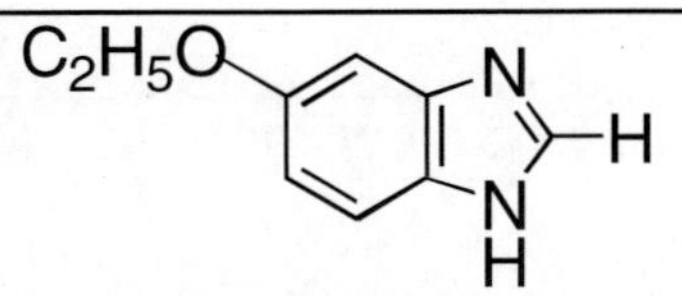
III	2-Benzyl- 5-hydroxybenzimidazole	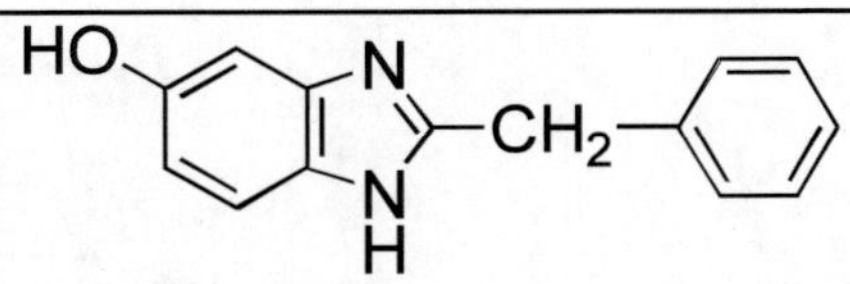
IV	2- Benzyl - 5-ethoxybenzimidazole	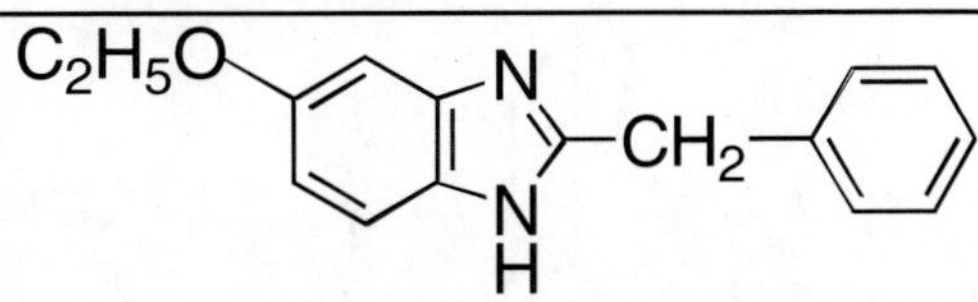

They are seen to differ from each other by substituents at positions 2 and 5. It is known that these substituents differ in hydrophobic properties, specifically, in Hansch hydrophobicity constant π for fragments of aromatic series [18]. The 5-hydroxybenzimidazole derivatives are arranged in order of increasing hydrophobic properties as I, II, III, and IV.

The effect of the 5-hydroxybenzimidazole derivatives on the erythrocyte membrane structure was studied by the spin probe method.

It was previously shown that changes in the kinetic characteristics of a hydrophobic spin probe during its intercalation into the biomembrane can be used to analyze the effect of various factors, including exogenous chemical substances, on the structural state of the membrane.

In this context, we studied the effect of the 5-hydroxybenzimidazole derivatives on the kinetics of insertion of spin probe 1 into the erythrocyte membrane.

The EPR spectra of this spin probe added to the suspension of erythrocyte membranes (ghosts) (Figure 1) are superpositions of the EPR signals of aggregates of the probe in the aqueous medium of the suspension and the probe molecules intercalated into the membrane structure. As a structure-sensitive parameter, we used the quantity Δ, which was determined as shown in Figure 1. Changes in Δ characterize the redistribution of the spin probe between the aqueous medium and the erythrocyte membrane in the erythrocyte ghost suspension. The larger Δ, the higher the spin probe content of the membrane.

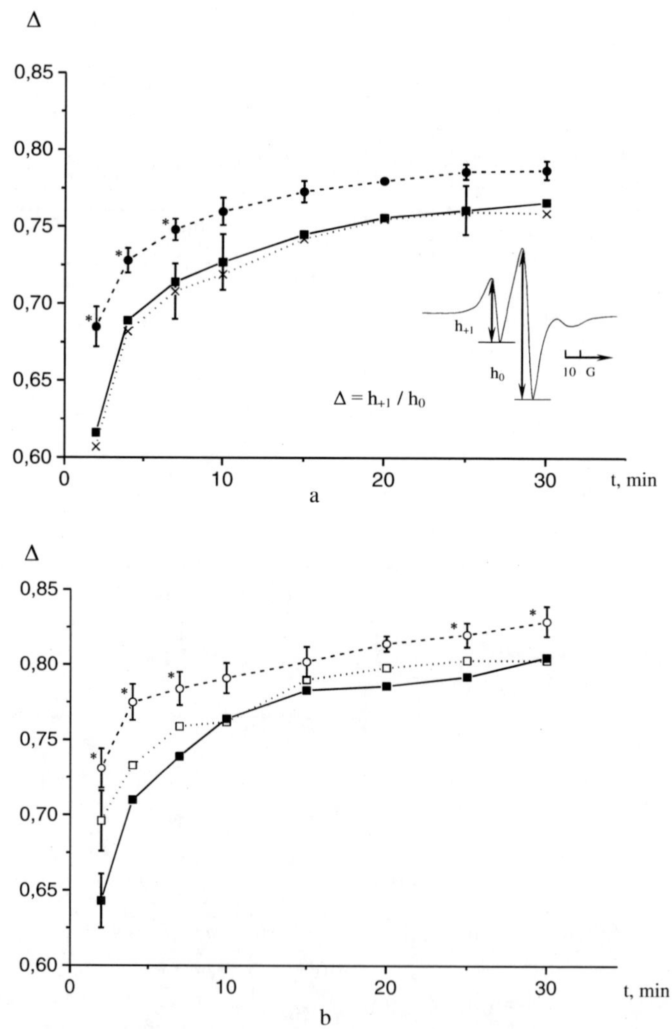

Figure 1. Effect of 5-hydroxybenzimidazole derivatives ($1{\cdot}10^{-3}$M) on the kinetics of insertion of spin probe 1 into the erythrocyte membrane (* $p < 0.05$) and the method for determining the parameter Δ : (a) (······×·····) derivative I, (——■——) reference sample, and (- - - - ●- - - -) derivative III; and (b) (——■——) reference sample, (······□·····) derivative II, and (- - - - ○- - -) derivative IV. The erythrocyte ghost suspension in phosphate buffer contained 7.5 mg protein per milliliter, and the spin probe concentration was $5{*}10^{-4}$ M.

Figure 1 presents typical kinetic curves of the parameter Δ. It is seen that 5-hydroxybenzimidazole derivative I hardly affects the kinetics of insertion of spin probe 1 into the erythrocyte membrane (Figure la). Derivative II, which contains an ethoxy group at position 5, somewhat accelerates the insertion (Figure lb). A pronounced effect on the process kinetics is exerted by derivatives III and IV, which contain benzyl substituents at positions 2; the most efficient substance is hydrophobic derivative IV, which also contains an ethoxy group at position 5 (Figure lb).

Derivatives III and IV increase not only the rate of insertion of the spin probe into the membrane but also the membrane capacity. The increase in the membrane capacity is indicated by the fact that the parameter Δ increases and the experimental curve lies above the reference curve. Note that statistically reliable effects were observed at a derivative concentration of 10^{-3} M.

The results of the experiments showed that, as the most hydrophobic 5-hydroxybenzimidazole derivatives are intercalated into the erythrocyte membrane and are distributed in it, the membrane structure is modified. As a result, the spin probe enters the intramembrane space more rapidly and in larger amounts.

The data obtained suggested that the structural modifications of the membrane that are caused by the distribution of 5-hydroxybenzimidazole derivatives in it can be accompanied by changes in the erythrocyte morphology. Using scanning electron microscopy, we experimentally confirmed these assumptions.

Figure 2a presents a typical electron micrograph of a reference sample of rat blood erythrocytes. It is seen that most of erythrocytes are discocytes. Along with these, the other known erythrocyte forms occurring in normalcy in small amounts were also detected (echinocytes, stomatocytes, etc.) [19].

The effect of 5-hydroxybenzimidazole derivatives on the erythrocyte morphology was studied over a wide derivative concentration range of 10^{-7} to 10^{-3} M.

It was found that all the derivatives with different efficiencies cause concentration-dependent transformation of some of the erythrocytes from discocytes into echinocytes (cells with outward projections—crenations) (Figure 2).

Derivatives III and IV, which contain hydrophobic substituents at positions 2, have a stronger crenating effect; derivative IV, which contains benzyl and ethoxy substituents, is the most efficient. Figures 2b-2d show typical electron micrographs of erythrocytes in the presence of derivative III at different concentrations.

It is seen that an increase in the derivative concentration in the erythrocyte suspension leads to a decrease in the concentration of discocytes and an increase in the concentration of echinocytes. Morphological changes were observed as early as the first 10 min of incubation with the derivatives.

According to Sheetz and Singer [8, 9], transformation of erythrocytes from discocytes into echinocytes under the action of xenobiotics are caused by the distribution of the agents in the outer monolayer of the erythrocyte membrane. This increases the monolayer area and gives rise to characteristic spine like projections on the erythrocyte surface.

In terms of these concepts, the character of changes in the erythrocyte morphology under the action of 5-hydroxybenzimidazole derivatives observed here suggests that these nonelectrolytes differing in hydrophobic properties are likely to be distributed in the outer monolayer of the erythrocyte membrane.

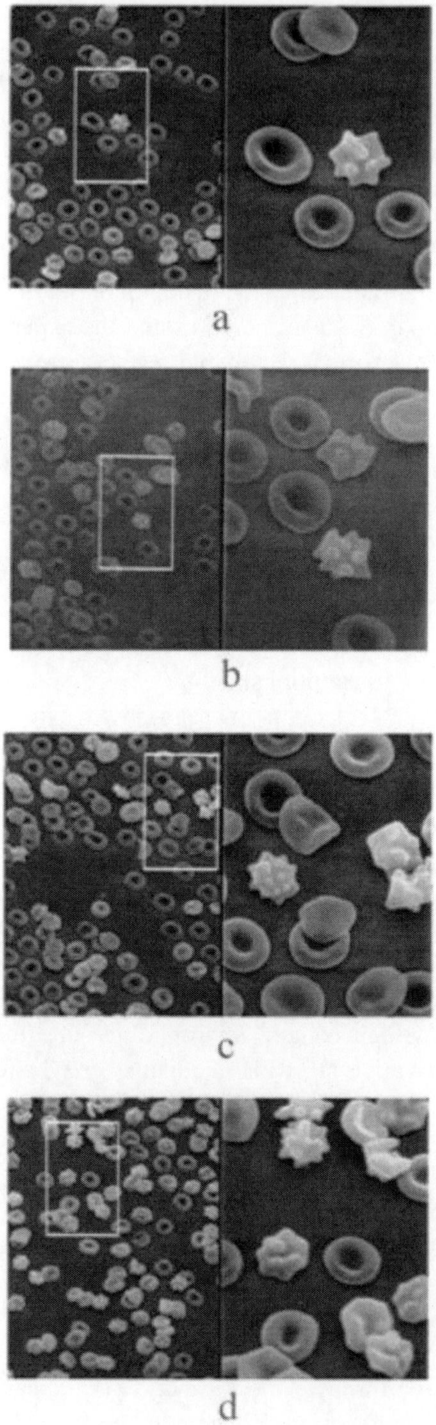

Figure 2. Electron micrographs of rat blood erythrocytes (a) in the absence of derivative III and in the presence of (b) $1 \cdot 10^{-6}$ M, (c) $1 \cdot 10^{-4}$ M, and (d) $1 \cdot 10^{-3}$ M derivative III. The enlargements for the left and right parts of each micrograph are x1000 and x3000, respectively.

The data obtained imply that the hydrophobicity of 5-hydroxybenzimidazole derivatives largely determines the efficiency of their modifying effect on the erythrocyte membrane structure and the erythrocyte morphology.

Scanning electron microscopy revealed structural changes in the erythrocyte membrane occurring at lower concentrations of 5-hydroxybenzimidazole derivatives than those at which changes are detected by the spin probe method. This is likely to be related to the distribution of the spin probe in a membrane structure region remote from the intramembrane regions modified during transport of 5-hydroxybenzimidazole derivatives.

CONCLUSION

Thus, it is established that the intercalation of organic nonelectrolytes such as 5-hydroxybenzimidazole derivatives into the erythrocyte membrane and their transport and distribution are accompanied by concentration-dependent modifications of the membrane structure and changes in the surface architectonics of erythrocytes. These effects are more pronounced for the more hydrophobic 5-hydroxybenzimidazole derivatives. The character of the revealed transformations of erythrocytes suggests that the regions of distribution and possible biological action of 5-hydroxybenzimidazole derivatives are probably in the outer monolayer of the erythrocyte membrane.

REFERENCES

[1] W.D. Stein, in *Membrane Transport,* Ed. by S. L. Bouting and J. A. M. de Pont (*Elsevier North Holland Biomedical,* New York, 1981), pp. 1-28.

[2] R. J. Hitzemann. Effects of non-electrolyte molecules with anesthetic activity on the physical properties of DMPC multilamellar liposomes. *Biochim. Biophys. Acta* 983, 205 (1989).

[3] J. Sikkema, J. A. M. de Pont, and B. Poolman. Interactions of cyclic hydrocarbons with biological membranes. *J. Biol. Chem*. 269 (11), 8022 (1994).

[4] L. Ya. Gendel', L. V. Kim, O. G. Luneva, V. A. Fedin, and K. E. Kruglyakova. Changes in the surface architectonics of erythrocytes under the influence of the synthetic antioxidant fenozan-1. Izv. *Akad. Nauk, Ser. Biol*., No.4, 508 (1996).

[5] S. C. McKarns, C. Hansen, W. S. Caldwell, *et al.* Correlation between hydrophobicity of short-chain aliphatic alcohols and their ability to alter plasma membrane integrity. Fundam. *Appl. Toxicol*. 36 (1), 62 (1997).

[6] V. D. Zinchenko, L. G. Kuleshova, Yu. V. Kucherenko, and M. I. Shchetinskii, *Proceedings of II Congress of Biophysicists of Russia, Moscow, Russia, August 23-27, 1999,* pp. 506-507.

[7] L. Ya. Gendel' and K. E. Kruglyakova, *Spin Labeling and Probing Method* (Nauka, Moscow, 1986), pp. 163-194 [In Russian].

[8] M. P. Sheetz and S. J. Singer. Biological membranes as bilayer couples. A molecular mechanism of drug-erythrocyte interactions. *Proc. Natl. Acad. Sci. USA* 71 4457 (1974).

[9] M. P. Sheetz and S. J. Singer. Biological membranes as bilayer couples. III. Compensatory shape changes induced in membranes. *J. Cell. Biol.* 70 247 (1976).

[10] B. Isomaa, Hagerstrand, and G. Paatero. Shape transformations induced by amphiphiles in erythrocytes. *Biochim. Biophys. Acta* 899, 93 (1987).

[11] O. G. Luneva, L. Ya. Gendel', and K. E. Kruglyakova. Features of organic nonelectrolyte binding to the erythrocyte membrane. *Biofizika* 47, 38 (2002).

[12] Parshina E.Yu., Gendel L.Ya., and Rubin A.B. Effect of ichphans on the structural properties of erythrocyte membranes. Izv. *RAN Ser.Biol.*, no.6, 645 (2007).

[13] F. Rusina, Yu. V. Kuznetsov, L. V. Gol'tsova, and L. D. Smirnov, *Proceedings of VI International Conference on Bioantioxidants, Moscow, Russia, April 16-19,* 2002, pp. 495-496.

[14] J. W. Wyse and D. A. Butterfield. Electron spin resonance investigation of the interaction of the anion and glucose transport inhibitor, p-azidobenzylphlorizin, with the human red cell membrane. *Biochim. Biophys. Acta* 979, 121 (1989).

[15] Yu. V. Kuznetsov, L. G. Stolyarova, V. P. Lezina, and L. D. Smirnov. Friedel-Crafts acylation of 5-hydroxybenzimidazole derivatives. *Izv. Akad. Nauk SSSR, Ser. Khim.*, No.8, 1888 (1990).

[16] E. G. Rozantsev and V. I. Suskina, *Izv. Akad. Nauk SSSR, Ser. Khim.*, No. 5, 1191 (1969).

[17] G. I. Kozinets and Yu. A. Simovart, *Surface Architectonics of Peripheral Blood Cells in Normalcy and in Pathologies of Blood System* (Valgus, Tallinn, 1984) [In Russian].

[18] A. Klesov and I. V. Berezin, *Enzyme Catalysis (*Moscow State Univ., Moscow, 1980) [In Russian].

[19] M. Bessis, *Corpuscles: Atlas of Red Blood Cell Shape* (Berlin Ect., 1974).

In: Biology and Communication in the Information Age
Editor: Rafiqul Islam
ISBN: 978-1-62808-252-4

Chapter 15

STUDY OF MEMBRANE PROTECTIVE PROPERTIES OF ISOBORNYLPHENOLS ON THE ERYTHROCYTE INDUCED HEMOLYSIS MODEL

L. N. Shishkina[1*], O. G. Shevchenko[2], I. J. Chukicheva[3] and A. V. Kutchin[3]

[1]Emanuel Institute of Biochemical Physics of the Russian Academy of Sciences, Moscow, Russia

[2]Institute of Biology of Komi Science Centre, Ural Branch of the Russian Academy of Sciences, Syktyvkar, Komi Republic, Russia

[3]Institute of Chemistry of Komi Science Centre. Ural Branch of the Russian Academy of Sciences, Syktyvkar, Komi Republic, Russia

ABSTRACT

Membrane protective properties of a number of isobornylphenols have been studied on model of the erythrocyte hemolysis induced by hydrogen peroxide. It has been shown that their ability to protect the erythrocyte membrane against oxidative hemolysis substantially depends both on the structure and the concentration of the compound. The highest membrane protective activity has been found for substance having isobornyl substitutes at both *ortho* positions. It has been found that all the isobornylphenols studied inhibit the accumulation of oxidation products, which indicates their ability to inhibit lipid peroxidation in cells. The cytotoxicity of a number of isobornylphenols at high concentrations and the absence of the linear dependence of "biological effect - dose" make it possible to suggest that the biological activity of isobornylphenols may be associated not only with their antiradical or antioxidant activity.

Keywords: Erythrocyte membrane, induced hemolysis, isobornylphenols, lipid peroxidation, antioxidant activity

* E-mail: shishkina@sky.chph.ras.ru.

AIM AND BACKGROUND

Obviously, the search for new biologically active substances synthesized on the basis of natural sources is important because they are generally less toxic for the organism. At the same time, the primary selection of the most promising biologically active substances requires the preliminary analysis of the mechanisms of their action in model biological systems, since the biological activity of most compounds is associated with their ability to have an effect both on the regulation of lipid peroxidation (LPO) processes and on the structural state of cell membranes [10]. It is known that the value of antioxidant and/or antiradical activity of biologically active substances substantially depends on the model system and is not always consistent with their biological activity when these compounds have administered to the mammalian organism [11, 37]. The model of the induced hemolysis of erythrocytes [1, 3, 4, 17, 24, 28. 38, 41, 42] is a convenient model for studying the membrane protective effects of the various substances of the herbal origin and their influence on the properties of biological membranes in vitro.

The aim of this work is to study the membrane protective and antioxidant properties of isobornylphenols of different structures with regard to their concentration using the model of the hydrogen peroxide-induced hemolysis of erythrocytes.

INTRODUCTION

Terpenephenols synthesized on the basis of green coniferous and containing an isobornyl group as alkyl substituents [13] represent promising antioxidants (AO) since one of them, as shown in [12, 30, 32, 33], has a wide spectrum of the biological activity. Antioxidant properties of some isobornylphenols were studied in chemical model systems [25, 30]. However, it should be noted that the dependence of the biological effectiveness for the majority of AO on their concentration is not, as a rule, linear [11]. This is the reason why a more detailed study of the biological properties of isobornylphenols in the biological model systems is needed.

EXPERIMENTAL PART

We studied isobornylphenols differing in their chemical structure and position of substituents in the phenolic ring: 2-(1,7,7-trimethylbicyclo [2,2,1] hept-exo-2-yl) oxyphenol (TP-1), 4-methyl -2 - (1,7,7- trimethylbicyclo [2,2,1] hept-exo-2-yl) phenol (TP-5); 2-(1,7,7-trimethylbicyclo [2,2,1] hept-exo-2-yl) phenol (TP-6); 4-methyl-2,6-di-(1,7,7-trimethylbicyclo [2,2,1] hept-exo-2-yl) phenol (TP-7, dibornol); 6-methyl-2-(1,7,7-trimethylbicyclo [2,2,1] hept-exo-2-yl) phenol (TP-8). The above compounds are obtained by means of a well-known method of the alkylation of phenols with a natural unsaturated monoterpene - camphene - using the corresponding aluminum phenolate as a catalyst [13 - 15]. The structural formulas of the compounds are shown in Figure 1. A standard low-toxic AO 4-methyl-2,6-di-tert-butylphenol (BHT) which is a sterically hindered phenol with a similar structure and has an expressed antiradical activity [11] was used as a standard of comparison.

The blood of white outbreed male mice aged 2 to 3 months and weighing 22 to 29 g was used to isolate red blood cells. The blood from the decapitated animals was collected into tubes treated with a 5% sodium citrate solution. Red blood cells were separated from plasma by centrifugation for 5 min at 3000 rev/min and washed twice with phosphate-buffered saline (PBS: 150 mM NaCl, 8.1 mM Na2HPO4, 1.9 mM NaH2PO4), pH 7.4. The resulting red cell mass from 5-9 animals was combined in each series of experiments. A 1% (v/v) erythrocyte suspension in PBS (the concentration was determined by hematocrit) was used in experiments. Since both isobornylphenol and BHT are hydrophobic compounds, all the AO were injected into the erythrocyte suspension in a 96% ethanol solution whose concentration in the incubation medium was always 0.1% (v/v). Solutions of the substances tested at concentrations of 0.001 to 0.1 M were placed in the erythrocyte suspension at a ratio of 1:1000 and incubated at 37°C for 10 min. Hemolysis was next initiated by hydrogen peroxide solution, chemically pure grade, whose concentration in the incubation medium was 0.006% (1.8 mM). The reaction mixture was incubated at 37°C in the thermostated shaker «Biosan» (Latvia) during 6 h. The aliquot was collected from the sample each hour and then centrifuged for 5 min at 3000 rev/min. The yield of hemoglobin was determined spectrophotometrically by measuring the optical density of the supernatant at λ 541 nm [42]. The percentage of hemolysis was calculated relatively to the complete hemolysis of the sample [38], which was achieved by a double freeze-thaw cycle.

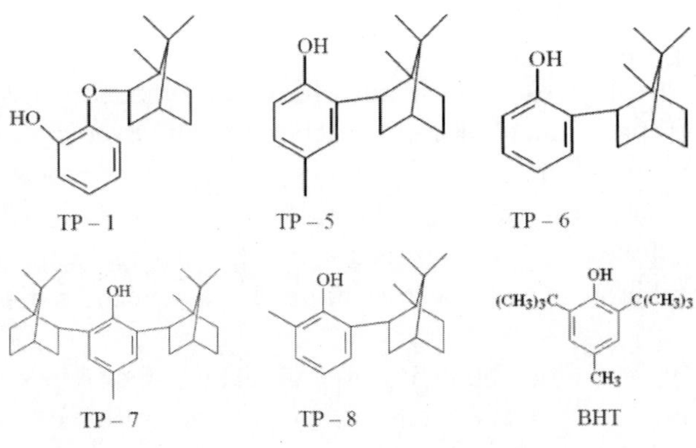

Figure 1. Structural formulas of the isobornylphenols studied.

The content of LPO secondary products (nmol/ml) reacting with 2-thiobarbituric acid (TBA-reactive substances, TBA-RS) in the incubation mixture was estimated spectrophotometrically [2]. To assess the accumulation of hemoglobin oxidation products, the aliquot of the erythrocyte suspension was subjected to the complete hemolysis following incubation and centrifuged to precipitate the erythrocyte ghosts; the absorption spectrum was analyzed using spectrofluorimeter "Fluorat-02-Panorama" (St. Petersburg, Russia). The content of various forms of hemoglobin (oxyHb, metHb and ferrylHb) was calculated taking into account the corresponding extinction coefficients [5]:

$$[oxyHb] = -472\ A560 + 536\ A577 - 104\ A630$$
$$[metHb] = -784\ A560 + 430\ A577 + 1500\ A630$$
$$[ferrylHb] = 976\ A560 - 536\ A577 - 476\ A630$$

Each experiment was performed 2-4 times. The statistical data processing and plotting were performed using Microsoft Office Excel 2007 and Statistica 6.0 software packages. The experimental data in figures and tables are represented as arithmetical average values with their mean square errors (M±m).

RESULTS AND DISCUSSION

It is known that hydrogen peroxide easily penetrates into red blood cells [18] and reacts with hemoglobin (oxyHb), the latter being oxidized to methemoglobin (metHb) and then to ferrylHb (globin-Fe(IV)=O) [5, 20, 21].

The ratio between these oxidation products depends on the reaction conditions and the concentration of hydrogen peroxide [7, 18, 20, 21, 26]. FerrylHb and its radicals forming as a result of the interaction between metHb and hydrogen peroxide (H2O2) are highly active, easily interact with different proteins, in particular, inactivate enzymes containing thiol groups at the active site, thus leading to the formation of cross-links between the chains of hemoglobin and its denaturation [21, 26].

The impact of radicals on membrane lipids results in the formation of peroxyradicals which initiate the development of oxidative reactions in membranes. Thus, the damage of red blood cells under oxidative stress is mainly the result of two processes - the oxidation of hemoglobin and a free radical damage of polyunsaturated fatty acids of their membrane lipids, as well as the thiol groups of membrane proteins [23, 26].

To adequately assess the antioxidant properties of isobornylphenols, it was necessary to determine whether the presence of alcohol in the reaction mixture even at low concentrations has an effect on the parameters of the induced hemolysis of erythrocytes and whether the AO studied can directly interact with hydrogen peroxide, thus inhibiting the oxidation of hemoglobin in the cell, as well as to investigate the dependence of the effectiveness of erythrocyte hemolysis inhibition by AO on the initial parameters of red blood cells from different groups of mice.

First of all, it should be noted that the incubation of the all isobornylphenols studied with hydrogen peroxide did not lead to alterations in the concentration of the latter, which indicates the absence of their interaction with H2O2. The analysis of numerous literature data leads to the conclusion about the absence of a similar reaction for BHT.

Table 1. Effect of BHT at a concentration of 0.1 mM on the ratio between different forms of hemoglobin in erythrocyte hemolysates 6 h after their incubation with hydrogen peroxide

Experimental conditions	metHb/oxyHb	ferrylHb/oxyHb
H_2O_2	0.277 ± 0,009	0.204 ± 0,002
H_2O_2+ethanol (control)	0.350 ± 0,032	0.285 ±0,005
H_2O_2+ethanol+BHT	0.269 ± 0,007*	0.272 ± 0,001

* - differences are statistically significant as compared with the control at $p < 0.05$.

Table 2. Effect of isobornylphenols and BHT at a concentration of 0.1 mM on the parameters of the blood erythrocyte hemolysis induced by hydrogen peroxide

Experimental conditions	Index			
	metHb/oxyHb, 6 h incubation	ferrylHb/oxy Hb, 6 h incubation	[TBA-RS], nmol/ml 3 h incubation	% hemolysis, 6 h incubation
H_2O_2	0.356±0.011	0.224±0	2.43±0.42	48.4±4.3
H_2O_2+ethanol (control)	0.467±0.043	0.210±0.004	2.79±0.11	61.7±5.0
H_2O_2+ ethanol + BHT	0.322±0.008*	0.207±0.001	1.43±0.05***	11.0±0.7**
H_2O_2+ ethanol + TP-1	0.253±0.036*	0.208±0.006	1.54±0.03***	9.9±0.4**
H_2O_2+ ethanol + TP-5	0.444±0.066	0.193±0.008	1.65±0.07*	62.4±11.2
H_2O_2+ethamol + TP-7	0.413±0.072	0.205±0.005	1.38±0.11**	9.6±1.3**

Hereinafter: the differences are statistically significant compared with the control at * -p≤0.05; **-p≤0.01; ***- p≤0.001.

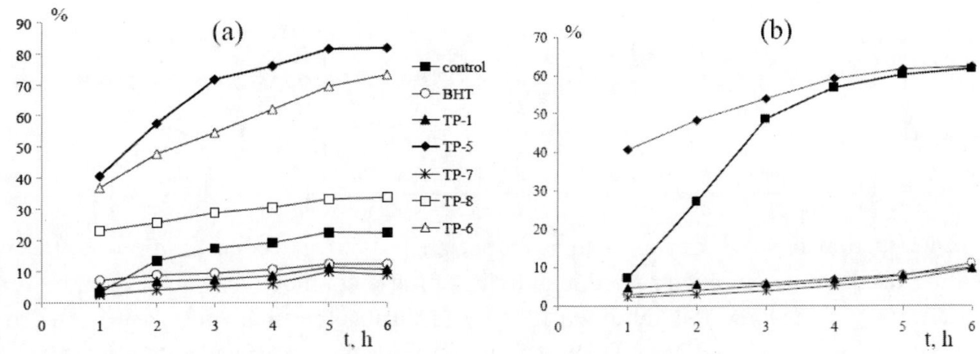

Figure 2. Effect of isobornylphenols and BHT at a concentration of 0.1 mM on the parameters of the blood erythrocyte hydrogen peroxide-induced hemolysis in different series of experiments.

The results represented in Tables 1 and 2 show that under conditions of the H2O2-induced hemolysis, ethanol even at low concentrations has an additional destabilizing effect on erythrocytes.

This is consistent with the published data and associated with the direct action of both ethanol and acetaldehyde forming as a result of its oxidation on erythrocyte membranes and other cellular components [40]. Thus, the samples of suspensions of the intact red blood cells incubated in the presence of 0.1% ethanol served as controls.

As known, the red blood cells of intact animals may have different initial characteristics: the activity of enzymes, the content of antioxidants, lipid composition and their physicochemical properties [6, 16, 31, 35, 39]. For example, it was shown earlier that the lipids of the red blood cells of laboratory animals can normally vary significantly in their supply of AO, the quantitative ratio between phospholipid fractions, the molar relationship [cholesterol]/ [phospholipids], the degree of unsaturation, and the ability to decompose peroxides [22, 34, 35]. Nevertheless, the data represented in Figures 2 a, b and Tables 1, 2 and 3 show that BHT significantly inhibited hemolysis of erythrocytes at the all studied concentrations. Besides, during incubation with hydrogen peroxide BHT prevented not only the destruction of cells, but significantly inhibited the oxidation of hemoglobin to metHb (Tables 1 and 2), which is most likely to be associated with the high antiradical activity of this compound. The findings of studies [20, 27] also support the ability of BHT to inhibit not only LPO but also the induced hemolysis of erythrocytes, although the plausible mechanisms of their membrane protective action are being debated [5, 20].

Table 3. The influence of isobornylphenols and BHT at different concentrations on the parameters of erythrocyte induced hemolysis

Variant	Concentration, μM	Parameters of induced hemolysis				
		initial %	max %	induction period, h	V max, % / h	Time of V max, h
Control	-	2.81	91.7	No	38.0	1
BHT	10	2.98	19.4	4	10.65	4.5
	50	4.32	8.5	5	0.9	-
TP-1	10	2.14	21.9	3	10.4	4.5
	50	3.65	7.1	5	1.8	-
TP-5	10	4.99	8.1	5	0.85	-
	50	9.39	13.4	5	1.55	-
TP-6	10	4.40	51.7	2	22.8	3.5
	50	9.94	15.3	5	2.2	-
TP-7	10	2.52	8.4	5	1.8	-
	50	2.56	9.4	5	1.7	-

The incubation of erythrocytes with isobornylphenols at the concentration of 0.1 mM showed the different membrane protection activity of these compounds (Figures 2 a, b, Table 2). Preparations TP-1 and TP-7 inhibited hemolysis during the incubation period, whereas TP-5, TP-6 and to a lesser degree TP-8 at this concentration exhibited a cytotoxic effect, substantially increasing the rate of hemolysis compared with controls. However, despite of differences in the parameters of the erythrocyte hemolysis in the control samples, the correlation between the biological activities of the isobornylphenols studied remained in both series of experiments (Figures 2 a, b).

The data in Table 2 show that all isobornylphenols inhibit, like BHT, the accumulation of the LPO secondary products in erythrocytes. Besides, TP-1 significantly decreases the oxidation of oxyhemoglobin to metHb. The presence in the TP-1 molecule of the oxygen atom at the ortho-position relative to the OH group is most likely to provide the increase of its antiradical activity with respect to hydroxyl radicals in the cell and/or its ability to penetrate into the erythrocyte. The effect of the hydrophobicity of the compound on its hemolytic activity is shown for a number of hybrid AO [29]. Interestingly, none of the compounds studied affected the value of the [ferrylHb]/[oxyHb] ratio (Table 2), although the reasons for this phenomenon require further investigation.

The study of membrane protective properties of isobornylphenols within the concentration range from 0.001 to 0.1 mM has shown that the effect substantially depends both on the concentration and the chemical structure of the compounds. For example, dibornol whose molecules have isobornyl substitutes at both ortho- positions has the same membrane protective activity within the whole range of the concentrations studied (Figure 3a). Such a high biological activity of dibornol at low concentrations can be associated with the highest value of its interaction rate constant with peroxyradicals (k7) among all the compounds studied [25]. In general, the membrane stabilizing action of TP-1 augmented with the increase of the substance concentration in the incubation medium (Figure 3b). As for TP-5 containing one isobornyl group at the ortho- position and one methyl group at the para position, its cytotoxic effect was clearly expressed only at a concentration of 0.1 mM. This substance exhibited its membrane protective properties at concentrations of 0.001-0.01 mM (Figure 3c).

The absence of the linear dependence of "biological effect – dose" has also been found for dibornol when analyzing its ability to inhibit the accumulation of TBA-reactive substances in erythrocyte hemolysates (Table 4). A statistically significant decrease in the content of oxidation products in the erythrocyte hemolysate was revealed during the incubation of erythrocytes with the TP-1 and TP-5 at certain concentrations (Table 4).

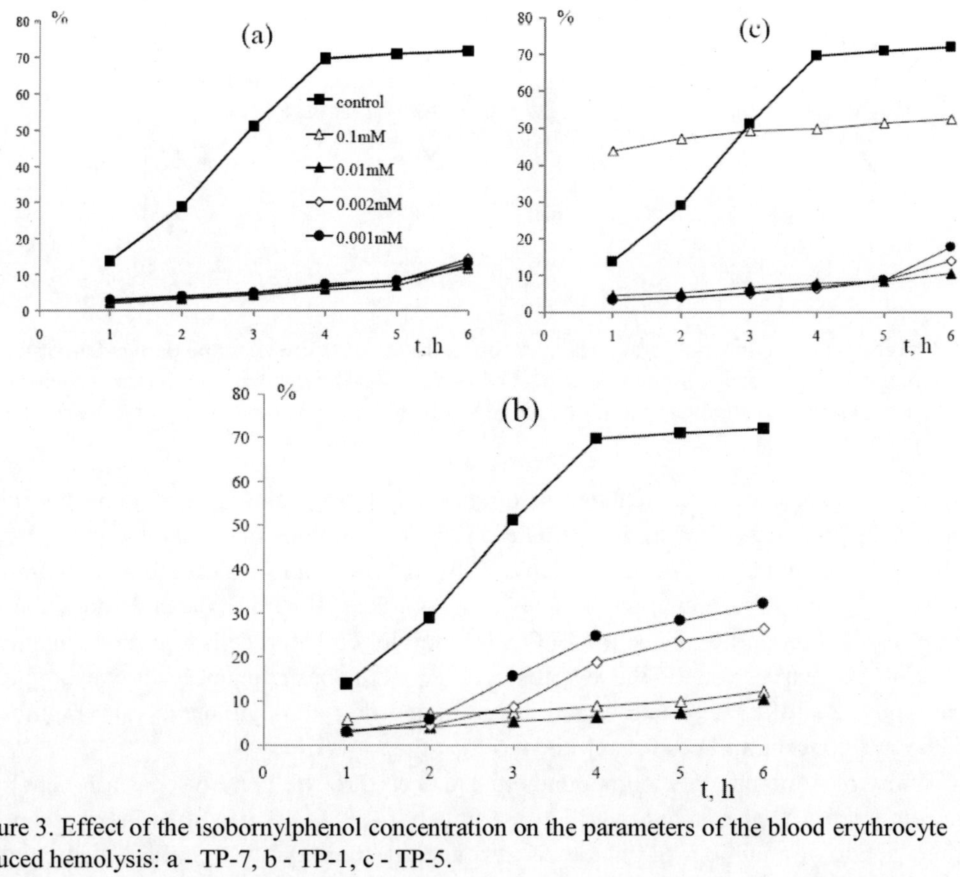

Figure 3. Effect of the isobornylphenol concentration on the parameters of the blood erythrocyte induced hemolysis: a - TP-7, b - TP-1, c - TP-5.

Table 4. The content of TBA-reactive substances (nmol/ml) in the erythrocyte hemolysates depending on the isobornylphenol concentration during 5 h incubation period

Variant	Substance concentration, мM			
	0.001	0.002	0.01	0.1
Control	3.08±0.29			
TP-1	2.06±0.11*	1.80±0.23*	2.39±0.09	1.62±0.21**
TP-5	2.70±0.65	1.61±0.13*	2.07±0.57	1.41±0.05**
TP-7	2.42±0.17	2.09±0.65	2.84±0.41	2.48±0.13

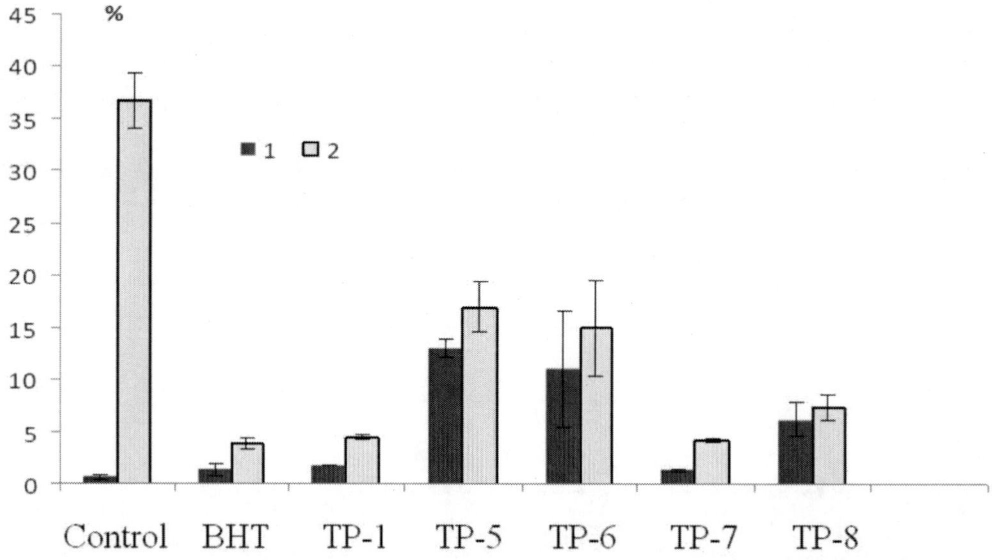

Figure 4. Effect of isobornylphenols and BHT at a concentration of 0.1 mM on the parameters of the blood erythrocyte hemolysis: 1 – initial percentage of hemolysis (the incubation of erythrocytes with compounds for 10 min.), 2 – the maximum percentage of hemolysis (the incubation for 3 h after the addition of hydrogen peroxide).

However, the presence in the reaction medium of TP-1, TP-5 and TP-7 at a concentration of 0.01 mM did not cause significant alterations in the content of oxidation products as compared with the controls (Table 4). Obviously, the mechanism for the interaction of isobornylphenols with erythrocyte membrane has a complex nature and depends not so much on their antioxidant properties, since the LPO inhibition in red blood cells does not sometimes correlate with the expression of the cytotoxic effect of isobornylphenols. For example, a significant inhibition of LPO (Table 4) and the pronounced erythrocyte hemolysis (Figures 2, 3c) by TP-5 at a concentration of 0.1 mM occurred at the same time.

The ability of some phenolic compounds to cause erythrocyte hemolysis as a function of their concentration was also noted by other researchers [4, 8, 9, 19, 29, 36]. The mechanism of the toxic action of a number of phenols is attributed to their ability to oxidize hemoglobin to metHb with a simultaneous diminution in the concentration of intracellular thiols, to intercalate into the membrane, interacting with phosphatidylcholine in the outer layer or changing the internal structure of the erythrocyte membrane, as well as to inactivate enzymes of the AO defense [6, 8, 9, 29].

The above findings have made it necessary to assess the membrane active properties of isobornylphenols investigated the initial parameters of erythrocyte hemolysis before the injection into the incubation medium of hydrogen peroxide (Figure 4). It has been found that the damage of cells and the release of hemoglobin into the extracellular medium can be observed 10 min after the incubation of erythrocytes with the preparations. According to the reduction of the cytotoxic effect at a concentration of 0.1 mM, the phenols studied range the following sequence: TP-5>TP-6>TP-8>TP-1>TP-7=BHT.

The findings of the study into membrane active properties of isobornylphenols and the parameters of the induced erythrocyte hemolysis using preparations at low concentrations are shown in Table 3. The incubation for 10 min reveals no toxicity of either dibornol at the studied concentrations or TP-1 and BHT at the concentration of 10 μM, whereas TP-5 and TP-6, especially at concentrations of 50 μM, had a remarkable cytotoxic effect. Nevertheless, during the whole incubation period of erythrocytes with hydrogen peroxide, TP-5 and TP-6 at the studied concentrations effectively protected cells against the hemolytic damage. The decrease in the concentration of these preparations to 10 μM significantly reduced the cytotoxic effect; however, only TP-5 at this concentration was able to effectively inhibit hemolysis during the whole period of incubation, whereas the reduction of the TP-6 protective effect was observed 2 hour after incubation. The decrease in the maximum rate and in the maximum hemolysis percentage with the increase in BHT and TP-1 concentrations indicate the existence of dependence of the membrane protection effect of these compounds on the concentration. A high membrane protective activity was found for TP-7 even at the lowest concentrations.

CONCLUSION

Thus, the model of hydrogen peroxide-induced hemolysis of red blood cells has demonstrated the ability of the isobornylphenols studied to inhibit lipid peroxidation and to hinder the oxidation of oxyHb to metHb. The experimental data obtained indicate that the membrane protective activity of some isobornylphenols is associated not only with their antioxidant properties. The highest membrane protective activity as compared with BHT within a wide range of concentrations was found for dibornol whose molecule contains an isobornyl group at both ortho positions. Compounds TP-5 and TP-6 containing one isobornyl group at the ortho position have cytotoxic effects on red blood cells at high concentrations and inhibit their hemolysis at concentrations less than 50 μM. The absence of a linear relationship "biological effect – dose", the dependence of the membrane protective activity on the chemical structure and kinetic characteristics of isobornylphenols bring out the need for both the detailed study into the mechanism of their effects on biological objects and the selection of the more worthwhile compounds for further research work. For the first select of these compounds obtained on the natural raw material base the model of the hydrogen peroxide induced hemolysis of the blood erythrocytes could be used.

References

[1] Ajila C.M., Rao P.U.J.S. 2008. Protection against hydrogen peroxide induced oxidative damage in rat erythrocytes by *Mangifera indica* L. peel extract. *Food Chem. Toxicol.* 46, 303–309.

[2] Asakawa T., Matsushita S. 1980. Coloring Conditions of Thiobarbituric Acid Test for Detecting Lipid Hydroperoxides. *Lipids*. 15(3), 137–140.

[3] Asghar Z., Masood Z. 2008. Evaluation of antioxidant properties of silymarin and its potential to inhibit peroxyl radicals in vitro. *Pak J. Pharm. Sci.* 21 (3), 249-254.

[4] Banerjee A., Kunwar A., Mishra B., Priyadarsini K.I. 2008. Concentration dependent antioxidant/pro-oxidant activity of curcumin studies from AAPH induced hemolysis of RBCs. *Chem. Biol. Interact.* 174(2), 134-139.

[5] Berg J.M., Kamp J.A.F., Lubin B.H., Roelofsen B., Kuypers F.A. 1992. . Kinetics and site specificity of hydroperoxide-induced oxidative damage in red blood cells. *Free Radic. Biol. Med.* 12(6), 487-498.

[6] Berg J.M., Kuypers F.A., Lubin B.H., Roelofsen B., Kamp J.A.F. 1991. Direct and continuous measurement of hydroperoxide-induced oxidative stress on the membrane of intact erythrocytes. *Free Radic. Biol. Med.* 11(3), 255-261.

[7] Blasa M., Candiracci M., Accorsi A., Piacentini M.P., Piatti E. 2007. Honey flavonoids as protection agents against oxidative damage to human red blood cells. *Food Chemistry*. 104, 1635–1640.

[8] Bukowska B. , Michałowicz J., Krokosz A. , Sicińska P. 2007. Comparison of the effect of phenol and its derivatives on protein and free radical formation in human erythrocytes (in vitro). *Blood Cells, Molecules, and Diseases*. 39, 238–244.

[9] Bukowska B., Kowalska S. 2004. Phenol and catechol induce prehemolytic and hemolytic changes in human erythrocytes. *Toxicology Letters*. 152, 73-84.

[10] Burlakova E.B. 2005 Bioantioxidants: yesterday, today, tomorrow: In: *Chemical and Biological Kinetics. New Horizons. Vol. 2. Biological Kinetics.* Eds. Burlakova E.B., Varfolomeev S.D.. Leden. Boston: VSP, p. 1-33.

[11] Burlakova E.B., Alesenko A.V., Molochkina E.M., Pal'mina N.P., Khrapova N.G. 1975. *Bioantioxidants in the radiation damage and malignant growth*. Moscow: Nauka. 214 p. (in Russian)

[12] Chukicheva I.Yu., Fedorova I.V., Buravlev E.V., Lumpov A.E., Vikharev Yu.B., Anikina L.V., Grishko V.V., Kutchin A.V. 2010. Anti-inflammatory activity of isobornylphenol derivatives. *Chemistry of Natural Compounds*. No 3, 402-403 (in Russian)

[13] Chukicheva I.Yu., Kutchin A.V. 2004. Natural and synthetic terpenephenols. *Russian Chemical Journal*. XLVIII (3), 21-38 (in Russian)

[14] Chukicheva I.Yu., Kutchin A.V., Spirihin L.V., Borbulevich O.Ya, Churakov A.V., Belokon' A.I. 2003. Alkylation of phenol by camphene in the presence of aluminum phenolate. *Chemistry and Computer Modeling. Butlerov's Commun.*. No 1, 9-13 (in Russian)

[15] Chukicheva I.Yu., Timusheva I.V., Spirikhin L.V., Kutchin A.V. 2007. Alkylation of pyrocatechol and resorcinol by camphene. *Chemistry of Natural Compounds*. 43(3), 245-249.

[16] Clemens M.R., Waller H.D. 1987. Lipid peroxidation in erythrocytes. *Chem. Phys. Lipids*. 45, 251-268.

[17] Deng S.L., Chen W.F., Zhou B., Yang L., Liu Z.L. 2006. Protective effects of curcumin and its analogues against free radical-induced oxidative hemolysis of human red blood cells. *Food Chemistry*. 98 (1), 112-119.

[18] Halliwell B., Clement M.V., Longa L.H. 2000. Hydrogen peroxide in the human body. *FEBS Letters*. 486, 10-13.

[19] He J., Lin J., Li J., Zhang J.H., Sun X.M., Zeng C.M. 2009. Dual effects of *Ginkgo biloba* leaf extract on human red blood cells. *Basic Clin. Pharmacol. Toxicol*. 104(2), 138-144.

[20] Ko, F.N., Hsiao G., Kuo Y.H. 1997. Protection of oxidative hemolysis bydemethyl diisoeugenol in normal and b-thalassemic red blood cells. *Free Radic. Biol. Med*. 22, 215–222.

[21] Kowalczyk A., Puchała M., Wesołowska K., Serafin E. 2007. Inactivation of alcohol dehydrogenase (ADH) by ferryl derivatives of human hemoglobin. *Biochim. Biophys. Acta*. 1774., 86–92.

[22] Kozlov M.V., Kushnireva E.V. Urnysheva V.V., Taran Yu.P., Shishkina L.N. 2007. Effect of lipid characteristics on the regulation of biochemical processes in liver and blood of animals. *Biophysics*. 52, 693-698 (in Russian)

[23] Lopez-Revuelta A., Sanchez-Gallego J.I., Hermandez-Hernandez A., Sanchez-Yague J., Llanillo M. 2006. Membrane cholesterol contents influence the protective effects of quercetin and rutin in erythrocytes damaged by oxidative stress. *Chem. Biol. Interact*. 161, 79–91.

[24] Magalhres A.S., Silva B.M., Pereira J.A., Andrade P.B., Valentro P., Carvalho M. 2009. Protective effect of quince (*Cydonia oblonga* Miller) fruit against oxidative hemolysis of human erythrocytes. *Food Chem. Toxicol*. 47, 1372–1377.

[25] Mazaletskaya L.I., Sheludchenko N.I. Shishkina, L.N., Kutchin A.V., Fedorova I.V., Chukicheva I.Yu. 2011. Kinetic Parameters of the Reaction of Isobornylphenols with Peroxy Radicals. *Petroleum Chemistry*. Vol. 51.No 5. P. 348-353.

[26] Murakami K., Mawatari S. 2003. Oxidation of hemoglobin to methemoglobin in intact erythrocyte by a hydroperoxide induces formation of glutathionyl hemoglobin and binding of α-hemoglobin to membrane. *Arch. Biochem. Biophys*. 417(2), 244-250.

[27] Nohl H., Stolze K. 1998. The effects of xenobiotics on erythrocytes. *General Pharmacology*. 31(3), 343-347.

[28] Paiva-Martins F., Fernandes J., Rocha S., Nascimento H., Vitorino R., Amado F., Borges F., Belo L., Santos-Silva A. 2009. Effects of olive oil polyphenols on erythrocyte oxidative damage. *Mol. Nutr. Food Res*. 53(5), 609-616.

[29] Parshina E.Yu., Hendel, .LYa., Rubin A.B. 2009. Hemolytic activity of new hybrid antioxidants - IHFANs. *Biophysics*. 54 (6), 1051-1054 (in Russian)

[30] Patent No. 2 347 661 C2 (RF) (Russia)

[31] Pieri C., Moroni F., Marra M. 1996. Food restriction increases the protection of erythrocytes against the hemolysis induced by peroxyl radicals. *Mechanisms of Ageing and Development*. 87, 15-23.

[32] Plotnikov M.B., Chernysheva G.A., Smolyakova V.I., Ivanov I.S., Kutchin A.V., Chukicheva I.Yu., Krasnov E.A. 2009. Neuroprotective effects of dibornol and

mechanisms of action in the case of cerebral ischemia. *Herald of Russ. AMS*. No 11, 12-17 (in Russian)

[33] Plotnikov M.B., Smol'yakova V.I., Ivanov I.S., Kutchin A.V., Chukicheva I.Yu., Krasnov E.A. 2008. Antithrombogenic and antithrombocyte activity of the derivative of the ortho-isobornylphenol. *Bull. Exp. Biol. and Med.* 145 (3), 296-299 (in Russian)

[34] Shevchenko O.G., Shishkina L.N. 2009. Erythrocyte lipids as a model for assessing biological consequences of exposure to physical and chemical factors. *Technologies of living systems. 6* (8), 21-32 (in Russian)

[35] Shishkina L.N. Khrustova N.V. 2006. Kinetic characteristics of lipids from mammalian tissues in autoxidation reactions. *Biophysics*. 51, 340-346 (in Russian)

[36] Suwalskya M., Vargasb P., Avellob M., Villenac F., Sotomayor C.P. 2008. Human erythrocytes are affected in vitro by flavonoids of *Aristotelia chilensis* (Maqui) leaves. *Intern. J. Pharmaceutics*. 363, 85–90.

[37] Tabart J., Kevers C., Pincemail J., Defraigne J.-O., Dommes J. 2009. Comparative antioxidant capacities of phenolic compounds measured by various tests. *Food Chemistry*. 113, 1226–1233.

[38] Takebayashi J., Chen J., Tai A. 2010. A Method for Evaluation of Antioxidant Activity Based on Inhibition of Free Radical-Induced Erythrocyte Hemolysis. *Advanced Protocols in Oxidative Stress, Methods in Molecular Biology*. 594, 287-296.

[39] Toussaint O., Houbion A., Remacle J. 1993. Relationship between the critical level of oxidative stresses and the glutathione peroxidase activity. *Toxicology*. 81(2), 89-101.

[40] Tyulina O.V., Huentelman M.J., Prokopieva V.D., Boldyrev A.A., Johnson P. 2000. Does ethanol metabolism affect erythrocyte hemolysis? *Biochim. Biophys. Acta. Molecular Basis of Disease*. 1535(1), 69-77.

[41] Wang J., Sun B., Cao Y., Tian Y. 2009. Protection of wheat bran feruloyl oligosaccharides against free radical-induced oxidative damage in normal human erythrocytes. *Food Chem. Toxicol*. 47(7), 1591-1599.

[42] Yang H.-L., Chen S.-C., Chang N.-W., Chang J.-M., Lee M.-L., Tsai P.-C., Fu H.-H., Kao W.-W., Chiang H.-C., Wang H.-H., Hseu Y.-C. 2006. Protection from oxidative damage using *Bidens pilosa* extracts in normal human erythrocytes. *Food Chem. Toxicol*. 44, 1513–1521.

In: Biology and Communication in the Information Age
Editor: Rafiqul Islam
ISBN: 978-1-62808-252-4

Chapter 16

EFFECT OF ALKYL SUBSTITUTE POSITION AND KIND ON PHENOL TOXICITY

I. S. Belostotskaja, E. B. Burlakova, V. M. Misin*, G. A. Nikiforov, N. G. Khrapova and V. N. Shtol'ko

Emanuel Institute of Biochemical Physics of Russian Academy of Sciences, Moscow, Russia

ABSTRACT

Toxicity of 27 phenols with different 2-, 6-positions of alkyl substitutes and functional groups NH_2 , CN, OH in 4-posicion was studied. The least toxicity has phenols with two tert-butyl substitutes of benzene nuclei, which increases with mowing p-off benzene nuclei. Obtained results can help to synthesize new nontoxic bioactive phenols.

Keywords: Antioxidant, substituted phenols, toxicity

AIM AND SCOPE

The definition of substitute phenol toxicity which are analogues of biologically active 2,6-di-tert-butyl-4-methylphenol.

INTRODUCTION

Notwithstanding general point of view, peroxide oxidation of lipids is a physiological process of normal metabolizing tissues and follows the same laws as lipids oxidation does *in vitro* in liquid phase. Fine regulation of lipid oxidation in cell takes place both by chemical ways and with the help of enzymes. The rate of lipid oxidation is supported by a few systems, each of which is working within strick limits on relating stages of the process, having its own limiting reactions [1].

* E-mail: misin@sky.chph.ras.ru.

The lipid oxidation rate depends mainly on concentration changes of the substances, breaking the oxidation chains. Only natural antioxidants (basically phenols) have the ability to react with peroxide radicals. There is no interaction with pyroxile radicals of lipids in any other system of defense. None of the enzymes with antioxidant action break oxidation chain. Only natural antioxidants (AO) are inhibitors of radical processes. They can destroy redundant pyroxile radicals of lipids. Lipid oxidation rate is significally affected by AO more than by other systems. It defines the unique role of natural AO in regulation the intensity of lipid oxidation [2].

The low level quantity of lipids peroxide in normal tissues is explained by well-balanced processes of peroxide formation and expenses. Breaking this balance results in changing the antioxidant status of the organism and may be the cause of some pathologies [2, 3]. In some studies it was established that AO therapy was necessary for treating many diseases, and free radical processes do play a certain role.

For inhibition reactions of chain oxidation *in vivo* as substances, phenol derivatives are widely used .They possess physiological activity in mammals. There appeared Biologically Active Supplements and medicines of the antioxidant effect, which are of great help to treat many diseases, for example: dibunoli, mexidol, emoxipin, trimexidini, probucoli [4-10]. In Russia the substances group of the antioxidant effect was chosen from Biologically Active Supplements [11].

For many substances toxicity was carefully studied; clinical, preclinical tests and experiments were carried out. However, in literature there is little information on toxicity of numerous synthesized phenols. They are joined into different groups and have various phenols, having substituents in position 2,4,6 to the group OH. For example, in [12], pulmonary toxicity of 24 phenols was studied on mice, which have various lipophil alkyl substituents in positions 2,4,6 and injected I.P in mammals.

In this study systematic comptive investigations were carried out for estimation toxicity of 27 phenol compounds, having in p- position substituents with different heteroatoms. They have the same formula:

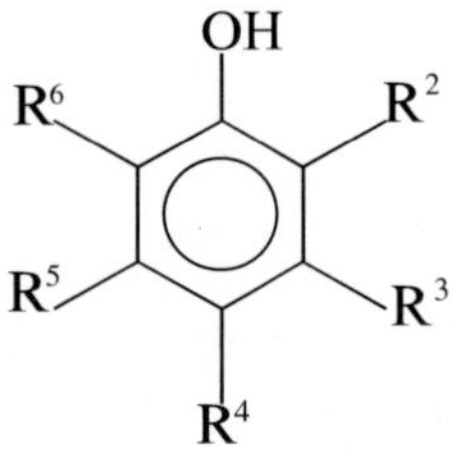

R^2, R^4, R^6 – various substituents; R^3=R^5=H

Experimental Part

Synthesis of studied phenols was described in [4, 13, 14]. Toxicity of all compounds was estimated by a value LD_{50}, measured in mg/kg weight of animal. For some compounds, values as much as possible transferable dose (MTD) and LD_{100} were additionally determined. Toxicity was determined on mice (males) of the line *Bulb* (mass18-22gr) with a single I.P. These mammals were housed in standard conditions of vivarium. Water-soluble preptions were injected as solutions of distilled water; liposoluble - 10% in solution Twin-80, used as solubilisator. Each doze off this preption was tested on not less than 4 mammals. For calculation, Beren's method was used (the method "frequency accumulation"), because it was simple and rather reliable [15].

RESULTS AND DISCUSSION

All investigated substances are resulted in table 1. For better comparison of the obtained results on toxicity, investigated compounds were grouped separately in tables 2-9. It was necessary to estimate toxicity of the substituted phenols depending on:

1. the type of p-substituents under identical 2,6-di-tert-butyl substituents ;
2. spacer length $-(CH_2)_n-$ between phenol ring and p-substituents under identical 2,6-di-tert-butyl substituents;
3. the type of both o-substituents under identical p-substituents.

Table 1. Toxicity of the phenols containing various substituents. $R^3 = R^5 = H$

№ patt.	substituents R^2	R^4	R^6	MTD, mg/kg	LD_{50}, mg/kg	LD_{100}, mg/kg
1	$C(CH_3)_3$	CH_3	$C(CH_3)_3$	250	375	650
2	$C(CH_3)_3$	NH_2	$C(CH_3)_3$	40	80	100
3	$C(CH_3)_3$	CH_2NH_2	$C(CH_3)_3$	30	70	130
4	$C(CH_3)_3$	$(CH_2)_2NH_2$	$C(CH_3)_3$	60	75	80
5	$C(CH_3)_3$	$(CH_2)_3NH_2$	$C(CH_3)_3$	50	60	-
6	$C(CH_3)_3$	$(CH_2)_4NH_2$	$C(CH_3)_3$	-	50	-
7	$C(CH_3)_3$	$CH(CH_3)NH_2$	$C(CH_3)_3$	80	105	130
8	$CH(CH_3)_2$	$CH(CH_3)NH_2$	$C(CH_3)_3$	30	80	100
9	$CH(CH_3)_2$	$CH(CH_3)NH_2$	$CH(CH_3)_2$	27	35	45
10	CH_3	$CH(CH_3)NH_2$	$C(CH_3)_3$	-	50	100
11	$C(CH_3)_3$	$CH(C_2H_5)NH_2$	$C(CH_3)_3$	40	70	110
12	H	$(CH_2)_3NH_2$	$C(CH_3)_3$	100	165	-
13	$C(CH_3)_3$	$CH_2NHCOCH_3$	$C(CH_3)_3$	350	425	-
14	$C(CH_3)_3$	$(CH_2)_2NHCOCH_3$	$C(CH_3)_3$	-	175	-
15	$C(CH_3)_3$	$(CH_2)_3NHCOCH_3$	$C(CH_3)_3$	75	125	-
16	$C(CH_3)_3$	CN	$C(CH_3)_3$	300	450	525
17	$C(CH_3)_3$	CH_2CN	$C(CH_3)_3$	50	95	180
18	$C(CH_3)_3$	$(CH_2)_2CN$	$C(CH_3)_3$	200	360	475
19	H	$(CH_2)_2CN$	$C(CH_3)_3$	250	285	350
20	H	$(CH_2)_2CN$	H	200	302	400
21	CH_3	CH_2CN	$C(CH_3)_3$	50	152	250
22	CH_3	CN	$C(CH_3)_3$	50	150	-
23	H	CN	$C(CH_3)_3$	250	185	-
24	H	CN	H	200	300	-
25	$C(CH_3)_3$	OH	$C(CH_3)_3$	-	390	-
26	$C(CH_3)_3$	$(CH_2)_2OH$	$C(CH_3)_3$	-	300	-
27	$C(CH_3)_3$	$(CH_2)_3OH$	$C(CH_3)_3$	-	225	-

First, phenol toxicity was compared for various p-substituents and identical 2,6-di-tert-butyl substituents, possessing maximum screening effect (table 2). Toxicity estimated by the value LD_{50} was found to decrease in a row (in brackets values LD_{50} were given) of 2,6-di-tert-butylphenols having next n- substituents.

$-NH_2$ (80) ≈ $-CH_2NH_2$ (70) > $-CH_3$ (375) > -OH (390) ≈ $-C(CH_3)_3$ (400) > -CN (450)

Table 2. Dependence of toxicity of the phenols having identical assistants R^2=R^6 = C $(CH_3)_3$, from type of p- substituents; R^3=R^5=H

substituents			MTD, mg/kg	LD_{50}, mg/kg	LD_{100}, mg/kg
R^2	R^4	R^6			
$C(CH_3)_3$	NH_2	$C(CH_3)_3$	40	80	100
$C(CH_3)_3$	CH_2NH_2	$C(CH_3)_3$	30	70	130
$C(CH_3)_3$	CH_3	$C(CH_3)_3$	250	375	650
$CH(CH_3)_2$	OH	$CH(CH_3)_2$	-	390	-
$C(CH_3)_3$	$C(CH_3)_3$	$C(CH_3)_3$	250	400	-
$C(CH_3)_3$	CN	$C(CH_3)_3$	300	450	525

Toxicity decreasing of some phenol compounds from this row was approved with increasing the value of another parameter toxicity-MTD (in brackets the value MTD is given).

$-NH_2$ (40) ≈ $-CH_2NH_2$ (30) >> $-CH_3$ (250) = $-C(CH_3)_3$ (250) > -CN (300-350)

The same decrease in toxicity was being observed for parameter LD_{100} in a row.

$-NH_2$ (110) > $-CH_2NH_2$ (130) > > -CN (525) > $-CH_3$ (650)

For phenols, having identical 2,6-di-tert-butyl substituents, dependence of their toxicity on the distance of the functional p-substituents from benzene ring was investigated (table 3). The tendency was found to increase phenol toxicity connected with the spacer length $-(CH_2)_n-$ for all investigated types of substituents ($-NH_2$, $-NHCOCH_3$, –OH , –CN) both electron donors and electron acceptors (table 3).

However, for 2,6-di-tert-butylphenol with p-substituent - CN, separated only by one group $-CH_2-$ (n=1) from benzene ring, a sharp toxicity increase, as compared with phenol toxicity, having n=0, was being observed.

Further, for substances with the increasing bridges $-(CH)_n-$ toxicity sharply diminishes. But it remained greater than the phenol toxicity did, having p-substituent - just near benzene ring (n=0).

It is interesting that the observed extreme dependence was repeating for all values, characterizing the toxicity of this compounds (MTD, LD_{50}, LD_{100}).

The observed occurrence of extreme phenol toxicity, having methylene (or methyl) group in p-position, could possibly be explained by greater lability of these phenols in comparison with other phenols.

Table 3. Dependence of toxicity of phenols on remoteness of p-substituents on a benzene ring
$R^4 = (CH_2)_nX$; $R^2=R^6= C(CH_3)_3$; $R^3=R^5=H$

n	X=CN			X=NH_2			X=$NHCOCH_3$			X=OH
	MTD, mg/kg	LD_{50}, mg/kg	LD_{100}, mg/kg	MTD, mg/kg	LD_{50}, mg/kg	LD_{100}, mg/kg	MTD, mg/kg	LD_{50}, mg/kg	LD_{100}, mg/kg	LD_{100}, mg/kg
0	300	450	525	40	80	100	-	-	-	390
1	50	95	180	30	70	130	350	425	500	-
2	200	360	475	60	75	80	-	175	-	300
3	-	-	-	50	60	-	75	125	200	225
4	-	-	-	-	50	-	-	-	-	-

Thus there can be reactions of appearing of phenol radicals with the following reaction of demerization or reaction of disproportionation with formation cyclohexadienone-2,5 [4, 16, 17].

For 2,6-di-tert-butylphenol with p-substituent $-NH_2$ the same tendency of toxicity increase, according to increase of spacer length $-(CH_2)_n-$ has been observed (table 3). Toxicity increase was observed for this phenol (n=1) with the help of values MTD and LD_{50} . It was interesting that sharper toxicity increase has been observed for similar phenol, but with CN substituent.

Identical dependence of sufficient toxicity increase with the spacer length $-(CH_2)_n-$ was observed for 2,6-di-tert-butylphenol having p-substituents $-(CH_2)_nNHCOCH_3$ (table 3).

Symmetric 2,6-di-tert-butylphenol (as an example), containing group $-CH_2NH_2$ in p-position, toxicity dependence on the type of α-substituents in p-methylene group was investigated (table 4).

It was found phenol toxicity to decrease after introduction CH_3 group in α-position of the substitute CH_2NH_2. The introduction of C_2H_5 group returns the substituted phenol toxicity to its initial meaning.

Table 4. Toxicity of the phenols containing various R^4 substituents $R^3 = R^5$ =H

№	substituents			MTD,	LD_{50},	LD_{100},
patt.	R^2	R^4	R^6	mg/kg	mg/kg	mg/kg
1	$C(CH_3)_3$	CH_2NH_2	$C(CH_3)_3$	30	70	130
2	$C(CH_3)_3$	$CH(CH_3)NH_2$	$C(CH_3)_3$	80	105	130
3	$C(CH_3)_3$	$CH(C_2H_5)NH_2$	$C(CH_3)_3$	40	70	110

It was noted that the introduction of acetyl group to N-position of symmetrical 4-aminomethyl-2,6-di-tert-butylphenol has decreased the substance toxicity according to the value LD_{50} in 6 times (table 5).

Table 5. Toxicity of the phenols containing various R^4 substituents. $R^3 = R^5$ =H

№	substituents			MTD,	LD_{50},	LD_{100},
patt.	R^2	R^4	R^6	mg/kg	mg/kg	mg/kg
1	$C(CH_3)_3$	CH_2NH_2	$C(CH_3)_3$	30	70	130
2	$C(CH_3)_3$	$CH_2NHCOCH_3$	$C(CH_3)_3$	350	425	500

It was interesting to study the dependence of phenol toxicity on the type of substituents in o-positions.

For phenols (table 6), having identical p-substituent $-CH(CH_3)NH_2$, it was investigated phenol toxicity influenced by the substituents of the types R_2 and R_6 (tert-butyl, iso-propyl, methyl).

Over all the values, describing toxicity (MTD, LD_{50}, LD_{100}), phenol toxicity increases with the quantity decrease of tert-butyl groups, which are in both o-positions of these phenol in a row

tert-butyl + tert-butyl < tert-butyl + iso-propyl < tert-butyl + methyl < iso-propyl + iso-propyl

Table 6. Toxicity of the phenols containing various R^2, R^6 substituents. $R^3 = R^5 = H$

№ patt.	substituents R^2	R^4	R^6	MTD, mg/kg	LD_{50}, mg/kg	LD_{100}, mg/kg
1	$C(CH_3)_3$	$CH(CH_3)NH_2$	$C(CH_3)_3$	80	105	130
2	$CH(CH_3)_2$	$CH(CH_3)NH_2$	$C(CH_3)_3$	30	80	100
3	CH_3	$CH(CH_3)NH_2$	$C(CH_3)_3$	-	50	100
4	$CH(CH_3)_2$	$CH(CH_3)NH_2$	$CH(CH_3)_2$	25	35	45

Such toxicity increase can be explained both by decrease of steric hindrance, created to electron donor group -OH by the volume tert-butyl substituents, and by hyper conjugation effect of o-substituents in OH-group [18].

For n-CN of the substituted phenols (table 7) after decreasing the number of tert-butyl substituents in o-position it was found the sharp toxicity increase with the following decrease of it. Nevertheless the toxicity of CN-substituted 2,6-di-tert-butylphenol is less than toxicity of others CN-substituted phenol derivatives.

Table 7. Toxicity of the phenols containing various R^2, R^6 substituents. $R^3 = R^5 = H$

№ patt.	substituents R^2	R^4	R^6	MTD, mg/kg	LD_{50}, mg/kg	LD_{100}, mg/kg
1	$C(CH_3)_3$	CN	$C(CH_3)_3$	300	450	525
2	CH_3	CN	$C(CH_3)_3$	50	150	-
3	H	CN	$C(CH_3)_3$	250	185	-
	H	CN	H	200	300	-

The analogous dependence has been observed under investigation according to values MTD, LD_{50} and LD_{100} of three phenols, having group-CH_2CH_2CN in p-position. They differ in quantity of tert-butyl substituents in o-position (table 8). Toxicity increasing as found when the quantity of tert-butyl substituents decrease. However, extreme dependence of mono-tert-butyl-substituted phenol toxicity has been observed.

Table 8. Toxicity of the phenols containing various R^2, R^6 substituents. $R^3 = R^5 = H$

№ patt.	substituents R^2	R^4	R^6	MTD, mg/kg	LD_{50}, mg/kg	LD_{100}, mg/kg
1	$C(CH_3)_3$	$(CH_2)_2CN$	$C(CH_3)_3$	200	360	475
2	H	$(CH_2)_2CN$	$C(CH_3)_3$	250	285	350
3	H	$(CH_2)_2CN$	H	200	300	400

So, the least toxicity of sterically hindred 2,6-di-tert-butilphenols has been observed for a few phenol groups with different p-substitutes. It can be explained by their worst bioaccessibility [4]. However, only for 2,6-di-tert-butyl-4-cyanophenol and 2-tert-butyl-6-methyl-4-cyanophenol (table 9) reverse dependence was being observed: the change of one tert-butyl group to methyl group resulted in essential toxicity decrease. This was proved by the growth of both measured values LD_{50} and LD_{100}.

Table 9. Toxicity of the phenols containing various R^2, R^6 substituents. $R^3 = R^5 = H$

№ patt.	substituents			MTD, mg/kg	LD_{50}, mg/kg	LD_{100}, mg/kg
	R^2	R^4	R^6			
1	$C(CH_3)_3$	CH_2CN	$C(CH_3)_3$	50	95	180
2	CH_3	CH_2CN	$C(CH_3)_3$	50	150	250

The observed property dependence on the structure for various o-substituted phenols with polar substitutes in p-position does not correlate with the same dependence observed for those with no polar alkyl substitutes [12]. Perhaps after introducing the substituents of various types to phenols, the competition of a few tendencies is being observed, each of them influencing on selected toxicity of the investigated substances [19].

First, the change of lipophility takes place, which results in the change of phenol transport through lipid membrane layers to the following receptors. For example, in [20] the influence of AO on the activity of lipodependent proteinkinase was noted not only through lipid membrane changes, but through direct interaction with enzyme. As the substituted phenols are accumulated in lipids, so lipophility dropping of substituted phenol should promote it to supply those fields of lipid membranes which are enriched with oxidized lipids and the products of their metabolism [19]. Second, the structure change results in the reaction ability change in substituted phenol groups, responsible for metabolizing with these phenols. OH - groups, particularly, are responsible for conjugation reactions glucuronic acid and sulfates [21, 22]. In turn heteroatom groups in p-position take place participate in the processes on the first stage of metabolism with the following stage of metabolism by conjugation with possible metabolites – 2,5-cyclohexadienones formation [4, 12]. Third, the structure change of substituted phenols results in constant of ionization change of OH-group. It is sufficiently influences upon selective toxicity, connected with substrate ionization [19].

So, the above stated factors make difficult to find the true phenol toxicity dependence on structure. Besides it is difficult to compare these results with those obtained for phenols with lipophilic alkyl substituents [12]. Indeed, from these results in [12], it follows that 2,6-di-tert-butyl-4-methylphenol is one of the strongest toxicants. Similar principal result difference may be explained that strong phenol toxicity has been studied on mice in this study, and in [12] specific mice organs (lungs) have been investigated there; its affects on lungs.

For full explanation obtained, it is desirable to investigate specific affect of these phenols on various targets in some organs with help of pharmacokinetics. Nevertheless, the obtained results in this study may be of help in pharmacological investigations and planning studies connected with synthesis of new, nontoxical, biologically active phenol compounds.

CONCLUSION

1. Phenols having two sterically hindred tert-butyl substitutes in o-position have the least toxicity.
2. Toxicity of 2,6-di-tert-butylphenol derivatives increases with removing the functional p-substituents of benzene ring.

3. The declinations from the law of nature of toxicity changes were observed for p-substituted phenols, in which functional group is selected from benzene ring only by one methylene group.
4. The obtained results may be of help in pharmacological investigations and planning studies connected with synthesis of new, nontoxical, biologically active phenol compounds.

REFERENCES

[1] Burlakova E.B., Khrapova N.G. Peroxide oxidation of lipids of membranes and natural antioxidants // *Russian Chemical Reviews.* 1985. V. 54. № 9. P. 1540-1558.

[2] Khrapova N.G. Peroxide oxidation of lipids of Biological membranes and Food Additives, In «Chemical and Biological Kinetics. New Horizons. V. 2. Biological Kinetics», Ed. Burlakova E.B., Khimiya, Moscow, 2005, P. 46-60.

[3] Burlakova E.B., Alesenko A.V., Molochkina E.M. etc., Bioantioxidants in Radiation Sickness and Malignant Disease, Ed. Burlakova E.B. Nauka, Moscow, 1975.

[4] Ershov V.V, Nikiforov G.A, Volodkin A.A. Sterically hindred phenols, Khimiya, Moscow, 1972.

[5] Zarudy F.S., Gilmutdinov G.Z., Zarudy R.F. etc. 2,6-di-tert-butyl-4- methyl phenol (Dibunoli, Ionoli, Tonafol) is a classical antioidant // *Khimiko-Farmacevtichesky Zh.* 2001. V. 35. № 3. P. 42-48.

[6] Zorkina A.V., Kostin J.V., Inchina V.I. etc. Antioxidative both hypolipidemical properties of mexidol and emoxipin at long immobili stress // *Khimiko-Farmacevtichesky Zh.* 1998. V. 32. № 5. P. 3-6.

[7] Kotljarov A.A., Smirnov L.D., Smirnova L.E. etc. Research of joint application mexidol with antiarrhytmic preparations // *Experimental and clinical pharmacology: Two-month scientific-theoretical magazine.* 2002. V. 65. № 5 P. 31-34.

[8] Zenkov N.K., Kandalintseva N.V., Lankin V.Z. etc. Phenolic bioantioxidants, SO RAMN, Novosibirsk, 2003.

[9] Burlakova E.B. Bioantioxidants: yesterday, today, tomorrow. In «Chemical and Biological Kinetics. New Horizons. V. 2. Biological Kinetics». V. 2. Ed. Burlakova E.B., Khimiya, Moscow, 2005, P. 10-45.

[10] Kravchuk E.A., Keselyova T.N., Ostrovsky M.A. etc. Influence an antioxidant preparation trimeksidin on a current of an early stage experimental uveitis // *Refraction surgery and ophthalmology* 2008. V. 8. № 1. P. 36-41.

[11] The Russian federal register of biologically active additives to food, 2th edition, processed and added. 2001.

[12] Mizutani T., Ishida I., Yamamoto K., Tajima K. Pulmonary Toxicity of butylated hydroxytoluene and related alkylphenols: structural requirements for toxic potency in mice // *Toxicology and Applied Pharmacology*, 62, 273-281 (1982).

[13] Ershov V.V, Belostotskaja I.S. Di-tert-butylspirocyclodienones // *Izv.Akad. Nauk SSSR, Ser. Khim.* 1965. № 7. P. 1301-1303.

[14] Belostotskaja I.S., Volodkin A.A., Ostapets-Sveshnikova G. D, Ershov V.V. Synthesis α-replaced hydroxyphenylacetic acids and phenylethylamines of some the sterically hindred phenols // *Izv.Akad. Nauk SSSR, Ser. Khim.* 1966. № 10. P. 1833-1835.

[15] Belen'ky M.L. Elements of a quantitative estimation of pharmacological effect, Ed. Acad. Sci. of Latv. SSR, 1959.

[16] Roginsky V.A. Phenolic antioxidants: Reactionary ability and efficiency, Nauka, Moscow, 1988.

[17] Takahashi O., Hiraga K. 2,6-Di-tert-butyl-4-methylene-2,5-cyclohexadienone: a hepatic metabolite of butylated hydroxytoluene in rats // *Fd. Cosmet. Toxicol* 1997, V. 17, P. 451-454.

[18] Temnikova T.I. Course of theoretical bases of organic chemistry, Khimiya, Leningrad, 1968.

[19] Albert A. Selective Toxicity, 7th edition. Chapman and Hall, London, 1985.

[20] Hohlov A.P. Mechanisms of regulation of activity of membrane receptors synthetic antioxidants from a class sterically hindred phenols. // *The bulletin of experimental biology and medicine* 1988. № 10. P. 440-444.

[21] Kabiev O.K., Balmuhanov S.B. Natural phenols – a perspective class of antineoplastic and radio protective connections, Meditsina, Moscow, 1975.

[22] Vergejchik T.H. Toxicological chemistry, MEDpress-inform, Moscow, 2009.

In: Biology and Communication in the Information Age
Editor: Rafiqul Islam
ISBN: 978-1-62808-252-4

Chapter 17

ASSESSMENT OF THE MEMBRANE ACTIVE AND PROTECTIVE PROPERTIES OF ISOBORNYLPHENOLS IN THE CHEMICAL AND BIOLOGICAL MODEL SYSTEMS

K. M. Marakulina*[1], *I. G. Plaschina*[1], *O. G. Shevchenko*[2], *S. N. Plyusnina*[3], *I. V. Fedorova*[3], *M. V. Kozlov*[1], *I. Yu. Chukicheva*[3], *A. V. Kutchin*[3] and *L. N. Shishkina*[1*]

[1]Emanuel Institute of Biochemical physics of Russian Academy of science, Moscow
[2]Institute of Biology Komi scientific center, Ural Branch, Russian Academy of Sciences, Syktyvkar, Russia
[3]Institute of Chemistry, Komi scientific center, Ural Branch, Russian Academy of Sciences, Syktyvkar, Russia

ABSTRACT

Semisynthetic isobornylphenols (IBP) was shown to exhibit the antioxidant properties in biological model systems reducing the intensity of lipid peroxidation. IBP can possess both membrane protective and membrane active properties in depend on their structure. Surface-active properties defined by their influence on the aggregation of lecithin in hexane are correlate with their effect on the architectonics of erythrocytes. The data obtained allow us to conclude that the biological activity of IBP is due to their antioxidant and surfactant properties, that is necessary to keep in mind under the analysis of their biological activity in systems *in vivo*.

Keywords: Isobornylphenols, BHT, phosphatidylcholine, sphingomyelin, dynamic light scattering, scanning electron microscopy

[*] E-mail: shishina@sky.chph.ras.ru.

AIM AND BACKGROUND

In current time the substituted phenols containing one or more isobornyl groups known as isobornylphenols (IBP) are actively studied. Earlier for this compounds in the chemical model the high antiradical activity was shown with inhibiting coefficient *f*=2 which is characteristic for active phenolic antioxidants (InH) [1]. It's important that IBP are synthesized from waste of timber-industry complex on the base of green conifers. Moreover, it was shown that one of the IBP has showed neuroprotective, antitrombogenic and antiplatelet activities in experiments *in vivo* [2, 3]. Thus, these compounds are perspective biologically active substances. The aim of this work was to study the surface active properties of IBP in chemical (phospholipid aggregation in a nonpolar solvent) and biological (induced and spontaneous hemolysis of the red blood cells) model systems.

INTRODUCTION

As known, biological activity of InH could rise from not only its inhibiting activity but their ability to interact with cell membrane components. The main components of any biomembranes – phospholipids (PL) in their structure contain polar heads and hydrophobic fatty acid chains. They are surface active substances which are capable to aggregation both polar and non-polar solvents [4, 5]. Due to the individual fractions of PL are asymmetric distributed between the membrane monolayers, the most saturated and therefore the more poorly oxidized choline-containing PL such as phosphatidylcholine (PC) and sphingomyelin (SM) are located in the outer membrane monolayer, and the less saturated amino-PL do in the inner. This arrangement of PC and SM makes them one of the first participants in the reactions with InH.

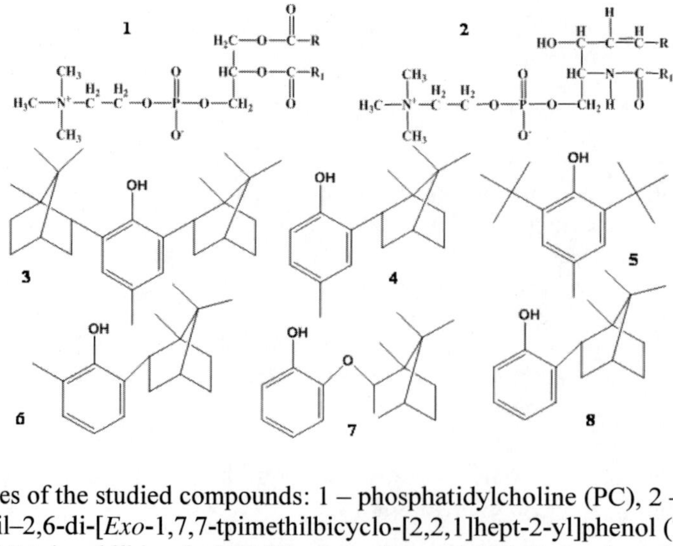

Figure 1. Structures of the studied compounds: 1 – phosphatidylcholine (PC), 2 – sphingomyelin (SM), 3 – 4-methil–2,6-di-[*Exo*-1,7,7-tpimethilbicyclo-[2,2,1]hept-2-yl]phenol (TP-7), 4 – 4-methil-2-[*Exo*-1,7,7-tpimethilbicyclo-[2,2,1]hept-2-yl]phenol (TP-5), 5 – 4-methyl-2,6-di*tert* butylphenol (BHT), 6 – 2-methil-6-[*Exo*-1,7,7-tpimethilbicyclo-[2,2,1]hept-2-yl]phenol (TP-8), 7 – 2-{[*Exo*-1,7,7-trimethylbicyclo-[2.2.1]hept-2-yl]hydroxy}phenol (TP-1), 8 – 2-[*Exo*-1,7,7-tpimethilbicyclo-[2,2,1]hept-2-yl]phenol (TP-6).

EXPERIMENTAL PART

The 10% alcoholic solution of soy bean lecithin (L) (Ukraine) and chloroform solution of sphingomyelin from bovine brain (Russia) were used in research. The IBP used were synthesized by the methods detailed in works [6, 7, 8]. BHT which is a standard antioxidant from the class of sterically hindered phenols was used for comparison. Structural formulas of the used InH, PC and SM are shown in Figure 1.

The purity and fractional composition of PL were controlled by thin layer chromatography [9]. The glass plates measuring 90 mm × 120 mm, silica gel type G (Sigma, USA) were used. Mixture of chloroform: methanol: glacial acetic acid: distilled water in a volumetric ratio 12,5:7,5:2,0:1,0 respectively was used as mobile phase. All the solvents were or specially or chemically pure grade. The development of chromatograms was performed by iodine vapor. For the color reaction for phosphorus we used ammonium molibdate and ascorbic acid and also perchloric acid of chemically pure grade. The amount of inorganic phosphorus was judged from the optical density of the solutions at the wavelength λ=800 nm with photoelectrocolorimeter KFK-3 (Russia).

After distilling the initial solvent the size of the PL aggregates in hexane was evaluated by dynamic light scattering method. The measurements were performed by the ZetaSizer Nano ZS (Malvern Instruments, UK) instrument equipped with the 4 mW He-Ne laser and automatic data processing program, at 25° C and fixed scattering angle 173°. The relative error in measuring of the aggregate size is less than 2%.

The red blood cells of laboratory mice were used. After decapitation blood were collected into tubes treated by 5% solution of sodium citrate. Red blood cells were separated from plasma by centrifugation and washed twice with phosphate-buffered saline (PBS), pH 7.4. The erythrocyte resulting was combined from several animals. The 1% erythrocyte suspension in PBS was used in the experiments. Since both the IBP and BHT are hydrophobic compounds, all InH at the concentration of 100 μM were injected in the erythrocyte suspension as the ethanol solution, concentration of which in the incubation medium was 0.1% in the experimental and control samples. Solutions of the investigated substances introduced into the erythrocyte suspension and incubated at 37 ° C in the thermostatic shaker (Biosan, Latvia). The content of secondary products of lipid peroxidation (LP) (nmol/ml) reacting with 2-thiobarbituric acid (TBA-reactive substances, TBA-RS), in the erythrocyte suspension was determined spectrophotometrically [10].

To study the surface architectonics the erythrocytes taken after 4 h incubation were fixed at 4° C for 1 h in 2.5% glutaraldehyde, prepared in phosphate buffer, pH 7.4, then washed twice in phosphate buffer (pH=7,4).

For electron microscopy studies the samples were dried, sprayed by gold and examined by the scanning electron microscope Tesla BS 300 (Czechoslovakia). Assessment of the quantitative distribution of the morphological forms of the red blood cells was performed by using of microscope Axiovert 200 M (Carl Zeiss, Germany). For each InH it was analyzed at least 1,000 cells.

The experimental data were proceeded by the common methods of variation statistics [11] and using the software package KINS [12]. Reliability of differences was calculated by t - Student's criterion. The results in figures are presented as arithmetic mean values with indication of its mean square error.

RESULTS AND DISCUSSION

The used PL samples were characterized by high content of the main fraction. So, phosphatidylcholine share in lecithin samples was range from 80,4% to 92,2%, and sphingomyeline share in their samples was 95,7-97,2%.

As known, the InH ability to participate in regulation of LP in complex biological systems are determined by the change of the TBA-reactive substances content as compared with control. As shown in Figure 2, all InH studied reduce the TBA-reactive substances content in the blood erythrocytes during hemolysis process, indicating a marked antioxidant properties of these compounds both in chemical [1], and in biological model systems.

Phenomenon of the complex formation between the molecules of PC, SM and IBP is earlier observed [13, 14]. It could have a significant effect on the ability of these compounds to interact with membranes. Besides, the complexation between PL and the different InH leads to change of the reactivity of the natural anf synthetic phenolic antioxidants, as shown in [15-17].

The assessment of membrane active and membrane protective properties of InH was carried out by means both chemical and biological model systems.

The aggregation parameters of L and SM in hexane in both presence and absence InH were studied by using of dynamic light scattering method to test the ability of InH to affect self-organization process of these PL in non-polar solvent. The diagram showing the ratio of the size of lecithin aggregates to the aggregates size obtained for the equimolar mixtures of L with InH is shown in Figure 3 a.

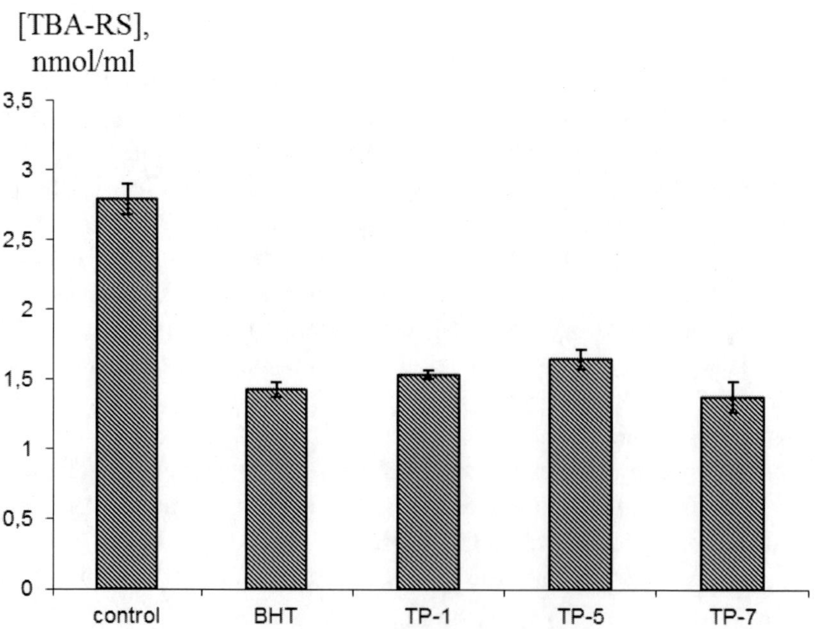

Figure 2. The TBA-reactive substances content in the blood erythrocyte suspension after 4 h incubation during their induced hemolysis in absence and presence of the phenolic antioxidants.

As seen, the significant changes are not observed for the L mixtures with TP-1, TP-7 and BHT, whereas for the L mixtures with TP-5, TP-6 and TP-8 the considerable reduction of the micelle sizes (20-40 times) are revealed, indicating that these compounds exhibit the membrane active properties. Acording to the reduction of the influence on the size of formed micelles in hexane all studied InH form the following sequence: TP-8 > TP-5 > TP-6 >> BHT ≥ TP-1 ≥ TP-7. The influence of studied InH on the size of the SM aggregates in hexane was shown in Figure 3 b. The comparison of the data presented in Figure 3 a and 3 b allows us to assume that the nature of PL plays a significant role in their interaction with InH. So, all investigated InH reliably increase the size of aggregates (20-40 %) at their equimolar mixtures with SM.

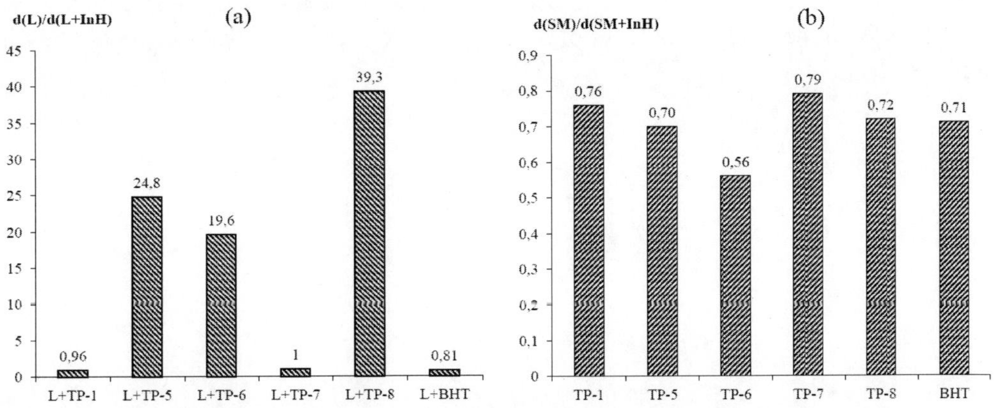

Figure 3. The effect of the studied phenolic antioxidants on the size of the lecithin (a) and SM (b) aggregates in hexane.

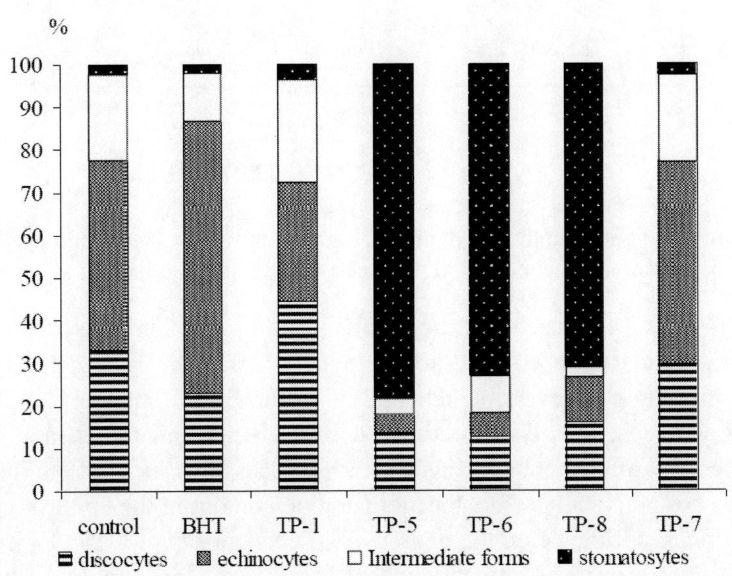

Figure 4. The ratio of the different forms of the blood erythrocytes after 4 h incubation in absence and presence of the studied phenolic antioxidants.

The data obtained about the InH influence both on the LP intensity and aggregation of PL in hexane indicate that the character of interaction of IBP with membrane lipids is complex. Although TP-5 reduces stationary level of the TBA-reactive substances content during the erythrocyte hemolysis, but the size of the lecithin aggregates in hexane decreases about 25 times in the presence of this InH. Consequently, on the one hand, the addition of this compound inhibits the LP process, and, on the other hand, leads to destruction of the lecithin aggregates in hexane.

The ratio of the different forms of erythocytes which are revealed after their incubation in the absence and presence of the studied InH is presented in Figure 4. As seen, the TF-1, TF-7 and BHT alter the ratio between discocytes (normal erythrocytes) and echinocytes (reversible cell form), but do not affect the total stomatocyte (inreversible pre-hemolytic cell form) level, which does not exceed the control value. This demonstrates the ability of these compounds to incorporate only in the outer monolayer of the erythrocyte membrane without penetrating into the inner.

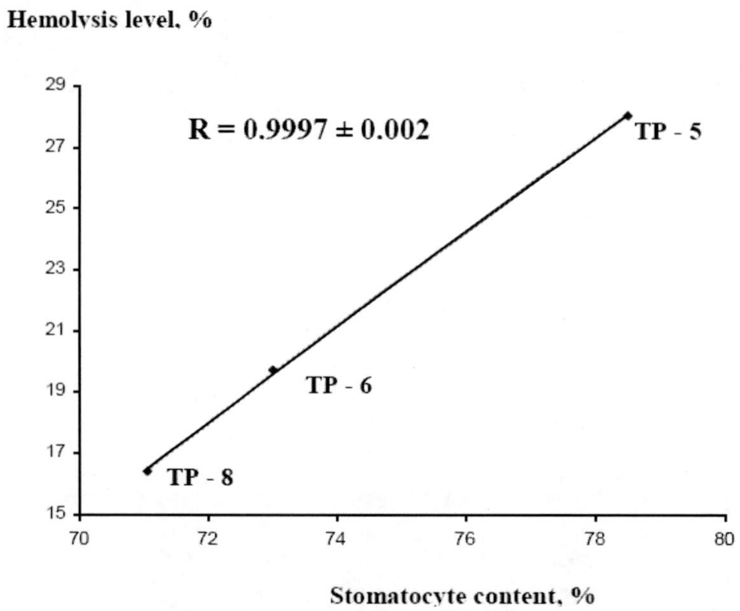

Figure 5. The correlation between the spontaneous hemolysis level of the blood erythrocytes and the stomatocyte share after 4 h incubation in presence of the studied phenolic antioxidants.

The TP-5, TP-6, TP-8 caused the more reliable dramatic increases of the stomatocyte proportion, indicating that these IBP are able to penetrate into the inner membrane of the red blood cells, resulting in the irreversible damage of erythrocytes. The correlation between the level of the spontaneous hemolysis of the red blood cells and the stomatocyte share in erythrocyte suspension after incubation with TP-5, TP-6, TP-8. In addition, TP-1, TP-7 and BHT, which does not practically affect the stomatocyte content in the erythrocyte suspension, have no the significant influence on the degree of spontaneous hemolysis. Whereas, the TP-5 TP-6 and TP-8 increasing the stomatocyte content also for certain influence on the spontaneous hemolysis level of the red blood cells as compared with control (Figure 5).

CONCLUSION

The analysis of the obtained and literature data, allow us to put up the series of conclusions.

Semisynthetic IBP exhibit antioxidant properties in biological model systems reducing the intensity of lipid peroxidation.

IBP could show both membrane protective (TP-1, TP-7) and membrane active (TP-5, TP-6, TP-8) properties in the dependence on their structure.

The study of surface active properties of IBP in the chemical model system allows us to assume that their membrane active and/or membrane protective properties were caused by their preferential interaction with PC.

Surface active properties determined by their influence on the aggregation of L in hexane are correlated with their effect on the architectonics of erythrocytes.

Thus, the biological activity of IBP is due to their antioxidant and surfactant properties, that is necessary to keep in mind under the analysis of their biological activity in systems *in vivo*.

REFERENCES

[1] Mazaletskaya L.I., Sheludchenko N.I. Shishkina, L.N., Kutchin A.V., Fedorova I.V., Chukicheva I.Yu. 2011. Kinetic Parameters of the Reaction of Isobornylphenols with Peroxy Radicals. *Petroleum Chemistry*. Vol. 51.No 5. pp. 348-353.

[2] M.B. Plotnikov, V.I. Smol'iakova, V.I. Ivanov, A.V. Kutchin, I.Yu. Chukicheva, E.A. Krasnov. Antithrombogenic and antiplatelet activity of isobornylphenols derivatives.// *Experimental Biology and Medicine Bulletin*. 2008, V. 145. № 3. pp. 296-299.

[3] M.B. Plotnikov, G.A. Chernysheva, V.I. Smol'iakova, V.A. Kutchin, I.Yu. Chukicheva, E.A. Krasnov. Neuroprotective effects of dibornol and mechanism of its action in rats with cerebral ischemia.// *Annals of the Russian Academy of Medical Sciences*. 2009, №11. pp.12-17.

[4] Ivkov V.G., Berestovskii T.N. *Dynamic Structure of the lipid bilayer*. Moscow: Nauka. 1981. P. 296.

[5] Barclay L.R.C., MacNeil J.M., VanKessel J., Forrest B.J., Porter N.A., Lehman L.S., Smith K.J., Ellington J.C.Jr. Autoxidation ant aggregation of phospholipids in organic solvents.// *J. Am. Chem. Soc*. 1984. Vol. 106. No. 48. P. 6740-6747.

[6] Chukicheva I.Yu., Kutchin A.V. Natural and synthesized terpenophenols.// *Russian Chemical journal*. 2004, V. 48. №3. pp. 21-38.

[7] Chukicheva I.Yu., Timusheva I.V., Spirikhin L.V., Kutchin A.V. Alkilation of pyrocatechol and resorcinol by camphen.// *Chemistry of natural compounds*. 2007, V. 43. № 3. pp. 245-249.

[8] I.Yu. Chukicheva, I.V. Fedorova, A.V. Kutchin. Investigation of alkylation of p-cresol by camphen in the presence acid catalysts.// *Chemistry of plant raw material*. 2009. № 3. pp. 63-68.

[9] *Biological membranes. A practice approach* (Eds J.B. Findlay, W.H. Evans). Moscow: Mir, 1990. (in Russian).

[10] T. Asakawa, S. Matsushita Coloring conditions of thiobarbituric acid test for detecting lipid hydroperoxides//Lipids. 1980. V. 15. № 3. P. 137-140.

[11] G.F. Lakin. *Biometrics*. — M.: Vyssh. shk., 1980. P. 293.

[12] E.F. Brin, S.O. Travin. Modeling the mechanisms of chemical reactions.// *Chem. Physics*. 1991. V. 10. № 6. pp. 830-837.

[13] K.M. Marakulina, R.V. Kramor, I.G. Plaschina, L.N. Shishkina "Evaluation of membrane active and membrane protective properties of the series of isobornylphenols in the model systems// *Proceedings of the International conference "Renewable wood and plant resources: Chemistry, technology, pharmacology, medicine"*, Saint-Petersburg: June 21-24, 2011. pp. 287-288.

[14] K.M. Marakulina, R.V. Kramor, Yu.K. Lukanina, L.N. Shishkina, A.V. Kutchin, I.Yu. Chukicheva. "Complexation of lecithin with phenolic antioxidants", In: *Morden problems in biochemical physics. New horizons* / Eds. G.E. Zaikov, S.D. Varfolomeev, E.B. Burlakova, A.A. Popov. N.Y: Nova Science Publishers. 2012.

[15] K. Xu, B. Liu, Yu. Ma, J. Du, G. Li, H. Gao, Yu. Zhang, Z. Ning. Physicochemical properties and antioxidant activities of luteolin-phospholipid complex.// *Molecules*. 2009, Vol. 14. pp. 3486-3493.

[16] Sharafutdinova R.R., Nasibullin R.S., Fahretdinova E.R. Investigation of the conformational state of quercetin upon complexation with lecithin // *Chemical physics and mesoscopic*. 2008. V. 10. № 4. pp.510-513.

[17] Mazaletskaya L.I., Sheludchenko N.I., Shishkin L.N. Lecithin influence antioxidant activity of flavonoids and tocopherol.// *Applied biochemistry and microbiology*. 2010. V. 46, № 2. pp. 148-152.

In: Biology and Communication in the Information Age
Editor: Rafiqul Islam
ISBN: 978-1-62808-252-4

Chapter 18

TRANSFORMATION OF HIGH-ENERGY BONDS IN ATP

G. A. Korablev*[1], N. V. Khokhriakov[1], G. E. Zaikov[2] and Yu. G. Vasiliev[1]

[1]Izhevsk State Agricultural Academy,
Basic Research and Educational Center of Chemical Physics and Mesoscopy,
Udmurt Scientific Center,
Ural Division, Russian Academy of Science, Russia, Izhevsk
[2]N.M. Emanuel Institute of Biochemical Physics, RAS, Russia, Moscow

ABSTRACT

With the help of spatial-energy concept it is demonstrated that the formation and change of high-energy bonds in ATP take place at the functional transitions of valence-active orbitals of the system "phosphorus-oxygen".

These values of energy bonds are in accord with experimental and quantum-mechanical data.

1. INTRODUCTION

During the interaction of oppositely charged heterogeneous systems the certain compensation of volume energy of interacting structures takes place which leads to the decrease in the resulting energy (for example, during the hybridization of atomic orbitals). But this is not the direct algebraic deduction of the corresponding energies. The comparison of multiple regularities of physical, chemical and biological processes allows assuming that in such and similar cases the principle of adding the reciprocals of volume energies or kinetic parameters of interacting structures is executed.

Lagrangian equation for the relative movement of the system of two interacting material points with the masses m_1 and m_2 in coordinate x is as follows:

* E-mail: korablev@udm.net.

$$M_{red}x'' = -\frac{\partial U}{\partial x}, \text{ where } \frac{1}{m_{red}} = \frac{1}{m_1} + \frac{1}{m_2} \quad (1), (1a)$$

here U – mutual potential energy of material points; m_{red} – reduced mass. Herein $x'' = a$ (system acceleration).

For the elementary areas of interactions Δx we can accept: $\frac{\partial U}{\partial x} \approx \frac{\Delta U}{\Delta x}$.

Then: $m_{red} a\Delta x = -\Delta U$; $\frac{1}{1/(a\Delta x)} \cdot \frac{1}{(1/m_1 + 1/m_2)} \approx -\Delta U$

or: $\frac{1}{1/(m_1 a\Delta x) + 1/(m_2 a\Delta x)} \approx -\Delta U$

Since the product $m_i a\Delta x$ by its physical sense equals the potential energy of each material point ($-\Delta U_i$), then: (2)

Thus the resulting energy characteristic of the system of two interacting material points is found by the principle of adding the reciprocals of initial energies of interacting subsystems.

"The electron with the mass m moving about the proton with the mass M is equivalent to the particle with the mass: $m_{red} = \frac{mM}{m+M}$" [1].

Therefore, modifying the equation (2) we can assume that the energy of atom valence orbitals (responsible for interatomic interactions) can be calculated [2] by the principle of adding the reciprocals of some initial energy components based on the equations:

$$\frac{1}{q^2/r_i} + \frac{1}{W_i n_i} = \frac{1}{P_E} \text{ or ;} \quad (3), (4), (5)$$

here: W_i – orbital energy of electrons [3]; r_i – orbital radius of i orbital [4]; $q=Z^*/n^*$ – by [5,6], n_i – number of electrons of the given orbital, Z^* and n^* – nucleus effective charge and effective main quantum number, r – bond dimensional characteristics.

The value P_O is called a spatial-energy parameter (SEP), and the value P_E – effective P–parameter (effective SEP). Effective SEP has a physical sense of some averaged energy of valence orbitals in the atom and is measured in energy units, e.g. in electron-volts (eV).

The values of P_0-parameter are tabulated constants for electrons of the given atom orbital.

For SEP dimensionality:

$$[P_0] = [q^2] = [E]\cdot[r] = [h]\cdot[\upsilon] = \frac{kgm^3}{s^2} = J\,m,$$

where [E], [h] and [υ] – dimensionalities of energy, Planck's constant and velocity.

The introduction of P-parameter should be considered as further development of quasi-classical concepts with quantum-mechanical data on atom structure to obtain the criteria of phase-formation energy conditions. For the systems of similarly charged (e.g. orbitals in the given atom) homogeneous systems the principle of algebraic addition of such parameters is preserved:

$$\sum P_E = \sum (P_0/r_i) \; ; \; \sum P_E = \frac{\sum P_0}{r} \qquad (6), (7)$$

$$\text{or: } \sum P_0 = P_0^{'} + P_0^{''} + P_0^{'''} + \ldots; \qquad (8), (9)$$

Here P-parameters are summed up by all atom valence orbitals.

To calculate the values of P_E-parameter at the given distance from the nucleus either the atomic radius (R) or ionic radius (r_I) can be used instead of r depending on the bond type.

Let us briefly explain the reliability of such an approach. As the calculations demonstrated the values of P_E-parameters equal numerically (in the range of 2%) the total energy of valence electrons (U) by the atom statistic model. Using the known correlation between the electron density (β) and intra-atomic potential by the atom statistic model [7], we can obtain the direct dependence of P_E-parameter on the electron density at the distance r_i from the nucleus. The rationality of such technique was proved by the calculation of electron density using wave functions by Clementi [8] and comparing it with the value of electron density calculated through the value of P_E-parameter.

The modules of maximum values of the radial part of Ψ-function were correlated with the values of P_0-parameter and the linear dependence between these values was found. Using some properties of wave function as applicable to P-parameter, the wave equation of P-parameter with the formal analogy with the equation of Ψ-function was obtained [9].

2. Wave Properties of P-Parameters and Principles of Their Addition

Since P-parameter has wave properties (similar to Ψ'-function), the regularities of the interference of the corresponding waves should be mainly fulfilled at structural interactions.

The interference minimum, weakening of oscillations (in antiphase) occurs if the difference of wave move (Δ) equals the odd number of semi-waves:

$$\Delta = (2n+1)\frac{\lambda}{2} = \lambda\left(n+\frac{1}{2}\right),$$

where n = 0, 1, 2, 3, … (10)

As applicable to P-parameters this rule means that the interaction minimum occurs if P-parameters of interacting structures are also "in antiphase" – either oppositely charged or heterogeneous atoms (for example, during the formation of valence-active radicals CH, CH_2, CH_3, NO_2 …, etc) are interacting.

In this case P-parameters are summed by the principle of adding the reciprocals of P-parameters – equations (3,4).

The difference of wave move (Δ) for P-parameters can be evaluated via their relative value of relative difference of P-parameters (coefficient α) which at the interaction minimum produce an odd number:

$$\gamma = \frac{P_2}{P_1} = \left(n+\frac{1}{2}\right) = \frac{3}{2};\frac{5}{2}\ldots\ldots \text{ When (main state) } \frac{P_2}{P_1}=\frac{1}{2} \qquad (11)$$

It should be pointed out that for stationary levels of one-dimensional harmonic oscillator the energy of these levels $\varepsilon=h\nu(n+\frac{1}{2})$, therefore in quantum oscillator, in contrast to the classical one, the least possible energy value does not equal zero.

In this model the interaction minimum does not provide zero energy corresponding to the principle of adding reciprocals of P-parameters – equations (3,4).

The interference maximum, strengthening of oscillations (in phase) occurs if the difference of wave move equals the even number of semi-waves:

$$\Delta=2n\frac{\lambda}{2}=\lambda n \text{ or } \Delta=\lambda(n+1).$$

As applicable to P-parameters the maximum interaction intensification in the phase corresponds to the interactions of similarly charged systems or systems homogeneous by their properties and functions (for example, between the fragments or blocks of complex inorganic structures, such as CH_2 and NNO_2 in octogene).

And then: (12)

By the analogy, for "degenerated" systems (with similar values of functions) of two-dimensional harmonic oscillator the energy of stationary states:

$$\varepsilon=h\nu(n+1)$$

By this model the interaction maximum corresponds to the principle of algebraic addition of P-parameters – equations (6-8). When n=0 (main state) we have $P_2=P_1$, or: the interaction maximum of structures occurs if their P-parameters are equal. This concept was used [2] as the main condition for isomorphic replacements and formation of stable systems.

3. Equilibrium-Exchange Spatial-Energy Interactions

During the formation of solid solutions and in other structural equilibrium-exchange interactions the unified electron density should be established in the contact spots between atoms-components. This process is accompanied by the re-distribution of electron density between valence areas of both particles and transition of a part of electrons from some external spheres into the neighboring ones.

It is obvious that with the proximity of electron densities in free atoms-components the transition processes between the boundary atoms of particles will be minimal thus contributing to the formation of a new structure. Thus the task of evaluating the degree of such structural interactions in many cases comes down to comparative assessment of electron density of valence electrons in free atoms (on the averaged orbitals) participating in the process.

Therefore the maximum total solubility evaluated via the structural interaction coefficient α is defined by the condition of minimal value of coefficient α which represents the relative difference of effective energies of external orbitals of interacting subsystems:

$$\alpha = \frac{P'_o/r_i' - P''_o/r_i''}{(P'_o/r_i' + P''_o/r_i'')/2}100\% \text{ or } \alpha = \frac{P_S' - P_S''}{P_S' + P_S''}200\% \qquad (13), (14),$$

where P_S – structural parameter is found by the equation:

$$\frac{1}{P_S} = \frac{1}{N_1 P_E'} + \frac{1}{N_2 P_E''} + \ldots \qquad (15),$$

here N_1 and N_2 – number of homogeneous atoms in subsystems.

The nomogram of the dependence of structural interaction degree (ρ) upon the coefficient α, the same for the wide range of structures, was prepared by the data obtained (the figure is not available).

Isomorphism as a phenomenon is usually considered as applicable to crystalline structures. But obviously the similar processes can also take place between molecular compounds where the bond energies can be assessed via the relative difference of electron densities of valence orbitals of interacting atoms. Therefore the molecular electronegativity is rather easily calculated via the values of corresponding P-parameters.

In complex organic structures the main role in intermolecular and intramolecular interactions can be played by separate "blocks" or fragments considered as "active" areas of the structures. Therefore it is necessary to identify these fragments and evaluate their spatial-energy parameters. Based on wave properties of P-parameter, the total P-parameter of each element should be found following the principle of adding the reciprocals of initial P-parameters of all the atoms. The resulting P-parameter of the fragment block or all the structure is calculated following the rule of algebraic addition of P-parameters of their constituent fragments.

Apparently, spatial-energy exchange interactions (SEI) based on leveling the electron densities of valence orbitals of atoms-components have in nature the same universal value as purely electrostatic coulomb interactions and complement each other. Isomorphism known from the time of E. Mitscherlich (1820) and D.I. Mendeleev (1856) is only a special demonstration of this general natural phenomenon.

The quantitative side of evaluating the isomorphic replacements both in complex and simple systems rationally fits into P-parameter methodology. More complicated is the problem of evaluating the degree of structural SEI for molecular structures, including organic ones. Such structures and their fragments are often not completely isomorphic to each other. Nevertheless, SEI is going on between them and its degree can be evaluated either semi-quantitatively numerically or qualitatively. By the degree of isomorphic similarity all the systems can be divided into three types:

I Systems mainly isomorphic to each other – systems with approximately the same number of *heterogeneous* atoms and cumulatively similar geometric shapes of interacting orbitals.

II Systems *with organic isomorphic similarity* – systems which:

1) either differ by the number of heterogeneous atoms but have cumulatively similar geometric shapes of interacting orbitals;
2) or have certain differences in the geometric shape of orbitals but have the same number of interacting heterogeneous atoms.

III Systems without isomorphic similarity – systems considerably different both by the number of heterogeneous atoms and geometric shape of their orbitals.

Taking into account the experimental data, all SEI types can be approximately classified as follows:

Systems I

1. α < (0-6)%; ρ = 100 %. Complete isomorphism, there is complete isomorphic replacement of atoms-components;

2. 6 % < α < (25-30)%; ρ = 98 – (0-3) %.

There is wide or unlimited isomorphism.

3. α > (25-30) %; no SEI.

Systems II

1. α < (0-6)%;

a) There is reconstruction of chemical bonds that can be accompanied by the formation of a new compound;

b) Cleavage of chemical bonds can be accompanied by a fragment separation from the initial structure but without adjoinings and replacements.

2. 6 % < α < (25-30)%; the limited internal reconstruction of chemical bonds is possible but without the formation of a new compound and replacements.

3. α > (20-30) %; no SEI.

Systems III

1. α < (0-6)%; a) The limited change in the type of chemical bonds in the given fragment is possible, there is an internal re-grouping of atoms without the cleavage from the main molecule part and replacements;

b) The change in some dimensional characteristics of the bond is possible;

2. 6 % < α < (25-30)%;

A very limited internal re-grouping of atoms is possible;

α > (25-30) %; no SEI.

When considering the above systems, it should be pointed out that they can be found in all cellular and tissue structures in some form but are not isolated and are found in spatial-time combinations.

The values of α and ρ calculated in such a way refer to a definite interaction type whose nomogram can be specified by fixed points of reference systems.

If we take into account the universality of spatial-energy interactions in nature, such evaluation can have the significant meaning for the analysis of structural shifts in complex bio-physical and chemical processes of biological systems.

Fermentative systems contribute a lot to the correlation of structural interaction degree. In this model the ferment role comes to the fact that active parts of its structure (fragments, atoms, ions) have such a value of P_E-parameter which equals the P_E-parameter of the reaction final product.

That is the ferment is structurally "tuned" via SEI to obtain the reaction final product, but it will not enter it due to the imperfect isomorphism of its structure (in accordance with III). The important characteristics of atom-structural interactions (mutual solubility of components, chemical bond energy, energy of free radicals, etc) for many systems were evaluated following this technique [10, 11].

4. Calculation of Initial Data and Bond Energies

Based on the equations (3-5) with the initial data calculated by quantum-mechanical methods [3-6] we calculate the values of P_0-parameters for the majority of elements being tabulated, constant values for each atom valence orbital. Mainly covalent radii – by the main type of the chemical bond of interaction considered were used as a dimensional characteristic for calculating P_E-parameter (Table 1).

The value of Bohr radius and the value of atomic ("metal") radius were also used for hydrogen atom. In some cases the bond repetition factor for carbon and oxygen atoms was taken into consideration [10].

For a number of elements the values of P_E-parameters were calculated using the ionic radii whose values are indicated in column 7. All the values of atomic, covalent and ionic radii were mainly taken by Belov-Bokiy, and crystalline ionic radii – by Batsanov [12]. The results of calculating structural P_S-parameters of free radicals by the equation (15) are given in Table 2. The calculations are done for the radicals contained in protein and amino acid molecules (CH, CH_2, CH_3, NH_2 etc), as well as for some free radicals formed in the process of radiolysis and dissociation of water molecules. The technique previously tested [10] on 68 binary and more complex compounds was applied to calculate the energy of coupled bond of molecules by the equations:

$$\frac{1}{E}=\frac{1}{P_S}=\frac{1}{\left(P_E\frac{n}{K}\right)_1}+\frac{1}{\left(P_E\frac{n}{K}\right)_2}; \quad P_E\frac{n}{K}=P \qquad (16), (17)$$

where n – bond average repetition factor, K – hybridization coefficient which usually equals the number of registered atom valence electrons. Here the P-parameter of energy characteristic of the given component structural interaction in the process of binary bond formation. "Non-valence, non-chemical weak forces act … inside biological molecules and between them apart from strong interactions" [13].

Table 1. P-parameters of atoms calculated via the bond energy of electrons

Atom	Valence electrons	W (eV)	r_i (Å)	q^2_0 (eVÅ)	P_0 (eVÅ)	R (Å)	P_0/R (eV)
H						0.5292	9.0644
	$1S^1$	13.595	0.5295	14.394	4.7985	0.28	17.137
						R^-_I=1.36	3.525
C	$2P^1$	11.792	0.596	35.395	5.8680	0.77	7.6208
						0.67	8.7582
	$2P^2$	11.792	0.596	35.395	10.061	0.77	13.066
						0.67	15.016
	$2S^2$				14.524	0.77	18.862
	$2S^2+2P^2$				24.585	0.77	31.929
					24.585	0.67	36.694
N	$2P^1$	15.445	0.4875	52.912	6.5916	0.70	9.4166
	$2P^2$				11.723	0.70	16.747
	$2P^3$				15.830	0.70	22.614
						0.55	28.782
	$2S^2$	25.724	0.521	53.283	17.833	0.70	25.476
	$2S^2+2P^3$				33.663	0.70	48.09
O	$2P^1$	17.195	0.4135	71.383	6.4663	0.66	9.7979
	$2P^1$					R_I=1.36	4.755
	$2P^1$					R_I=1.40	4.6188
	$2P^2$	17.195	0.4135	71.383	11.858	0.66	17.967
						0.59	20.048
						R_I=1.36	8.7191
						R_I=1.40	8.470
	$2P^4$	17.195	0.4135	71.383	20.338	0.66	30.815
						0.59	34.471
	$2S^2$	33.859	0.450	72.620	21.466	0.66	32.524
	$2S^2+2P^4$				41.804	0.66	63.339
						0.59	70.854
Ca	$4S^1$	5.3212	1.690	17.406	5.929	1.97	3.0096
	$4S^2$				8.8456	1.97	4.4902
	$4S^2$					R^{2+}=1.00	8.8456
	$4S^2$					R^{2+}=1.26	7.0203
P	$3P^1$	10.659	0.9175	38.199	7.7864	1.10	7.0785
	$3P^1$					R^{3-}=1.86	$P_Э$=4.1862
	$3P^3$	10.659	0.9175	38.199	16.594	1.10	15.085
	$3P^3$					R^{3-}=1.86	8.9215
	$3S^2+3P^3$				35.644	1.10	32.403
Mg	$3S^1$	6.8859	1.279	17.501	5.8568	1.60	3.6618
	$3S^2$				8.7787	1.60	5.4867
						R^{2+}=1.02	8.6066
Mn	$4S^1$	6.7451	1.278	25.118	6.4180	1.30	4.9369
	$4S^1+3d^1$				12.924	1.30	9.9414
	$4S^2+3d^2$				22.774	1.30	17.518
Na		4.9552	1.713	10.058	4.6034	1.89	2.4357
	$3S^1$					R^{1+}_I=1.18	3.901
						R^{1+}_I=0.98	4.6973
K	$4S^1$	4.0130	2.612	10.993	4.8490	2.36	2.0547
						R^{1+}_I=1.45	3.344

Table 2. Structural P_S-parameters calculated via the bond energy of electrons

Radicals, molecule fragments	$P_i'(eV)$	$P_i''(eV)$	$P_S(eV)$	Orbitals
OH	9.7979 17.967	9.0644 17.138	4.7080 8.7712	O ($2P^1$) O ($2P^2$)
H_2O	2·9.0644	17.967	9.0227	O ($2P^2$)
CH_2	17.160 31.929	2·9.0644 2·17.138	8.8156 16.528	C ($2S^1 2P^3_r$) C ($2S^2 2P^2$)
CH_3	15.016 40.975	3·9.0644 3·9.0644	9.6740 16.345	C ($2P^2$) C ($2S^2 2P^2$)
CH	31.929	12.792	9.1330	C ($2S^2 2P^2$)
NH	16.747 19.538	17.138 17.132	8.4687 9.1281	N($2P^2$) N($2P^2$)
NH_2	19.538 28.782	2·9.0644 2·17.132	9.4036 18.450	N($2P^2$) N($2P^3$)
CO–OH	8.4405	8.7710	4.3013	C($2P^2$)
C=O	15.016	20.048	8.4405	C($2P^2$)
C=O	31.929	34.471	16.576	O($2P^4$)
CO=O	36.694	34.471	17.775	O($2P^4$)
C–CH_3	17.435	19.694	9.2479	–
C–NH_2	17.435	18.450	8.8844	–
CO–OH	12.315	8.7712	5.1226	C($2S^2 2P^2$)
(HP)O_3	23.122	23.716	11.708	O($2P^2$) P($3S^2 3P^3$)
(H_3P)O_4	17.185	17.244	8.6072	O($2P^1$) P($3P^1$)
(H_3P)O_4	31.847	31.612	15.865	O($2P^2$) P($3S^2 3P^3$)
H_2O	2·4.3623	8.7191	4.3609	O($2P^2$) r =1.36 Å
H_2O	2·4.3623	4.2350	2.8511	O($2P^2$) r=1.40
C-H_2O	2.959	2.8511	1.4520	–
(C-H_2O)$_3$ Lactic acid	–	–	1.4520·3= 4.3563	–
(C-H_2O)$_6$ Glucose	–	–	1.4520·6= 8.7121	–

At the same time, the orientation, induction and dispersion interactions are used to be called Van der Waals. For three main biological atoms (nitrogen, phosphorus and oxygen) Van der Waals radii numerically equal approximately the corresponding ionic radii (Table 3).

It is known that one of the reasons of relative instability of phosphorus anhydrite bonds in ATP is the strong repulsion of negatively charged oxygen atoms. Therefore it is advisable to use the values of P-parameters calculated via Van der Waals radii as the energy characteristic of weak structural interactions of biomolecules (Table 3).

Bond energies for P and O atoms were calculated taking into account Van der Waals distances for atomic orbitals: $3P^1$ (phosphorus)-$2P^1$ (oxygen) and for $3P^3$ (phosphorus)-$2P^2$ (oxygen). The values of E obtained slightly exceeded the experimental, reference ones (Table 4). But for the actual energy physiological processes, e.g. during photosynthesis, the efficiency is below the theoretical one, being about 83 %, in some cases [14,15].

Perhaps the electrostatic component of resulting interactions at anion-anionic distances is considered in such a way. Actually the calculated value of 0.83E practically corresponds to the experimental values of bond energy during the phospholyration and free energy of ATP in chloroplasts. Table 4 contains the calculations of bond energy following the same technique but for stronger interactions at covalent distances of atoms for the free molecule P$\overset{...}{-}$O (sesquialteral bond) and for the molecule P=O (double bond).

The sesquialteral bond was evaluated by introducing the coefficient n = 1.5 with the average value of oxygen P_E-parameter for single and double bonds.

The average breaking energy of the corresponding chemical bonds in ATP molecule obtained in the frameworks of semi-empirical method PM3 with the help of software GAMESS [16] are given in column 11 of Table 4 for comparison. The calculation technique is detailed in [17].

The calculated values of bond energies in the system K-C-N being close to the values of high-energy bond P$\sim$O in ATP demonstrate that such structure can prevent the ATP synthesis. When evaluating the possibility of hydrogen bond formation, we take into account such value of in which K =1, and the value $n = 3.525/17.037$ characterizes the change in the bond repetition factor when transiting from the covalent bond to the ionic one.

5. Formation of Stable Biostructures

At equilibrium-exchange spatial-energy interactions similar to isomorphism the electrically neutral components do not repulse but approach each other and form a new composition whose in the equations (13,14).

This is the first stage of stable system formation by the given interaction type which is carried out under the condition of approximate equality of component P-parameters: $P_1 \approx P_2$.

Hydrogen atom, element No 1 with the orbital $1S^1$ determines the main criteria of possible structural interactions. Four main values of its P-parameters can be taken from Tables 1 and 3:

1) for strong interactions: = 9.0644 eV with the orbital radius 0.5292 Å and = 17.137 eV with the covalent radius 0.28 Å.

2) for weaker interactions: = 4.3623 eV and P_E = 3.6352 eV with Van der Waals radii 1.10 Å and 1.32 Å. The values of P-parameters $P' : P'' : P'''$ relates as 1:2:4. In accordance with the concepts in Section 2, such values of interaction P-parameters define the normative functional states of biosystems, and the intermediary can produce pathologic formations by their values.

The series with approximately similar values of P-parameters of atoms or radicals can be extracted from the large pool of possible combinations of structural interactions (Table 5). The deviations from the initial, primary values of P-parameters of hydrogen atom are in the range $\pm$ 7%.

The values of P-parameters of atoms and radicals given in the Table define their approximate equality in the directions of interatomic bonds in polypeptide, polymeric and other multi-atom biological systems.

In ATP molecule these are phosphorus, oxygen and carbon atoms, polypeptide chains – CO, NH and CH radicals. In Table 5 you can also see the additional calculation of their bond energy taking into account the sesquialteral bond repetition factor in radicals C⋯O and N⋯H.

On the example of phosphorus acids it can be demonstrated that this approach is not in contradiction with the method of valence bonds which explains the formation peculiarities of ordinary chemical compounds. It is demonstrated in Table 6 that this electrostatic equilibrium between the oppositely charged components of these acids can correspond to the structural interaction for H_3PO_4 $3P^1$ orbitals of phosphorus and $2P^1$ of oxygen, and for HPO_3 $3S^23P^3$ orbitals of phosphorus and $2P^2$ of oxygen. Here it is stated that P-parameters for phosphorus and hydrogen subsystems are added algebraically. It is also known that the ionized phosphate groups are transferred in the process of ATP formation that is apparently defined for phosphorus atoms by the transition from valence-active $3P^1$ orbitals to $3S^23P^3$ ones, i.e. 4 additional electrons will become valence-active. According to the experimental data the synthesis of one ATP molecule is connected with the transition of four protons and when the fourth proton is being transited the energy accumulated by the ferment reaches its threshold [18,19]. It can be assumed that such proton transitions in ferments initiate similar changes in valence-active states in the system P-O. In the process of oxidating phospholyration the transporting ATP-synthase uses the energy of gradient potential due to $2H^+$-protons which, in the given model for such a process, corresponds to the initiation of valence-active transitions of phosphorus atoms from $3P^1$ to $3P^3$-state.

In accordance with the equation (17) we can assume that in stable molecular structures the condition of the equality of corresponding effective interaction energies of the components by the couple bond line is fulfilled by the following equations:

$$\left(P_E \frac{n}{K}\right)_1 \approx \left(P_E \frac{n}{K}\right)_2 \rightarrow P_1 \approx P_2 \tag{18}$$

And for heterogeneous atoms (when $n_1 = n_2$):

$$\left(P_E/K\right)_1 \approx \left(P_E/K\right)_2 \tag{18a}$$

In phosphate groups of ATP molecule the bond main line comprises phosphorus and oxygen molecules. The effective energies of these atoms by the bond line calculated by the equation (18) are given in Tables 4 and 5, from which it is seen that the best equality of P_1 and P_2 parameters is fulfilled for the interactions $P(3P^3)$ – 8.7337 eV and $O(2P^2)$ – 8.470 eV that is defined by the transition from the covalent bond to Van der Waals ones in these structures. The resulting bond energy of the system P–O for such valence orbitals and the weakest interactions (maximum values of coefficient K) is 0.781 eV (Table 4). Similar calculations for the interactions $P(2P^1)$ – 4.0981 eV and $O(2P^1)$ – 4.6188 eV produce the resulting bond energy 0.397 eV.

Table 3. Ionic and Van der Waals radii (Å)

Atom	Ionic radii			Van der Waals radii		
	Orbital	R_I	P_E/K (eV)	R_B	Orbital	P_E/K (eV)
H	$1S^1$	R^-=1.36 r =0.5292	3.525 9.0644	1.10 1.32	$1S^1$	4.3623 3.6352
N	$2P^3$	R^{3-}=1.48	10.696/3=3.5653	1.50 1.50 1.50	$2P^1$ $2P^3$ $2S^22P^3$	4.3944/1 10.553/3=3.5178 22.442/5=4.4884
P	$3P^3$	R^{3-}= 1.86	8.9215/3=2.9738	1.9 1.9 1.9	$3P^1$ $3P^3$ $3S^23P^3$	4.0981/1 8.7337/3=2.9112 18.760/5=3.752
O	$2P^2$	R^{2-}=1.40 R^{2-}=1.36	8.470/2=4.2350 8.7191/2=4.3596	1.40 1.50 1.40 1.50	$2P^1$ $2P^1$ $2P^2$ $2P^2$	4.6188/1 4.3109/1 8.470/2=4.2350 7.9053/2=3.9527
C	$2S^22P^2$	d*/2=3.2/2=1.6	15.365/4=3.841	1.7 1.7 1.7	$2P^1$ $2P^2$ $2S^22P^2$	3.4518/1 5.9182/2=2.9591 14.462/4=3.6154

d* – contact distance between C-C atoms in polypeptide chains [13].

The difference in these values of bond energies is defined by different functional states of phosphorous acids HPO_3 and H_3PO_4 in glycolysis processes and equals 0.384 eV that is close to the phospholyration value (0.34 – 0.35 eV) obtained experimentally.

Such ATP synthesis is carried out in anaerobic conditions and is based on the transfer of phosphate residues onto ATP via the metabolite. For example: ATP formation from creatine phosphate is accompanied by the transition of its NH group at ADP to NH_2 group of creatine at ATP.

From Table 4 it is seen that the change in the bond energy of these two main radicals of metabolite is 5.3238 – 4.9654 = 0.3584 eV – taking the sesquialteral bond N⋯H into account (as in polypeptides) and 5.3238 – 4.6186 = 0.7052 eV – for the single bond N-H. This is one of the intermediary results of the high-energy bond transformation process in ATP through the metabolite.

Table 4. Bond energy (eV)

Atoms, structures, orbitals	Bond	Component 1		Component 2		Component 3		Calculation	E [13,14,15]	E [16] [17]	Remarks
		P_E (eV)	n/K	P_E (eV)	n/K	P_E (eV)	n/K	E			
1	2	3	4	5	6	7	8	9	10	11	12
P⋯O $3S^23P^3$- $2S^22P^4$	cov.	32.403	1.5/5	70.854 63.339	1.5/6 1.5/6	6.14		6.277 6.024 <6.15>	6.1385 6.14		PO free molecule
H_2O $1S^1$-$2P^2$	cov. cov.	2 × 9.0624 2 × 9.0624	1/1 1/1	17.967 20.048	1/6 2/2			2.570 9.520	2.476	10.04	Decay of one molecule
H_3PO_4	cov.	3 × 9.0624	1/1	32.405	1/5	4 × 17.967	1/2	4.8779	4.708		
1	2	3	4	5	6	7	8	9	10	11	12
C⋯O ($2P^1$-$1S^1$)	cov.	7.6208	1.125/2	9.7979	1/1			4.2867			
C-N $2P^1$-$2P^1$	cov.	7.6208	1/4	9.4166	1/5			0.9471			
C⋯N $2P^1$-$2P^1$	cov.	7.6208	1.125/4	9.4166	1.1667/5			1.0898	0.870		
K-C-N $4S^2$-$2P^1$-$2P^1$	cov.	2.0547	1/1	7.6208	1/4	9.4166	1/5	0.648			
(C-H_2O)-(C-H_2O)	VdW	1.4520	1/1	1.4520	1/1			0.726			
C⋯O $2S^22P^2$-$2P^2$	cov.	31.929 31.929	1.125/4 1/4	20.048 20.042	1/2 1/2			4.7367 4.4437			
N⋯H $2P^1$-$1S^1$	cov.	9.4166 9.4166	1.1667/1 1/1	9.0644 9.0644	1/1 1/1			4.9654 4.6186			

Table 4. (Continued)

Atoms, structures, orbitals	Bond	Component 1		Component 2		Component 3		Calculation	E [13,14,15]	E [16] [17]	Remarks
		P_E (eV)	n/K	P_E (eV)	n/K	P_E (eV)	n/K	E			
C-H $2P^1$-$1S^1$	cov.	13.066	1/2	9.0644	1/1			3.797	3.772		
C-H $2P^2$-$1S^1$		13.066	1/2	17.137	1/1			4.7295			
N-H_2 $2P^3$-$1S^1$	cov.	22.614	1/3	2×9.0644	1/1			5.3238			
-H···O		3.525	3.525/17.037	4.6188	1/6			0.3730	0.3742		Hydrogen bond
P=O $3P^3$-$2P^2$	cov.	15.085	2/3	20.042	2/2			6.6970	6.504	6.138 5	Free molecule
P-O $3P^3$-$2P^2$	VdW	8.7337	1/5	8.470	1/6			0.781	0.670		ΔG ATP
P-O $3P^1$-$2P^1$	cov.	7.0785	1/1	9.7979	1/1			4.1096	4.2059	4.293 1	
P-O $3P^1$-$2P^1$	VdW	4.0981	1/5	4.6188	1/6			0.3970	0.34-0.35		Phospholy ration

Table 5. Bio-structural spatial-energy parameters (eV)

Series No	H	C	N	O	P	CH	CO	NH	Glucose	Lactic acid	OH	Remarks
I	9.0644 ($1S^1$)	8.7582 ($2P^1$) 9.780 ($2P^1$)	9.4166 ($2P^1$)	9.7979 ($2P^1$)	8.7337 ($3P^3$)	9.1330 ($2S^22P^2$–$1S^1$)	8.4405 ($2P^2$–$2P^2$)	8.4687 ($2P^2$–$1S^1$) 9.1281 ($2P^2$–$1S^1$)	8.7121 $2P^2$– –($1S^1$–$2P^2$)		8.7710	Strong interaction
II	17.132 ($1S^1$)	17.435 ($2S^12P^1$)	16.747 ($2P^2$)	17.967 ($2P^2$)	18.760 ($3S^23P^3$)	C and H blocks	16.576 ($2S^22P^2$–$2P^4$)	N and H blocks				Strong interaction
III	(4.3623) ($1S^1$)	3.8696 ($2P^2$)	4.3944 ($2P^1$)	4.3109 ($2P^1$) 4.6188 ($2P^1$)	4.0981 ($3P^1$)	4.7295	4.4437 4.7367	4.6186 4.9654		4.3563 $2P^2$– ($1S^1$–$2P^2$)	4.7084	Weak interaction
IV	3.6352 ($1S^1$)	3.4518 ($2P^1$) 3.6154 ($2S^22P^2$)	3.5178 ($2P^3$)	4.2350 ($2P^2$) 3.6318 ($2P^4$)	4.0981 ($3P^1$) 3.752 ($3S^23P^3$)	4.7295	4.4437 4.7367	4.6186 4.9654				Effective bond energy

Table 6. Structural interactions in phosphorus acids

Molecule	Component 1			Component 2			$\alpha=(\Delta P/\langle P\rangle)*100\%$
	Atom	Orbitals	$P=P_1+P_2$ (eV)	Atom	Orbitals	P (eV)	
$(H_3P)O_4$	H_3P	$1S^1 - 3P^1$ $1S^1$-($3S^23P^3$)	4.3623*3+ 4.0981=17.185 4.3623*3+18.760=31.847	O_4 at r=1.50Å	$2P^1$ $2P^2$	4.3109*4=17.244 7.9053*4=31.612	0.34 0.74
$(HP)O_3$	HP	$1S^1$ – ($3S^23P^3$)	4.3623+18.760=23.122	O_3 at r=1.50Å	$2P^2$	7.9053*3=23.716	2.54

From Tables 4 and 6 we can conclude that the phosphorous acid H_3PO_4 can have two stationary valence-active states during the interactions in the system P-O for the orbitals with the values of P-parameters of weak and strong interactions, respectively. This defines the possibility for the glycolysis process to flow in two stages. At the first stage, the glucose and H_3PO_4 molecules approach each other due to similar values of their P-parameters of strong interactions (Table 2).

At the second stage, H_3PO_4 P-parameter in weak interactions 4.8779 eV (Table 4) in the presence of ferments provokes the bond (H_2O-C)-(C-H_2O) breakage in the glucose molecule with the formation of two molecules of lactic acid whose P-parameters are equal by 4.3563 eV. The energy of this bond breakage process equalled to 0.726 eV (Table 4) is realized as the energy of high-energy bond it ATP.

According to the reference data about 40% of the glycolysis total energy, i.e. about 0.83 eV, remains in ATP.

By the hydrolysis reaction in ATP in the presence of ferments

$$(HPO_3 + H_2O \rightarrow H_3PO_4 + E)$$

for structural P_S-parameters (Table 2) E = 11.708 + 4.3609 – 15.865 = 0.276 eV.

It is known that the change in the free energy (ΔG) of hydrolysis of phosphorous anhydrite bond of ATP at pH = 7 under standard conditions is 0.311 – 0.363 eV. But in the cell the ΔG value can be much higher as the ATP and ADP concentration in it is lower than under standard conditions. Besides, the ΔG value is influenced by the concentration of magnesium ions which is the acting co-ferment in the complex with ATP. Actually Mg^{2+} ion has the P_E-parameter equalled to 8.6066 eV (Table 1) which is very similar to the corresponding values of P-parameters of phosphorous and oxygen atoms.

The quantitative evaluation of this factor requires additional calculations.

CONCLUSION

1. Bond energies of some biostructures have been calculated following P-parameter and quantum-mechanical techniques.
2. High-energy bonds in ATP are formed in the system P-O under functional transitions of their valence-active states.
3. The data obtained agree with the experimental ones.

REFERENCES

[1] Eyring H., Walter J., Kimball G.E. *Quantum chemistry* // I. L., M., 1948, 528p.

[2] Korablev G.A. *Spatial-Energy Principles of Complex Structures Formation* // Brill Academic Publishers and VSP, Netherlands, 2005, 426pp. (Monograph).

[3] Fischer C.F. Average-Energy of Configuration Hartree-Fock Results for the Atoms Helium to Radon //*Atomic Data*, 1972, № 4, pp. 301-399.

[4] Waber J.T., Cromer D.T. Orbital Radii of Atoms and Ions // *J. Chem. Phys.*, 1965, V 42, №12, pp. 4116-4123.

[5] Clementi E., Raimondi D.L. Atomic Screening constants from S.C.F. Functions, 1 // *J.Chem. Phys.*, 1963, V.38, №11, pp. 2686-2689.

[6] Clementi E., Raimondi D.L. Atomic Screening constants from S.C.F. Functions, 1 // *J.Chem. Phys.*, 1967, V.47, №14, pp. 1300-1307.

[7] Gombash P. *Atom statistical model and its application* // M.: I. L., 1951, 398 p.

[8] Clementi E.Tables of atomic functions // J.B.M. S. Re./ *Develop. Suppl.*, 1965, V. 9, №2, 76.

[9] Korablev G.A., Zaikov G.E. Spatial-Energy Parameter as a Materialised Analog of Wafe Function // *Progress on Chemistry and Biochemistry*, Nova Science Publishers, Inc. New York, 2009, 355-376.

[10] Korablev G.A., Zaikov G.E. Energy of chemical bond and spatial-energy principles of hybridization of atom orbitalls // *J. of Applied Polymer Science*, V.101, №3, Ang.5, 2006, pp. 2101-2107.

[11] Korablev G.A., Zaikov G.E. Formation of carbon nanostructures and spatial-energy criterion of stabilization // *Mechanics of composite materials and structures*, 2009, RAS, v.15, №1, 106-118.

[12] Batsanov S.S. *Structural chemistry. Facts and dependencies*. M.: MSU, 2009.

[13] Volkenshtein. *Biophysics.* M.: Nauka, 1988, 598 p.

[14] Photosynthesis / Ed. by Govindzhi. M.: Mir, v.1, 1987, 728 p; v.2, 1987, 460 p.

[15] Clayton R. Photosynthesis. *Physical mechanisms and chemical models*. M.: Mir, 1984, 350 p.

[16] Schmidt M.W., Baldridge K.K., Boatz J.A. et al. General atomic and molecular electronic structure system // *J.Comput. Chem.*, 1993, v.14, p. 1347-1363.

[17] Khokhriakov N.V., Kodolov V.I. Influence of active nanoparticles on the structure of polar liquids // *Chemical physics and mesoscopy*, 2009, v. 11. № 3, p. 388-402.

[18] Feniouk B.A. Study of the conjugation mechanism of ATP synthesis and ATP proton transport // *Referun - Biology and natural science*, 1998, 108 p.

[19] Feniouk B.A., Junge W., Mulkidjanian A. Tracking of proton flow across the active ATP-synthase of Rhodobacter capsulatus in response to a series of light flashes. *EBEC Reports,* Volume 10, p. 112, 1998.

In: Biology and Communication in the Information Age
Editor: Rafiqul Islam
ISBN: 978-1-62808-252-4

Chapter 19

QUANTUM MODELING OF THE KINETICS AND CHEMICAL MECHANISM OF THE METHYL ACETATE HYDROLYSIS CATALYZED BY CHYMOTRYPSIN AND SERINE

A. A. Turovsky[1]*, A. R. Kytsya[2], L. I. Bazylyak[2] and G. E. Zaikov[3]
[1]Chemistry and Biotechnology of Combustible Minerals Division,
Physical Chemistry of Combustible Minerals Department,
Institute of Physical–Organic Chemistry and Coal Chemistry named after
L. M. Lytvynenko; National Academy of Science of Ukraine, Lviv, Ukraine
[2]Chemistry of Oxidizing Processes Division,
Physical Chemistry of Combustible Minerals Department,
Institute of Physical–Organic Chemistry and Coal Chemistry named after
L. M. Lytvynenko; National Academy of Science of Ukraine, Ukraine
[3]Kinetics of Chemical andBiological Processes Division,
Institute of Biochemical Physics named after N. N. Emanuel;
Russian Academy of Sciences, Moscow, Russia

ABSTRACT

It has been done the quantum−chemical modeling of the methyl acetate hydrolysis reaction in the processes of enzymatic and homogeneous catalysis by Chemotrypsin and Serine respectively It was shown, that the structure of an active center of the Chemotrypsin taken in references is proved by our calculations. It were estimated the kinetic parameters for the reactions of acyl compounds formation and also of their hydrolysis for both types of the catalysis. It was shown, that the formation of acylchemotrypsin proceeds via a series of the intermediate complexes, caused by the conformation transition, which, in a great measure, caused by the hydrogen bonds of the Chemotrypsin and depend on the journey of Serine active group hydrogen atom. It was established, that the stabilization of the activated complexes of the acidulating and

* E–mail: hop_vfh@ukr.net.

deacidulating reactions in enzymatic and homogeneous catalysis in a great measure depends on a value of the charges of the reactive center, that is indicated on the values of the kinetic parameters of such reactions. It has been calculated the thermodynamics of the elementary stages for the acidulating and deacidulating reactions.

Keywords: Quantum–chemical modeling, enzymatic catalysis, homogeneous catalysis, methyl acetate hydrolysis, acidulating and deacidulating reactions

1. INTRODUCTION

Especially great successes under investigation of the driving forces of the enzymatic catalysis were achieved in a case of the *Chemotrypsin*. *Chemotrypsin* – this is endopeptidase, which cleaves the peptide bonds into peptides. The most important information about the structure of the *Chemotrypsin's* molecule has been obtained with the use of the roentgen investigations [**1–4**]. It was found, that the all charged groups into the molecule of the ferment are directed sideways to the aqueous solution (with the except for the three, which have the special functions into a mechanism action of the active center). The successes in kinetic investigations in the most cases were caused by the works of *M. Bergmann*, *D. Frugonn* and *H. Neyrag*, who determined that the *Chemotrypsin* can hydrolyze also the simple low–molecular products (amides, esters).

The force of the *Chemotrypsin* catalytic action under the esters hydrolysis approximately in 10^6 times exceeds the catalytic action both of OH^-, and H_3O^+.

Hydrolysis of the substrate (amides, esters) on the active center of the *Chemotrypsin* proceeds in some stages. The first stage of the enzymatic process includes the sorption (so–called formation of the *Michaels's* complex *ES*). The next stages include the chemical transformation of the sorbed molecule with the formation of the intermediate compound of the acyl–ferment *EA* in accordance with the following kinetic scheme:

$$E + S \underset{k_{-1}}{\overset{k_1}{\Leftrightarrow}} ES \underset{\downarrow P_1}{\overset{k_2}{\rightarrow}} EA \underset{\uparrow H_2O}{\overset{k_3}{\rightarrow}} E + P_2 \qquad (1)$$

$$K_S = \frac{k_{-1}}{k_1}$$

P_1 and P_2 are products of the hydrolysis.

The equilibrium position is determined only by the non–valence interactions with the protein of side chemically inert fragments of the substrate's molecule. The intermediate product represents by itself the acyl–ferment, which is unstable compound (lifetime ~ 0,01 *s*. [**5**]).

Finally, the enzymatic hydrolysis can be presented in accordance with the following scheme:

$$\underset{(E)}{EH} + \underset{(S)}{RC(O)OCH_3} \overset{K_S}{\Leftrightarrow} \underset{(E \cdot S)}{HE \cdot RC(O)OCH_3} \tag{2}$$

$$E \cdot S \overset{k_2}{\rightarrow} \underset{(EA)}{E - C(O)R} + \underset{(P_1)}{CH_3OH} \tag{3}$$

$$EA \underset{H_2O}{\overset{k_3}{\rightarrow}} E + RCOOH \tag{4}$$

However, the stages (3) and (4) are not elementary [6–9] and include the quick (and equilibrium) formation of the intermediate position, which corresponds to a new conformational state of the ferment.

Under hydrolysis of the molecule of a substrate which is sorbing on an active center, the *OH*–group of the *Serine* stands out as the attacking nucleophile [6, 10–12]. It is assumed that the high activity of the *Serine* related with its surroundings into the active center. Along with the *Serine,* the imidazole group of *His* also takes part in its activity [6, 10–11, 13]. At this, the *Nitrogen* atom of the *Histidine* forms the hydrogen bond with *Oxygen* of the *Serine* hydroxyl. Accordingly to *Blow* [14] the second hydrogen bond exists between the atoms *N* and *Histidine* and carbonyl group of the remains of *Asp*, which is located into the depths of the ferment globule. The system of the hydrogen bonds leads to increase of the negative charge on the *OH* group of the *Serin* that promotes to its nucleophilicity.

On the acidulating stage the nucleophilic attack of the carboxyl *Carbon* of the substrate by generalized nucleophile of active center: *Ser*, *His*, *Asp* proceeds. As a result of the active center acidulating, the turn of the *Ser* remains around the bonds C_α–C_β which is accompanying by the displacement of the *Oxygen* atom on ~ 2,5 *A,* takes place. At this, imidazole group of the *His* displaces sideways to the solvent [3]. As a result, the imidazole group of the *His* is included into a free ferment (and into the *Michaels's* complex) in hydrogen bond. *Ser* in acyl–ferment contributes one's own atom *N* for the formation of the hydrogen bond with water. As a result, the activated molecule of water has the ability to effectively attack the carbonyl *Carbon* of the substrate on the deacidulating stage. At this, it is formed the product of the hydrolysis and it is regenerated a free ferment. The above–said chemical mechanism of the hydrolytic action of *Chemotrypsin* is described in such a way in references. We were interested into the analysis of such approach to the enzymatic catalysis with the use of the quantum–chemical calculations.

2. Experimental Techniques

The aim of the presented work was the modeling of the kinetics and chemical mechanism of the enzymatic catalysis process for reaction of hydrolysis of methyl acetate by *Chemotrypsin* with the use of the quantum–chemical method.

Firstly, it was interesting to consider the stage–by–stage modeling of the process in accordance with the well–known schemes accepted in references and, as far as possible, to append them.

Secondly, it was interesting to estimate the kinetic parameters of the enzymatic catalysis reaction.

Thirdly, it was interesting to compare the kinetics of processes of enzymatic and homogene-ous catalysis of methyl acetate.

For quantum–chemical calculations the semiempirical method *PM*6 was used.

Modeling Objects

- methyl acetate;
- generally accepted active form of *Chemotrypsin* containing of three fragments (Figure 1).

On Figure 1 presented also the geometry and electronic characteristics of *Chemotrypsin*.

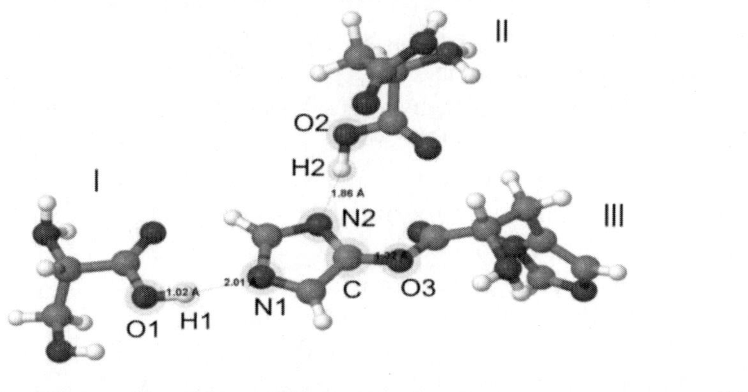

Charge on	H1	O1	N1	H2	O2	N2	C	O3
	+0,385	−0,546	−0,363	+0,403	−0,515	−0,474	+0,338	−0,429

Distance	O1 – H1	N1 – H1	O2 – H2	N2 – H2	C – O3
A	1,02	2,01	1,05	1,86	1,32

Figure 1. Geometry and electronic characteristics of the Chemotrypsin molecule: I – Ser; II – Asp; III – His. Ionization potential 9,605 eV; electron affinity: −2,601. Angle O2−N2−O3 104,8^0 eV.

3. RESULTS AND DISCUSSION

It can be seen from the optimal geometry of the active fragment of *Chemotrypsin* (Figure 1), that the hydrogen bonds correspondingly to the distances exist between the atoms H1 and *N*1, and also between atoms *H*2 and *N*2. Probably, the second bond is stronger (the distance is shorter).

Let's note, that the electron affinity in the *Chemotrypsin* fragment is enough high. At the fitting of the methyl acetate molecule to the *Chemotrypsin* along the coordinate *C*1−*O*1 (*Figure* 2*a*) at the distance 1,60 *A* a change of the complex's geometry takes place, and also the elongation of hydrogen bond *O*1…*H*…*N*1 takes place, that is caused by the conformational transformations of the *Chemotrypsin* molecule. Such conformational transformation proceeds with enough great energy charge (see peak № 1 on Figure 3).

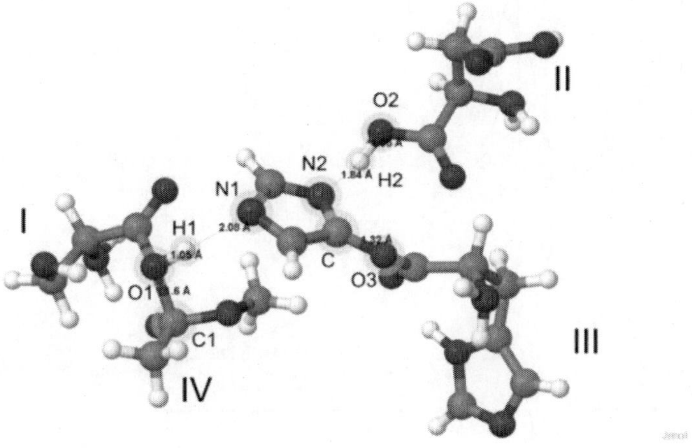

(a)

Distance *C1 – O1*: 1,60 *A*. Angle *O2 – N2 – O3*: $104,7^0$

Charge on	H1	O1	N1	C1	H2	O2	N2	C	O3
	+0,399	−0,531	−0,358	+0,758	+0,401	−0,519	−0,476	+0,339	−0,432

Distance	O1 – H1	N1 – H1	O2 – H2	N2 – H2	C – O3
A	1,05	2,08	1,05	1,84	1,32

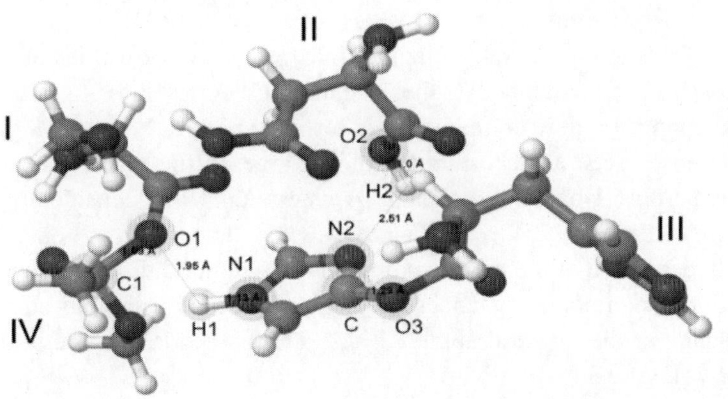

(b)

Distance *C1 – O1*: 1,59 *A*. Angle *O2 – N2 – O3*: $122,9^0$

Charge on	H1	O1	N1	C1	H2	O2	N2	C	O3
	+0,405	−0,706	−0,212	+0,811	+0,360	−0,500	−0,457	+0,423	−0,420

Distance	O1 – H1	N1 – H1	O2 – H2	N2 – H2	C – O3
A	1,95	1,13	1,00	2,51	1,29

Figure 2. Change of the geometry of complex in point 1 (correspondingly to Figure 3). **I** – Ser, **II** – Asp, **III** – His, **IV** – methyl acetate.

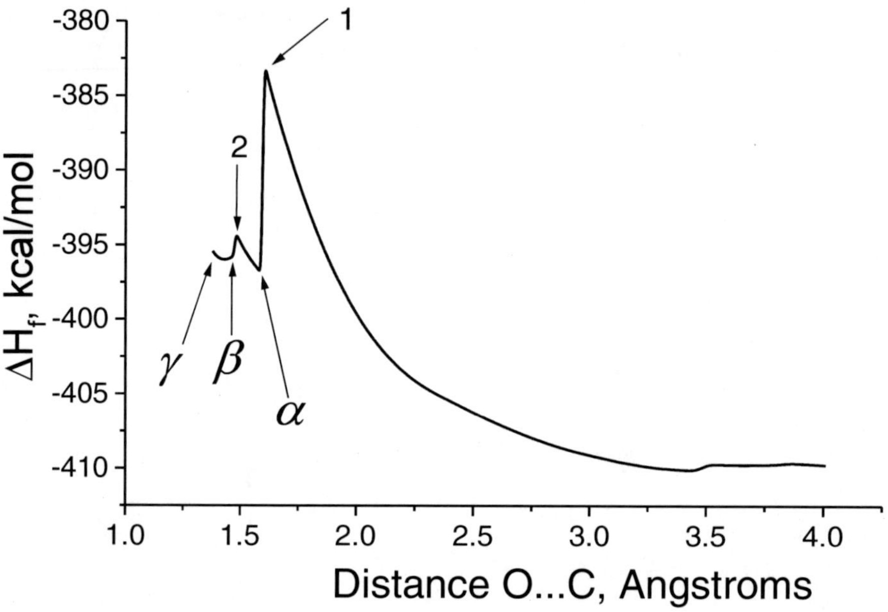

Figure 3. Formation of complex [Kh–CH_3CO]. **1** – complex coordination [Kh–CH_3COOCH_3] (1,60 A), transfer of Hydrogen atom from Serin to Nitrogen atom; **2** – detach of methanol (1,48 A).

It can be seen from Figure 2 (***a***), that at the distance *C*1−*O*1 1,60 *A* the charges on *H*1 and *O*1 are some increased; the charge on *N*1 is decreased. The change of angle *O*2−*N*2−*O*3 is insignificant. The interatomic distance *O*1−*H*1 is also enhanced. Insignificant change of a distance *C*1−*O*1 leads to considerable changes into geometry and values of charges on atoms of the *Chemotrypsin* fragment. At the distance *C*1−*O*1 1,59 *A* as a result of new conformational changes a new complex (α) is formed (Figure 2 (b) and Figure 3 (α)). From Figure 2 (b) it can be seen a great increase of a charge on the reactive center *O*1. It is also increased a charge on *C*1 of methyl acetate. A charge on *N*1 is considerably decreased as a result of a great decrease of a distance *N*1−*H*1. The charges on *N*2 and *H*2 are correspondingly decreased at the expense of the hydrogen bond distance increasing as a result of the change of the interatomic distances and angles of the surroundings atoms. It is observed an essential change of the dihedral angle *O*2−*N*2−*O*3 that leads to more stable conformation of *Chemotrypsin* (Figure 3 (α)).

At the fitting of atom *C*1 to *O*1 the atom O_4 (Figure 4 (a); Figure 3, peak № 2) is neared to the atom *H*1, and at the distance 1,48 *A* the transition of atom *H*1 to *O*4 and detach of methanol molecule from the complex take place. Such process is illustrated by a peak № 2 on Figure 3. For complex (Figure 4 (a)), which represents by itself the activated complex for the formation of the intermediate unstable compound, the charges on atoms H1 and N1 are increased the most essentially. The charges on C1 and O1 are decreased. As a result of the hydrogen atom to O4 of methyl acetate addition (Figure 4 (a)) the distances for N1−H1 are increased, and for N2−H2 are decreased correspondingly. In complex (β) (Figure 3) at distance C1−O1 = 1, 47 A the distances O4−H1, N2−H2 are considerably decreased and the interatomic distances C1−O4, N1−H1, C−O3 are increased as a result of the essential changes of the general geometry of complex.

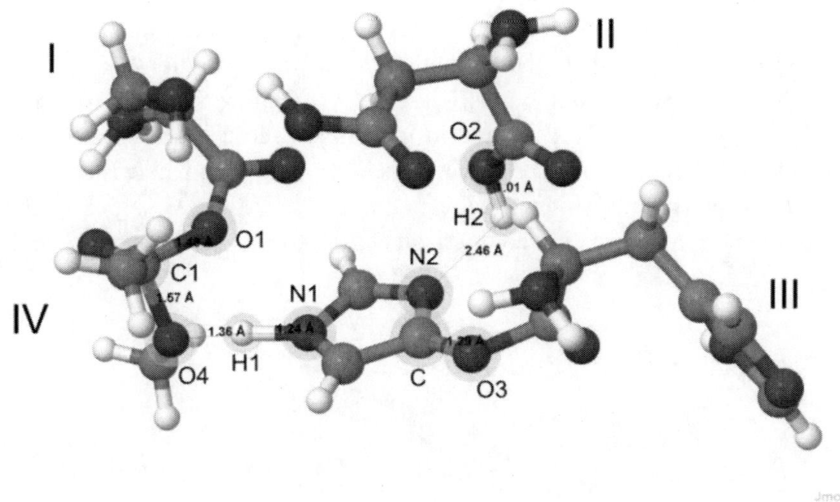

(a)

Distance $C1 – O1$: 1,48 A. Angle $O2 – N2 – O3$: 123,3^0

Charge on	H1	O1	N1	C1	O4	H2	O2	C	O3
	+0,420	−0,648	−0,280	+0,801	−0,550	+0,363	−0,502	+0,399	−0,425

Distance	O4 – H1	C1 – O4	N2 – H1	O2 – H2	N2 – H2	C – O3
A	1,36	1,57	1,24	1,01	2,46	1,29

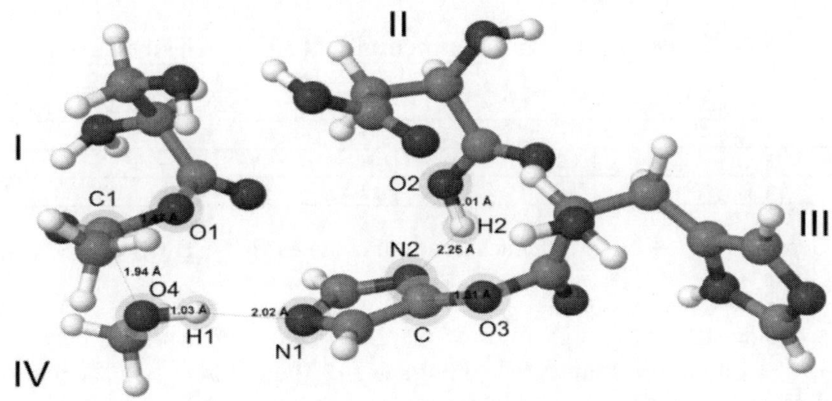

(b)

Distance $C1 – O1$: 1,47 A. Angle $O2 – N2 – O3$: 117,8^0

Charge on	H1	O1	N1	C1	O4	H2	O2	C	O3	N2
	+0,376	−0,623	−0,374	+0,794	−0,523	+0,378	−0,513	+0,349	−0,434	−0,481

Distance	O1 – H1	C1 – O4	N1 – H1	O2 – H2	N2 – H2	C – O3
A	1,03	1,94	2,02	1,01	2,25	1,31

Figure 4. The change of the geometry of complex in point 2. (a) I – Ser, II – Asp, III – His, IV – methyl acetate; (b) I – Ser + CH_3CO, II – Asp, III – His, IV – methanol.

At this, the charges on N4, N2 and H2 atoms are increased and on H1, O1, C1 and O4 atoms are decreased. The angle O2–N2–O3 is essentially changed. The complex (β) (Figure 3) transfers into the intermediate compound via the conformation (γ) (Figure 3). The formation of the intermediate compound [Kh–CH_3CO] (Figure 5) leads to decrease of the charges on atoms C1, O1, N1, N2, O3 and to increase of the charges on atoms N2, O2, H2 and O1. Interatomic distance C1–O1 for intermediate compound consists of 1,39 *A*. The distance *N2…H2* is considerably increased. The angle *O2 – N2 – O3* is greatly changed. The electron affinity is decreased in comparison with the starting active center of the *Chemotrypsin*.

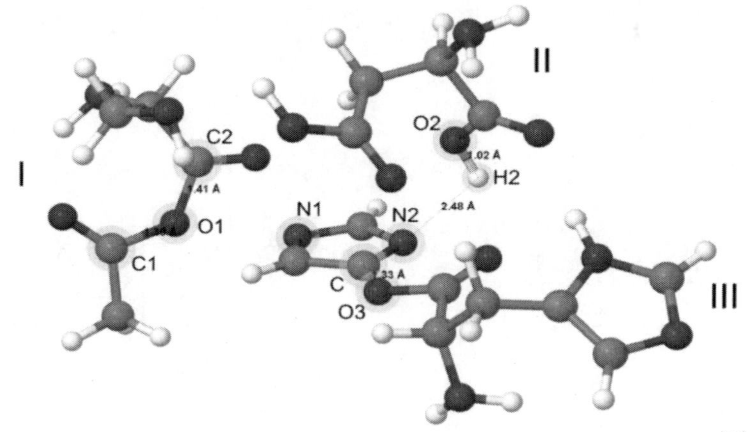

Angle O2 – N2 – O3: 83,20. Ionization potential: 9,050 eV. Electron affinity: –2,120 eV

Charge on	C2	O1	N1	C1	H2	O2	C	O3	N2
	+0,595	−0,559	−0,324	+0,717	+0,388	−0,520	+0,302	−0,412	−0,419

Distance	O1 – C1	C2 – O1	O2 – H2	N2 – H2	C – O3
A	1,39	1,41	1,02	2,48	1,33

Figure 5. Geometry and electronic characteristics of complex [Kh–CH_3CO]. I – Ser + CH_3CO, II – Asp, III – His.

Let's consider the thermodynamics of the enzymatic catalysis by *Chemotrypsin* for the reaction of *C–O* bond breakdown in methyl acetate. Below there is a scheme of a process, thermodynamics of which calculated without taking into account the interactions of complexes with the methanol (1) and acetic acid (2):

$$Kh + CH_3COOCH_3 \rightarrow [Kh{-}CH_3CO] + CH_3OH \quad (1)$$

(ΔH_p = + 12,9 *kilocalorie/mole*; ΔS_p = + 6,0 *calorie/mole·K*; ΔF_p = + 14,7 *kilocalorie/mole*)

$$[Kh{-}CH_3CO] + H_2O \rightarrow Kh + CH_3COOH \quad (2)$$

(ΔH_p = − 10,6 *kilocalorie/mole*; ΔS_p = + 10,1 *calorie/mole·K*; ΔF_p = − 13,6 *kilocalorie/mole*)

Here: Kh – Chemotrypsin; Kh–CH_3CO – Acylchemotrypsin
$\Sigma\Delta F$ = + 1,1 *kilocalorie/mole; K = exp(−ΔF/RT)* = 0,2.

Thermodynamic characteristics of the starting substances, of the reaction products and of the intermediate compound (*Acylchemotrypsin*) are represented in Table 1. As we can see from the calculations, the *Acylchemotrypsin* formation process is endothermic. Such intermediate compound thermodynamically is not stable and is easy transformed into the final products of the deacidulating reaction.

Table 1. Thermodynamic characteristics of the reagents and of the products of the methyl acetate hydrolysis reaction

№	Formula / substance	ΔH, kilocalorie/mole	ΔS, calorie/mole·K	ΔF, kilocalorie/mole
1	Chemotrypsin	−306,6	238,9	−377,8
2	$CH_3C(O)OCH_3$	−97,5	75,8	−120,1
3	H_2O	−54,3	45,0	−67,7
4	CH_3OH	−48,3	55,9	−65,0
5	CH_3COOH	−101,2	69,0	−121,8
6	[Kh−CH_3CO]	−342,9	252,8	−418,2

Let's consider the kinetics of the acidulating and deacidulating reactions of the fermentative part of *Chemotrypsin*. Kinetic data for the acylferment formation reaction and for the stages of its deacidulating are represented in Table 2. On Figure 3 presented the potential curve of the reaction of acylferment formation. It can be seen from the presented curve that as a result of the interaction of Chemotrypsin and methyl acetate it is formed a series of intermediate complexes, which are caused by the conformational changes of the ferment under the action of the methyl acetate.

Energy of the process of the substrate adsorption on the ferment is sufficient for the formation of Michael's complex (conformation on Figure 3 (1)). Complex (1), which is characterized by enough high free energy, is easy transformed into the complex (α).

Table 2. Kinetic parameters of the enzymatic catalysis of the acylferment formation

Stage	Point of calculation	*ΔH, kilocalorie/mole*	*ΔS, calorie/mole·K*	*ΔH*≠*, kilocalorie/mole*	*ΔS*≠*, calorie/mole·K*
1	Starting substances	−410,1	275,0	15,2	24,5
	Activated complex	−394,9	250,0		
	Products of the reaction	398,9	262,8		
2	Starting substances	406,4	212,8	11,2	37,6
	Activated complex	395,2	250,4		
	Products of the reaction	433,2	242,3		

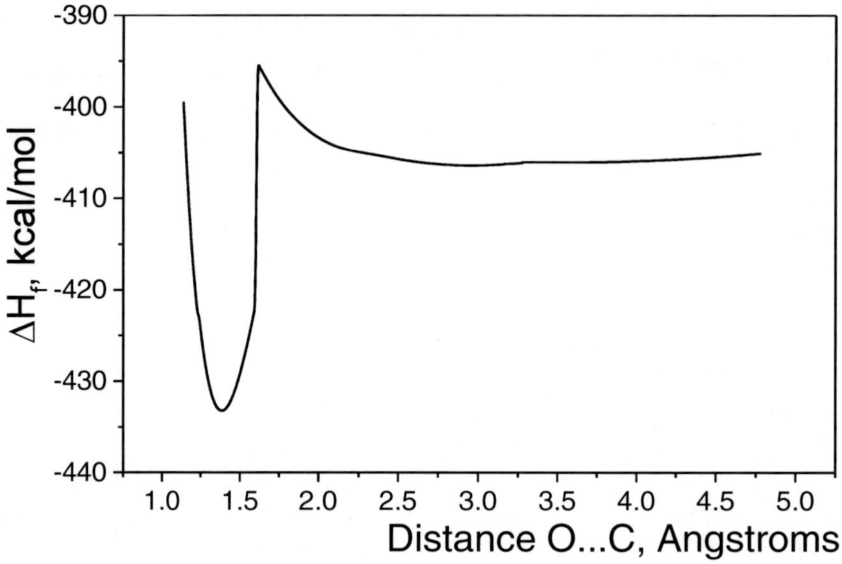

Figure 6. Potential curve of the deacidulating for Chemotrypsin complex.

At this, the transition from the complex (1) into the complex (α) is the exothermal process with the heat effect ~ 10 *kilocalorie/mole*. The complex (α) is easy transformed into the activated complex (2) (see Figure 3, peak № 2) with the activation energy ~ 2 *kilocalorie/mole*. Activated complex (2) transfers into the conformation (β), and (β) is transformed into the conformation (γ), which with a little barrier is transformed into the intermediate compound – acylferment. Activation enthalpies of the acylferment formation are some overestimate as a result of the inaccuracy of the semiempirical calculation methods. However, kinetic characteristics of the process may be trusted qualitatively (Table 2). Free activation energy of the acidulating reaction consists of 7,8 *kilocalorie/mole*, and of the deacidulating reaction this value consists of $\Delta F^{\neq}$ = 4,0 *kilocalorie/mole*, in other words the deacidulating reaction rate is considerably great. The acidulating reaction is the limiting stage of the methyl acetate hydrolysis process.

It can be seen from the Figure 6, that the activated complex of the intermediate product deacidulating reaction is formed at the interatomic distance $C{-}O$ 1,75 *A*. The deacidulating process is exothermal and proceeds with a heat effect 35 *kilocalorie/mole*. Probably, the deacidulating process is one−stage, in other words it can be considered as elementary one. A great difference in activation entropies for acidulating and deacidulating reactions is explained by more distended structure of the activated complex for the deacidulating reaction.

Generally, it is necessary to note, that a great contribution into the values of the rate constants in *Chemotrypsin* acidulating and deacidulating reactions is characterized by the entropic factors; this is connected with a great contribution of the activation oscillating constituents of entropy, which caused by the low−frequency vibrations of the bonds, in $\Delta S^{\neq}$.

It was interesting for comparison to study the modeling of the reaction of the methyl acetate acidic homogeneous catalysis.

The Objects of Investigation

- methyl acetate;
- *Serin.*

It was assumed, that the process of homogeneous catalysis proceeds also via the stage of acidulating and deacidulating. Calculations of the thermodynamic parameters of the reagents, of the intermediate products and of the final products are represented in Table 3.

Table 3. Thermodynamic characteristics of the reagents

№	Formula / substance	ΔH, kilocalorie/mole	ΔS, calorie/mole·K	ΔF, kilocalorie/mole
1	Ser–H	−138,7	87,8	−164,9
2	$CH_3C(O)OCH_3$	−97,5	75,8	−120,1
3	H_2O	−54,3	45,0	−67,7
4	CH_3OH	−48,3	55,9	−65,0
5	CH_3COOH	−101,2	69,0	−121,8
6	CH_3CO–O–Ser	−175,6	108,9	−208,0

Scheme of the Process

CH_3COOCH_3 + Ser–H –> CH_3CO–O–Ser + CH_3OH –$^{(+ H2O)}$–> CH_3COOH + Ser–H

CH_3COOCH_3 + Ser–H –> CH_3CO–O–Ser + CH_3OH (1)

(ΔH_p = + 12,3 kilocalorie/mole, ΔS_p = + 1,2 calorie/mole·K, ΔF_p = + 11,9 kilocalorie/mole)

CH_3CO–O–Ser + H_2O –> CH_3COOH + Ser–H (2)

ΔH_p = – 10,0 kilocalorie/mole,

ΔS_p = + 2,9 calorie/mole·K,

ΔF_p = – 10,8 kilocalorie/mole

$$\Sigma\Delta F = +\ 1{,}1 \text{ kilocalorie/mole}, \quad K = \exp\left(-\frac{\Delta F}{RT}\right) = 0{,}16.$$

It can be seen from the calculations, that the stage of the intermediate product formation is endothermic and its formation is not thermodynamically stable. Probably that the time of the intermediate product existing is very little. The hydrolysis reaction of the intermediate compound is thermodynamically efficient. Generally, free energy of the hydrolysis process consists of + 1,1 kilocalorie/mole. Approximately within the limits of experimental error the methyl acetate hydrolysis reaction is thermodynamically allowed.

Let's consider the kinetics of the catalysis process. On Figure 7 there are interatomic distances and electronic characteristics of the *Serin*. Ionization potential is some higher, than in a case of the *Chemotrypsin*. Electron affinity is sufficient low. In activated complex of the acylserin (see Figure 8) it is observed a great change of the charges on the reactive center *C−O*, in comparison with the starting system. The final intermediate product – acylserin – is formed at the distance C−O ~ 1,5 A (Figure 9).

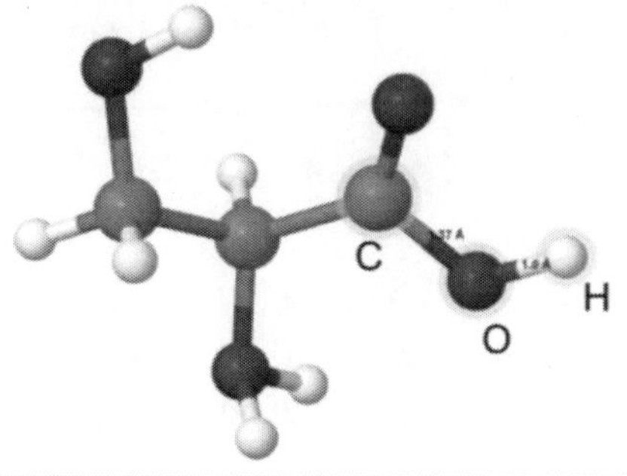

Angle *C* – O – *H*: 113,0^0. Ionization potential: 10,232 *eV*. Electron affinity: –0,066 *eV*

Charge on	H	O	C	Distance	O–H	C–O
	+0,348	–0,536	+0,561	A	1,00	1,37

Figure 7. Geometry and electronic characteristics of the Serin.

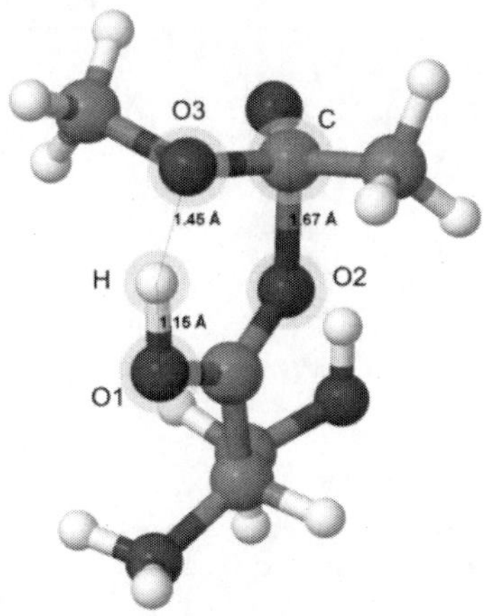

Charge on	H	O1	O2	C	O3	Distance	C–O2	O1–H	H–O3
	+0,418	–0,537	–0,570	+0,771	–0,585	A	1,67	1,15	1,45

Figure 8. Geometry and the values of charges for activated complex of the acylserin formation.

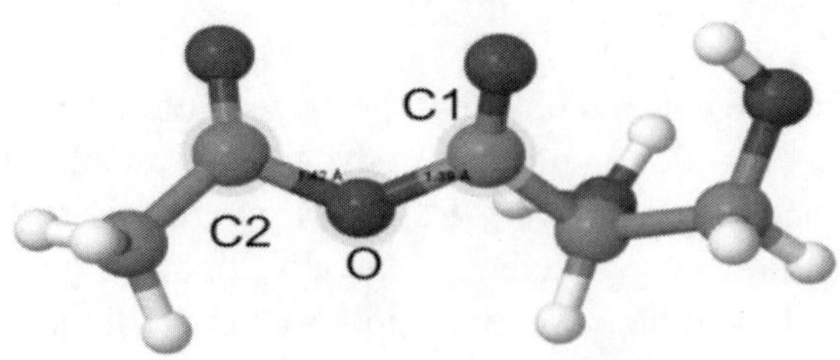

Angle *C* – O – *C*: 126,1^0. Ionization potential: 10,012 *eV*. Electron affinity: ––0,401 *eV*

Charge on	C1	O	C2	Distance	C1–O1	O–C2
	+0,584	–0,570	+0,703	A	1,390	1,42

Figure 9. Geometry and electronic characteristics of the complex *CH_3CO–O–Ser* (acidulating reaction).

The process of the intermediate product formation is endothermic (Figure 10) with the reaction heat ~ 11 *kilocalorie/mole*. Transformation of the activated complex into the intermediate product proceeds exothermally with heat ~ 6 *kilocalorie/mole*. Activated complex of the hydrolysis reaction (Figure 11) of the intermediate product is formed at the internuclear distance *C–O* ~ 1,6 *A*. Reaction of the hydrolysis is exothermal and proceeds with heat ~ 6,5 *kilocalorie/mole*.

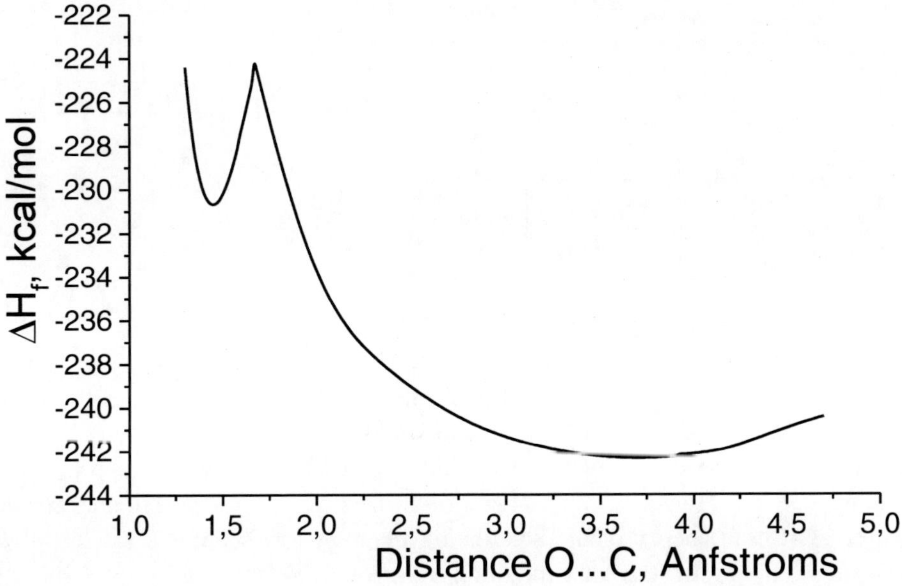

Figure 10. Potential curve of the intermediate product acylserin CH_3CO–O–Ser formation.

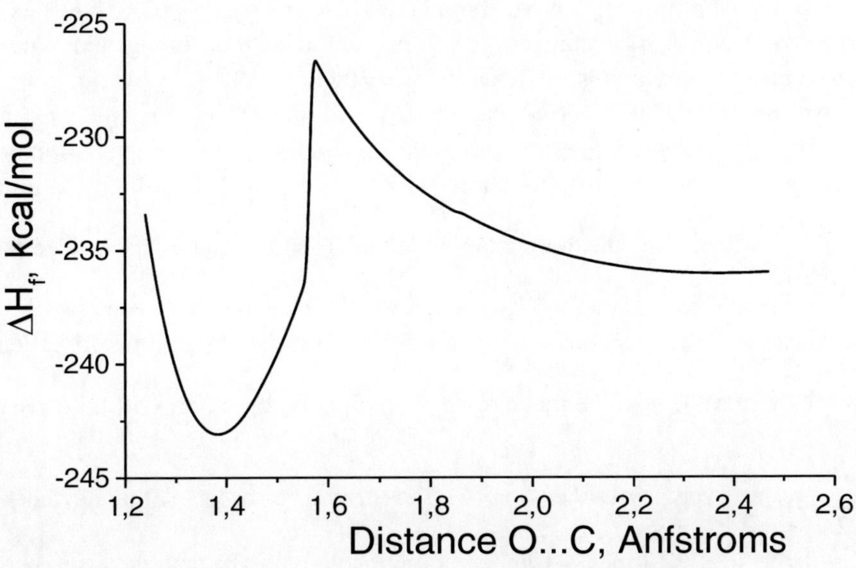

Figure 11. Potential curve of the deacidulating CH3CO–O–Ser reaction.

Table 4. Kinetic parameters of the acidulating and deacidulating reactions of Serin

Stage	Point of calculation	ΔH, kilocalorie/mole	ΔS, calorie/mole·K	E_A, kilocalorie/mole	$\Delta S^{\#}$, calorie/moleK
1	Starting substances	–242,3	137,1	17,9	31,0
	Activated complex	–224,4	106,1		
	Products of reaction	–230,7	136,6		
2	Starting substances	–236,0	122,7	9,7	8,3
	Activated complex	–226,3	114,4		
	Products of reaction	–243,1	103,7		

Kinetic parameters of the reaction of intermediate product formation (acidulating) and final products (hydrolysis) are represented in Table 4. The potential curvers of the stages are represented on Figure 10 and Figure 11 correspondingly above.

Heat of the activation of acidulating reaction is considerably greater, than the $\Delta H^{\neq}$ of the deacidulating reaction. However, the activation entropy of the first reaction is considerably greater. Free activation energy of the acidulating reaction consists of − 0,7 *kilocalorie/mole*. Free activation energy of the deacidulating reaction at $T = 298\ K$ consists of + 4,7 *kilocalorie/mole*. In other words, the second stage is the limiting one, whereas for the enzymatic reaction the limiting stage is the acidulating stage.

More electron affinity of the active fragment of *Chemotrypsin* (−2,601 *eV*) explains the acceleration of the enzymatic catalysis in comparison with the homogeneous catalysis by *Serin*, electron affinity of which is sufficiently low (−0,0066 *eV*).

If to assume, that the reactive centers $C1-O1$ for *Chemotrypsin* and *Serin* are characterized by approximately the same resonance integrals and by the overlapping integrals, then the orbital energy of these both reactions will be determined accordingly to formula **[15]**

$E_{orbit} \approx \dfrac{const}{I-E}$, where I is the ionization potential (donor), and E in an electron affinity (acceptor). For the methyl acetate $I \approx 11{,}66\ eV$; E for the *Chemotrypsin* consists of − 2,66 *eV*, whereas for *Serin* this value is equal to − 0,0066 *eV*. Correspondingly, the orbital energies of interaction for the reaction of intermadiare acylcompounds formation will be consist $\dfrac{const}{14{,}32}$ and $\dfrac{const}{11{,}67}$ respectively; *in other words*, the orbital energy of the acylchemotrypsin formation will be in 1,22 times lesser, in comparison with the intermediate energy of the acylcompound formation in a case of the homogeneous catalysis by *Serin*.

It is necessary to note, that the electron structure of the reagents has an essential influence on the hydrolysis reaction of $C-O$ bond in methyl acetate both via the enzymatic and via the homogeneous (acidic) catalysis. *For example*, the values of charges on the active atoms of the *Oxygen* of hydroxyl groups of the serinic fragment of *Chemotrypsin* and free *Serin* itself arc respectively equal to −0,546 i −0,536 that in some way leads to higher nucleophilicity of the *Chemotrypsin*.

During the enzymatic catalysis the electronic characteristics of the reagents are greatly changed, that leads to the essential changes of the nucleophilicity of the reactive center of *Chemotrypsin* and of the reactive center of the substrate (methyl acetate). If to assume the respective charges as the characteristics of such reactive centers, then we will obtain the following picture. The values of the atoms charges along the reaction way $\overset{+}{C}1-\overset{-}{O}1$ (equilibrium state) is changed from + 0,666, − 0,559 to + 0,801, − 0,648 into the activated complex. For the reaction of homogeneous catalysis of methyl acetate by *Serin* into the equilibrium state the charges on the reactive centers *C* and *O* are equal to + 0,666 and – 0,536 respectively, and for the activated complex + 0,771 and – 0,537.

It is following from the all above–said, that the nucleophilicity of the reactive center (atom of *Oxygen* of *Serin*) of *Chemotrypsin* is some more than the same characteristic for *Serin*. Generally, it can be concluded, that the stabilization of the activated complex at the expense of the charges of reactive center *C*−*O* will be higher in comparison with the homogeneous catalysis by *Serin* molecule. Accordingly to this also heat of the activated complex formation will less in a case of the enzymatic catalysis that leads to the less activation heat of the acylferment formation.

Let's note, that the activated complex of the acylferment in a case of the enzymatic catalysis is stabilized also at the expense of the formation of hydrogen bond of *His* and atom of oxygen of methyl acetate (finally, atom of hydrogen transfers on the oxygen with the formation of methanol).

Let's consider rather more detailed the reaction of *Chemotrypsin* deacidulating. The atom of *Carbon C*1 (Figure 12) is attacked by the atom of Oxygen O3 of water molecule with the simultaneous attack of Hydrogen of water the Nitrogen atom of the imidazole group of the His. Geometry and the charges distribution into the activated complex of acylchemotrypsin are represented on Figure 12.

The values of charges on the reactive center for activated complex of the acylchemotrypsin *C*1−*O*3 are respectively equal to + 794 i – 0,554, whereas for the activated complex of the acylserin the values of such charges are equal to *C*2 (+0,757) – *O*2 (−0,549) (see Figure 13).

From the comparison of both reactions of acylchemotrypsin and acylserin hydrolysis it can be concluded that most stabilization of the activated complex is observed in a case of the acylchemotrypsin deacidulating stage.

Due to such higher stabilization of the activated complex of the deacidulating reaction it is observed some less activation energy for the enzymatic reaction. In a case of the acylcompounds hydrolysis a molecule of water plays the role of the nucleophile. Electronic characteristics of water's molecule are represented in legends to the Figure 13.

Taking into account, that the ionization potential of the acylchemotrypsin is equal to 9,05 *eV*, and for the acylserin ~ 10 *eV* (the difference is 23 *kilocalorie/mole*), it can be concluded that the deacidulating reaction will be proceed with the less activation energy for the acylchemotrypsin. It possible that the deacidulating reactions will took place with the participation of the hydronium ion which is characterized with a high electron affinity.

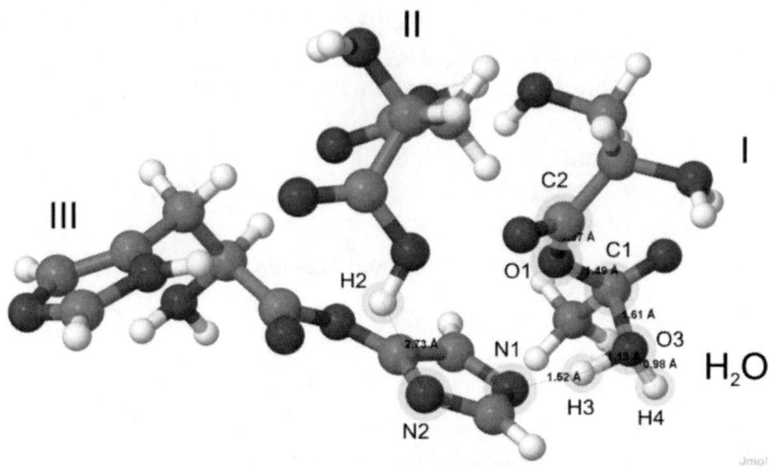

Charge on	C2	O1	N1	C1	H2	H3	H4	O3	N2
	+0,633	–0,651	–0,420	+0,794	+0,374	+0,412	+0,336	–0,554	–0,401

Distance	O1 – C1	C2 – O1	O3 – C1	O3 – H3	O3 – H4	N1 – H3	N2 – H2
A	1,49	1,37	1,61	1,13	0,98	1,52	2,73

Figure 12. Geometry of activated complex and the charges of some atoms *[Kh–CH*$_3$*CO]–H*$_2$*O*. I – *Ser + CH*$_3$*CO*, II – *Asp*, III – *His*.

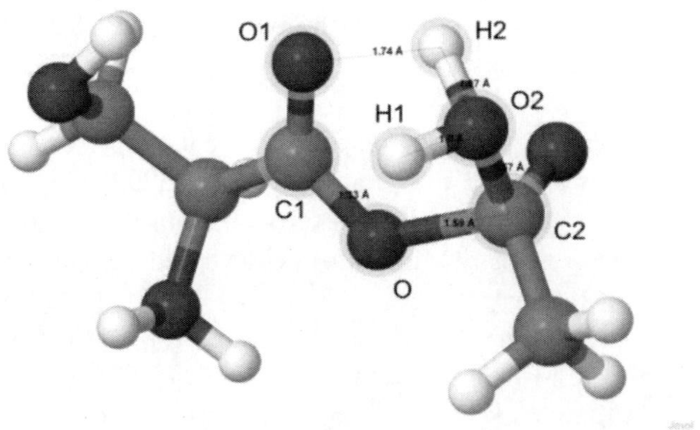

Charge on	C1	O	C2	O1	O2	H1	H2
	+0,634	–0,636	+0,757	–0,645	–0,459	+0,346	+0,404

Distance	C1 – O	O – C2	O2 – C2	O2 – H1	O2 – H2	O1 – H2
A	1,33	1,58	1,57	1,00	1,07	1,74

H_2O	H_3O^+
Charge on O: –0,6188	Charge on O: –0,122
Charge on H: +0,3094	Charge on H: +0,374
Distance O – H: 0,95 A	Distance O – H: 1,04 A
Ionization potential: 11,906 eV	Ionization potential: 22,869 eV
Electron affinity: 4,068 eV	Electron affinity: –6,346 eV

Figure 13. Geometry of the activated complex *[CH*$_3$*CO–O–Ser] – H*$_2$*O*.

Also it was considered the reaction of the methyl acetate hydrolysis in the presence of the sulphuric acid. It was shown, that the first acidulating stage proceeds with $\Delta H_{реакц.} = +\ 1{,}4$ *kilocalorie/mole*, $\Delta S_{реакц.} = -\ 2{,}9$ *calorie/mole·K*, $\Delta F_{реакц.} = +\ 2{,}3$ *kilocalorie/mole*. The second deacidulating stage is characterized by the following thermodynamic parameters: $\Delta H_{реакц.} = +\ 0{,}9$ *kilocalorie/mole*, $\Delta S_{реакц.} = 7{,}0$ *calorie/mole·K*, $\Delta F_{реакц.} = -\ 1{,}2$ *kilocalorie/mole*. Total reaction: $\Sigma\Delta F_{реакц.} = +\ 1{,}1$ *kilocalorie/mole*; $K = \exp\left(-\dfrac{\Delta F}{RT}\right) = 0{,}16$.

Activation parameters of the acidulating reaction are: $\Delta H^{\neq} = 34{,}2$ *kilocalorie/mole*, $\Delta S^{\neq} = 10{,}1$ *calorie/mole·K*, $\Delta F^{\neq} = \sim 31$ *kilocalorie/mole*.

Activation parameters of the deacidulating reaction are: $\Delta H^{\neq} = 10{,}3$ *kilocalorie/mole*, $\Delta S^{\neq} = 8{,}3$ *calorie/mole·K*, $\Delta F^{\neq} = 7{,}8$ *kilocalorie/mole*.

At the comparison of the methyl acetate hydrolysis kinetics in the presence of the sulphuric acid, it is necessary to note, that the deacidulating process proceeds with the same kinetic parameters as in a case of *Serin*. However, the activation energy for H_2SO_4 is practically in two times higher, and the activation entropy is too lesser.

It is seen from the value $\Delta F^{\neq}$ that the limiting stage in the methyl acetate hydrolysis with the sulphuric acid reaction is the stage of the acylproduct formation whereas in a case of the *Serin* it is observed the opposite situation. Probably, it is not be unambiguously stated, that for the all reactions of the ethers hydrolysis by different substrates the reaction of the intermediate product (*namely*, acylferment) formation is the limiting stage.

Quantum−chemical calculations show, that, really, into the activated state of the ferment there are hydrogen bonds, at the expense of which the nucleophilicity of the reactive atom of *Oxygen* of *Serin* is increased. These bonds substantially determine the chemical mechanism of the acylferment formation and have an influence on the values of their kinetic parameters for the enzymatic catalysis and on the geometry of conformers on the reaction way.

The following conclusions can be done from the all above−said:

A reaction of the intermediate product acylchemotrypsin formation is not elementary (as it is described in *ref.* [16]), but probably consists of the following stages:

1) Hydrogen bond «serinic group – imidazole group» *O−H…N* is some elongated; the *Michael's* complex is endothermicly formed;
2) Exothermal transition of the *Hydrogen* atom to the *Nitrogen* atom (imidazol group of *His*) (conformer α);
3) Formation of the activated complex of the acidulating reaction (2);
4) Transformation of the activated complex into the complex of the acylferment (conformer β), which transfers into the conformer γ, and finally into the intermediate product (acylchemotrypsin).

In other words, the reaction of the *Chemotrypsin* acidulation can be presented accordingly to the following scheme:

ferment + methyl acetate ↔ Michael's complex ↔ conformer α ↔ conformer β ↔ conformer γ ↔ acylferment.

Thereby, a reaction of the acylchemotrypsin formation consists of a series of the conformational transformations of the reagents themselves on the reaction way.

However, we have some doubts as to generally accepted positions of the *Michael's* complex formation; such doubts cannot be solved with the use of the semi–empirical calculations. It was accepted that the energy of the substrate adsorption process on the active center of ferment is enough for the *Michael's* complex formation, which requires the high energy of the formation. Ambiguously, that the physical adsorption can provides such energy.

As the practice of the quantum–chemical calculations for such systems shows, the interaction in them proceeds with enough low activation energies, however also with insufficient heat effects, which cannot able initiate the *Michael's* complex formation.

Chemical adsorption or some other approach can lead to such complex formation. *For example*, the substrate is adsorbed on active part and is overbalanced.

Each ferment has the essential protein part besides the reactive centers. Free energy of the ferment is into equilibrium nonworking (nonactivated) state and is relatively stable; that is, the ferment is characterized with minimal free energy. At the adsorption of the substrate on the active center the conformation of the ferment is greatly changed as a result of the interaction. The result of this there are changes of the non–valence and valence bonds into molecule that leads to the increase of the free energy of a system, which is distributed between the bonds of the system, including also the reactive center. Such «excited» system of the ferment leads to the short–term chemical act with the formation of the products plus free ferment which has the initial conformation which again is excited by the substrate.

Theoretically such hypotheses can be in some ways confirmed by quantum–chemical calculations based on the excitation theory.

Maybe, positive result about the distribution of the energy of an «excited system» can be obtained accordingly to *PPKM* theory [17]. In accordance with this theory the transformation of the complex «ferment–substrate», as it's known, connected with the energy redistribution between different freedom degrees. However, zero energy of the excited complex and the activated complex, and also the potential energy of bonds which form the active complex, are not ordered to such distribution. Residual energies of the excited complex and the activated complex are called by disconnected energies. Different freedom degrees play a different role into the energy change. Adiabatic freedom degrees practically not transfer the energies of bonds which are broken as a result of a low probability of the energy transfer from the external freedom degrees on the internal degrees, which correspond to three rotary and to three transitional movements of a system as a whole.

Internal degrees of freedom will be consisting of the active and non–active degrees of freedom. Active degrees of freedom are the degrees without limitations of the energy of bonds which are formed or broke; and the non–active degrees of freedom can transfer an energy of the bonds only in a case when the active complex is the activated complex.

Thereby, the reaction of the ferment initiation by the substrate can be schematically represented in accordance with the following scheme:

ferment (domain) + H_2O → solvate – ferment + substrate → «excited ferment – substrate» (new conformation) → ferment – substrate with distributed energy on degrees of freedom into *Michael's* complex → activated complex → products + ferment

CONCLUSION

In references the most essential reasons of the enzymatic reaction acceleration in comparison with the homogeneous catalytic processes are as follows:

1) sorption interaction with protein of the side substrate groups accelerates the reaction in 10^7 times;
2) polyfunctional catalysis (general acidic–base catalysis) can lead to the reaction acceleration in 10 times;
3) micro effects of medium of the active center can change the rate of catalysis on dozens orders.

It is possible, that the very important role into the processes of the enzymatic catalysis play the micro effects of medium of the active center. The rate of catalysis in such cases can be changed on dozens orders. However, unfortunately now it has done nothing in the presented direction.

REFERENCES

[1] Blow D. M., Steitz T. A. // *Ann. Rev. Biochem.*, 39, 63 (1970).
[2] Matthews B. W. *et al.* // *Nature*, 214, 652 (1967).
[3] Henderson R. J. // *Mol. Biol.*, 54, 341 (1970).
[4] Birktoft J. J. *et al.* // *Phil. Trans. Roy. Soc. (London)*, B 257, 67 (1970).
[5] Miller Ch. G., Bender M. L. // *J. Amer. Chem. Soc.*, 90, 6850 (1968).
[6] Bernhard S. *Structure and Function of the Enzymes* // M.: «Myr» (1971) *(in Russian).*
[7] Hess G. P. *et al.* // *Phil. Trans. Roy. Soc. (London)*, B 257, 89 (1970)
[8] Bernhard S. A., Gutfreund H. // *Trans. Phil. Trans. Roy. Soc. (London)*, B 257, 105 (1970),
[9] Himoe A., Brandt K., Hess G. P. // *J. Mol. Biol.*, 55, 215 (1971).
[10] Mosolov V. V. *Proteolytic Enzymes* // M.: «Nauka» (1971) *(in Russian)*
[11] Cunningham L. // *Comprehensive Biochem.*, 16, 85 (1965).
[12] Bender M. L., Kezdy F. J. // *J. Amer. Chem. Soc.,* 86, 3704 (1964).
[13] Berezin I. V., Martynek K. // *In Colln.: «Structure and Functions of Enzymes»* / Moscow State University Edition (1972) *(in Russian)*
[14] Blow D. M., Birktoft J. J., Hartley B. S. // *Nature*, 221, 337 (1969).
[15] *Reactive Ability And The Reaction Ways* // Under Edition: Klopmann G. / M.: «Myr» (1977). – 383 p. *(in Russian).*
[16] Berezin I. V., Martynek K. *The Principles Of The Physical Chemistry Of The Enzymatic Catalysis* / M.: «Vysshaya Shkola» (1978). – 350 p. *(in Russian).*
[17] Stiepukhovich A. D., Ulitsky V. A. *Kinetics And Thermodynamics Of Cracking Radical Reactions* // M.: «Khimiya» (1975). – 255 p. *(in Russian).*

In: Biology and Communication in the Information Age
Editor: Rafiqul Islam
ISBN: 978-1-62808-252-4

Chapter 20

ATOMIC HYDROGEN TRAPPING AS A PROBE OF INTRAMOLECULAR PHOTOELECTRON TRANSFER MEDIATED BY OXYACIDS

O. S. Nedelina**, *O. N. Brzevskaya and E. N. Degtyarev
Emanuel Institute of Biochemical Physics, Russian Academy of Sciences, Moscow, Russia

ABSTRACT

pH-dependent detection of the EPR spectra of atomic hydrogen $H^{\bullet}$ (g = 2.0023, A = 50.6 mT) in frozen aqueous solutions of oxyacids AOH initiated by photoinjected low energy electrons upon energy impact below the excitation thresholds of water or oxyacids (photon energy of not more than 4.5 eV) was crucial for identification of the dissociative electron capture by bound acidic proton. The data on direct electron attachment to an oxyacid that can create an acidic bound [H+] network under certain conditions allows one to consider direct participation of the acidic bound proton in the intramolecular ET as one electron relay being highly effective in mediating electron transport in acid catalysis of redox reactions.

Keywords: Photoinjected electrons, photoinduced intramolecular electron transfer, one electron relay, acid catalysis, EPR studies, atomic hydrogen, highly reactive inter-mediates, chemical trapping, low-energy electrons, dissociative electron attachment, polarized oxyacids,bound acidic proton

INTRODUCTION

Artificial photoredox systems involving an electron donor and an electron acceptor are potentially applicable for converting solar energy into chemical. Many different host systems

* E-mail: olga.nedelina@gmail.com.

have been tried to improve the efficiency of the energy storage by preventing the rapid back electron transfer reaction.

The most prevalent approach to photocatalysis is the construction of three-component systems comprising a sensitizer to absorb light, a proton reduction catalyst, and an electron relay to shuttle reducing equivalents from the sensitizer to the catalyst [1-3].

The rates of electron transfer (ET) can be finely controlled by the fundamental electron transfer properties of electron donor and acceptors (the redox potentials and the reorganization energy) and also by the electron donor-acceptor distance and the environment surrounding them. Various lines of experimental evidence have led to the conclusion that both thermal and photoinduced' electron transfer may occur between species separated by a distance significantly exceeding the sum of their van der Waals radii. In fact many biological electron transport processes including the primary steps of photosynthesis involve such long-range electron-transfer events.

Electron transfer rate constant falls off expotentially with increasing distance between the donor and acceptor molecules [4]. In contrast introduction of an appropriate third component acting as an effective catalyst or relay that control the direction and dynamics of ET by through-bond-coupling effects has been a central issue in designing modern chemical reactions which involve cleavage and formation of chemical bonds. ET over long distances doesn't proceed in a single step, but in multistep process reaction occurs in several faster steps over shorter distances each of them is a superexchange reaction. A condition for such a mechanism is that certain subunits or functional groups can play the role of relay stations where the charge resides for a short time before the ET continious by next hopping step. This configuration enables the photounduced ET process to take place in a sequence of a short rapid "hops" between adjacent relays the final result being efficient formation of a long-lived giant charge-separated state $^{+}D \rightarrow$[relay]$\rightarrow A^{-}$ [5-7].

The data presented repeatedly provide unambiguous evidence for the important influence of the intervening medium on the rate of electron transfer. Solvent molecules may play an important role as they actively take part in the reaction, being reactant, relay or product.

APPROACH

Here we report a simple approach to the problem of electron-transfer reaction distances based on EPR studies of photoinduced electron transfer in a condensed frozen solution (scheme 1), that would allow the underlying chemical and photochemical events to be probed with greater precision. This is the reductive electron transfer mechanism where an excited sensitizer or fluorophore acts as an electron-injecting species. The classical photochemical system (Scheme 1) we used in combination with control of the stepwise transformation of photoinjected electrons by EPR, nanosecond laser photolysis, and fluorescence quenching can serve as an adequate method for identification of one electron relays in the intramolecular electron transfer [8-10]. A discussion of photoionization in condensed rigid media must involve some consideration of the photoelectron transport and trapping processes.

It has been known for some time that a number of specific photochemical processes occur in frozen rigid solutions [11]. In this project the light energy is used to overcome the thermodynamic barrier associated with electron removal or attachment. Photon energy range corresponding to the near ultraviolet that is from 3 to 4,5 eV. The photon range is limited at low energies by the absorption threshold of the aromatic solute molecule (fluorophore) and at high energies by the water absorption. Depending on the nature of the solute and the solvent, the photochemical events that can occur subsequent to illumination of such a rigid solutions are a) solute (or fluorophore) photoionization that lead to the formation of an aromatic radical cation which is physically or chemically trapped in the solvent.

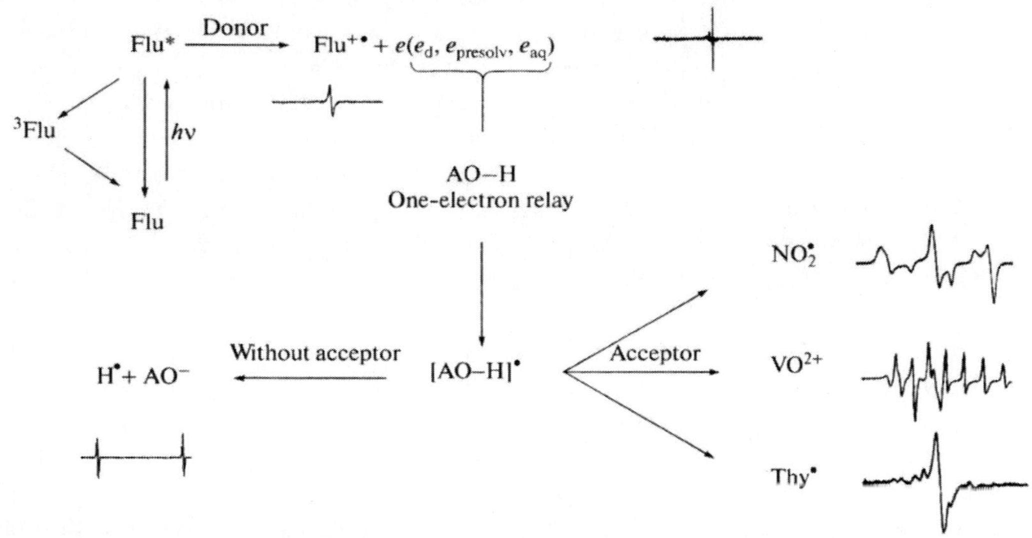

Scheme1. Capture of photoinjected low-energy electrons by bound acidic proton of oxyacids AOH as a a model of elementary steps of donor(e)-[H+]-acceptor(R) intramolecular transport.

The product obtained upon irradiation of fluorophore is identical to that produced by chemical oxidation. The details of these processes are not our major concern however and they will be only considered as factors which in many instances influence the primary processes of photoelectron transport.

Studies in frozen solvents are particularly amenable to systematic investigations of the electronic, structural and distance factors. In frozen solvents at liquid-nitrogen temperature, excited-state ET reactions must be highly optimized in order to compete with other radiative and nonradiative decay processes.

Condensed rigid systems related to those studied here appear to be of great value in such studies not only providing more insight into the thermodynamics of ET and its solvent dependence but especially also into the role of the nature of the coupling of donor and acceptor. Frozen matrices are known to be so rigid that highly reactive intermediates such as dry or solvated electrons and radicals may be trapped indefinitely, the generation of an initially random distribution of electrons and radicals alters to the steady state concentration gradients when the electrons diffuse to the scavenger ions.

Generally two types of trapping can be distinguished: a) physical trapping: the electron is trapped in a "self-created hole". The trapped or solvated electron is a specific entity with characteristic optical and ESR absorption b) chemical trapping: in the presence of an electron acceptor or scavenger or relay the electron attaches to a molecule or some other species and it is absorbed into the acceptor orbitals, which can in certain cases result in chemical change such as dissociation-fragmentation [11].

The trapping of electrons in frozen water is an extremely broad problem and it is currently the subject of extensive studies. This discussion is not intended to examine the details of electron trapping but rather to consider how it can be linked to photoelectron long-range transfer events between donor and distant acceptor in low concentrations.

The ability to unravel the more intricate details of electron transfer (ET) reactions has greatly benefited from the widespread interest that the chemistry of reactive intermediates including radicals, radical ions and metastable or resonant states has attracted in recent decades. The interaction of a molecule with an electron can provide, apart from its activation in chemical reaction or molecule fragmentation, also intermolecular electron transport from donor to acceptor, thus giving rise to intermediate electron relays. The considerable amount of experimental and theoretical data concerning these reactive intermediates that has become available has allowed a more accurate analysis of the dynamics of such ET reactions. Investigations in this area aim at an understanding of elementary reactions involved in such ET process.

Electron-induced chemistry at subionization energies can proceed either *via* electronic excitation of the target or, particularly at energies too low for electronic excitation, by way of dissociative electron attachment (DEA). The low-energy electrons can resonantly attach themselves to molecular targets and yield formation of temporary negative ions (TNI) [12]. The energized TNI in turn can either evolve by autodetaching the extra electron or could dispose of the excess energy within its molecular framework a path which may ultimately lead to a rupture of a chemical bond within the target: this process yielding a stable anion and a radical fragment, is known as dissociative electron attachment (DEA).

In modern terms, this reaction proceeds via the formation of short_lived electron adducts (negative ions, metastable anions, resonants) followed by energy and electron density transfer toward the weakened bond being cleaved to give stable products (free radical and anion): e + M–ABC → (M—ABC)– → M– + ABC• [13-14], in particular, e + MH → (MH)• → (M–H)– + H•. Moreover, in the dissociating transient negative ion any translational excess energy will nearly exclusively be transferred to the light hydrogen atom (via the principle of linear momentum conservation more than 99%) thus facilitating the removal of the hydrogen radical in this DEA reaction.

Most often the anion M–H detected by mass spectroscopy serves as the marker of this reaction (light $H^{\bullet}$ and $D^{\bullet}$ atoms are undetectable by this method). The formation of the (M–H) $^{-}$ anion, which has given off atomic hydrogen, was demonstrated experimentally by resonance electron capture mass spectroscopy for many compounds containing OH groups [15-18].

The energy of the electron at which the resonance increase in the capture is observed is different for different substances. For example, for compounds that easily dissociate and have high electron affinity, the probability of capture is the highest for nearly zero electron energy. If the affinity of the electron for an atom incorporated into a polyatomic molecule exceeds the energy for abstraction of this atom from the molecule, the electron attachment to this atom in a polyatomic molecule may occur at indefinitely low energy of the electron even at thermal energy.

Functional group dependence is observed in the dissociative electron attachment (DEA) to various organic molecules in which the DEA features seen in the precursor molecules of the groups are retained in the bigger molecules. This functional group dependence is seen to lead to site-selective fragmentation of these molecules at the hydrogen sites [19].

Here we show that the DEA process can also be used to probe the distance dependence of the electronic coupling in frozen solutions of oxyacids AOH. We have studied the photooxidation of different fluorophores and photoreduction of some biological acceptors in frozen oxyacids (AOH) diluted solution.

It was found that the AOH has a strong effect on both the photooxidation of fluorophores and photoreduction of remote acceptors. Detection of atomic hydrogen as a probe of low energy electron attachment to an oxyacid provides grounds for considering direct participation of the bound acidic proton of AOH as a one_electron relay in the intramolecular ET donor → [AOH] → acceptor [8-10].

Acid catalysis in redox reactions has been described once and again [20-22].

Quite a few facts have been accumulated indicating capture of electrons by Brønsted acids HX, which is detected by appearance of atomic hydrogen [23]. It has been suggested that acids HX can add an additional electron to give HX^- as the primary product: $e_{aq} + HX \rightarrow HX^- \rightarrow H + X^-$.This was shown experimentally (NMR) for the radiation hydroxonium: $e_{aq} + H_3O^+ \rightarrow H_3O \rightarrow H^\bullet + H_2O$ [24].

In radiation experiments, the irradiation energies are very high and the determined atomic hydrogen is the overall product of several reactions including decomposition of water ($H_2O^* \rightarrow H^\bullet + OH^\bullet$ or $H^\bullet + O^\bullet + H^\bullet$, the threshold energy is not lower than 5.5 eV), homolytic bond cleavage $X–H \rightarrow X\bullet + H\bullet$,and dissociative electron capture $e + HX \rightarrow [HX]^- \rightarrow X^- + H^\bullet$ [24]. Detection of atomic hydrogen occurring in subionization energy range is unambiguous evidence for interaction of the low energy electron with oxyacids AOH in the low energy dissociative electron capture, $e + AOH \rightarrow [AO–H]^{\bullet -} \rightarrow AO^- + H^\bullet$. DEA is in fact the only mechanism to induce a bond cleavage at such low electron energies.

Investigation of low energy electron addition and transfer to acids AOH containing a polarized $O^{\delta-}—H\delta^+$ bond, that is if there is a large accumulation of positive charge on the H atoms in a bond OH (charge delocalization usually determines the degree of acidity p*K*α) can be useful for elucidation of mechanisms of acid catalysis. The p*K*a value of an oxyacid is ultimately determined by the difference in the free energy between the dissociated and undissociated forms of the acid in the aqueous medium, characterized by the dissociation of the OH covalent bond.

As a result, the acidity of an oxyacid is directly reflected in the OH bond strength, exhibited by the equilibrium OH bond length and stretching frequency υ(OH). This explains the origin of correlations between the molecular properties (OH bond length, stretching frequency and positive charge on the bound proton) and the acidity. The larger the intrinsic acidity is of the proton concerned, the larger the polarization of the OH bond and the resulting OH dipole moment [25-26]. When acidity is considered as being related to the elongation of the O–H bond, prior to transfer to a base, we find that the relative acid strength are related to the vibrational properties ν of the acid site.

A vibrational fine structure in the effective ion yield curves at threshold energies for formation of $(M-H)^-$ ions through OH-bond cleavage, has been assigned to the υ(OH) stretching mode. It has been supposed that the formation of these ions occurs predissociation of the vibrationally excited state of the parent anion [18] being due to electron capture, followed by direct dissociation.

In this context, particular attention has been focused on pH dependence of atomic hydrogen detection as a probe of dissociative attachment of low energy electron to polarized oxyacids AOH.

RESULTS AND DISCUSSION

The main principle for instrumental identification of a one electron relay in the donor → relay → acceptor chain is detecting the product of its interaction with the mobile electron in the absence and in the presence of the remote acceptor. The photochemical system (Scheme 1) with appropriate fluorophores donating electrons with specified energy (Flu (fluorophore) + *h*ν → (Flu)* → Flu•+ + *e*–) to aqueous solutions of oxyacids makes it possible, by varying the composition of the medium, to elucidate the sequence and the nature of electron traps on the way of the photoinjected electron.

Electron migrates a considerable distance from the fluorophore cation radical prior to recombination during which time it is free to react with accessible scavengers or relays. The distance between the donor and relay as an acceptor is obviously very important. The pH-dependent survival probability of the donor mobile electron depends on the concentration of relay, i.e. acidic bound proton of hydroxyl groups which can serve as the electron captor to form hydrogen atoms as shown by the appearance of the hydrogen atom ESR signal. In the present work, we have varied the distance between the probe and relay by changing the relay (captor) concentration values.

The pH-dependent dissociative capture of a low energy electron by oxyacids AOH tracked by the behavior of the EPR spectrum of atomic hydrogen was observed in frozen (77 K) dilute (0.1 to 0.5 M) aqueous solutions of inorganic oxyacids AOH, in particular, HCl (H_3O^+), H_2SO_4 (H_3O^+, HSO_4^-), $H_2PO_4^-$, HPO_4^{2-}, H_3BO_3 on excitation of different fluorophores (tryptophan, NADH, phenothiazine, tetrasulfopyrene, quercetin), donating photoinjected electrons in their absorption band by exposure to near_UV radiation ($240 \leq \lambda \leq 400$ nm) whose energy is below the excitation threshold of both water and acceptor acids.

Thus, samples were irradiated at 77 K and examined by ESR at 77 K to detect hydrogen atoms. Figure 1 does show hydrogen atom spectra at 77 K characterized by two peaks separated by 506 G with g = 2.014.

Proton spinflip satellite lines are also seen on each side of the main hydrogen atom lines. The pattern of the EPR spectrum of atomic hydrogen, characterizing its proton environment was the same for oxyacids AOH with different p*K*α. For deuterated samples, the intensity of the EPR signals of atomic hydrogen sharply decreased in parallel with appearance of the deuterium spectrum, the EPR spectrum being narrowed down (Figure 2). H-D exchange occurs more easily on terminal OH group which brings about the highest OH vibrational band at around 3700cm-1 and the lowest value in proton signal in NMR spectrum.

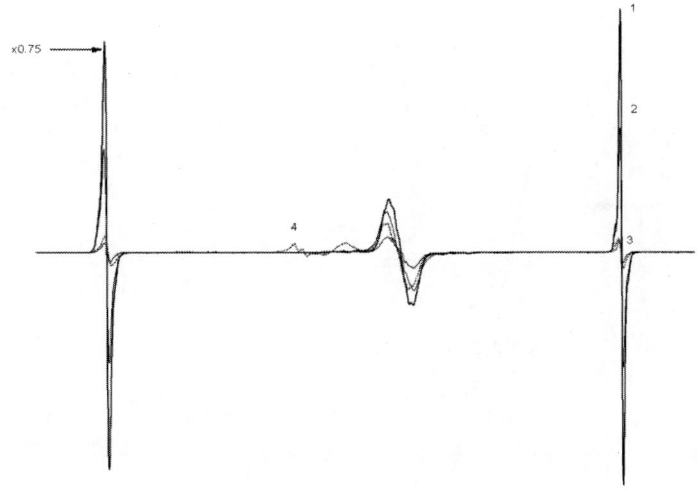

Figure 1. EPR spectra of atomic hydrogen obtained upon UV irradiation ($\lambda \geq 240$ nm) of frozen solutions of inorganic oxyacids in the presence of tryptophan (Trp) as the donor of photoinjected electrons at 77 K and a 2 μW microwave irradiation: (*1*) 0.5 M HSO_4^- (pH =0.8, p*K*α = 1.96); (*2*) 0.5 M $H_2PO_4^-$ (pH = 4.1, p*K*α = 7.2); (*3*) 0.5 M H_3BO_3 (pH = 9.0, p*K*α = 9.2); (*4*) 0.5 M HCl (pH = 0.6).

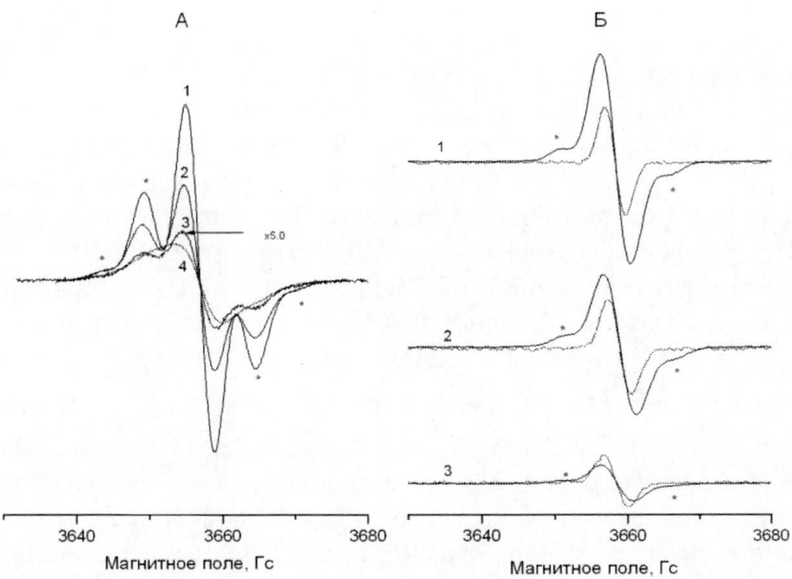

Figure 2. High_field component of the EPR spectrum of H atoms obtained upon UV irradiation ($\lambda \geq 240$ nm) of frozen aqueous solutions of inorganic oxyacids in the presence of tryptophan as a sensitizer at 77 K: (1) 0.5 M HSO_4^- (pH =0.8, p*K*α = 1.96); (*2*) 0.5 M $H_2PO_4^-$ (pH = 4.1, p*K*α = 7.2); (*3*) 0.5 M H_3BO_3 (pH = 9.0, p*K*α = 9.2); (*4*) 0.5 M HCl (pH = 0.6). The continuous curves on the right are satellite lines arising upon simultaneous flip_flop change in the spin state of adjacent protons (microwave radiation power 200 and 2 μW). The dotted lines on the right show the spectral components of the H atom in solutions of the corresponding acids in D_2O.

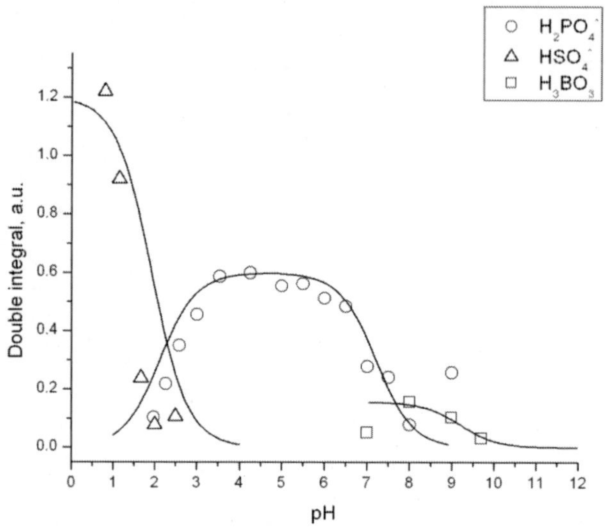

Figure 3. pH dependence of atomic hydrogen intensity for oxiacids with different pKα. The curves are calculated pH dependences of mole fractions of AO-H (1triangles) HSO4- , (2circles) H2PO4- and (3squares) H3BO3 ions.

The H/D isotope effect for the formation of H atoms is more than 7 for the phosphate monoanion and more than 30 for boric acid. This high H/D effect can be interpreted only assuming that the reaction first gives the intermediate electron adduct $[H_2PO_4\ ]^{\bullet}$ and/or $[D_2PO_4]^{\bullet}$, which then decomposes to give H or D.

It was found that attachment of low energy electrons with oxyacids, which we observed, is a pH dependent reaction; i.e. the intensity of the EPR spectrum of atomic hydrogen is proportional to the acidic bound proton concentration in AOH, and the highest amount of detectable atomic hydrogen is accumulated around the highest molar concentration of the acid in the predissociated state with the polarized $AO^{\sigma-}–H^{\sigma+}$ bond near its p*K*α.

Indeed, for oxyacids with p*K*α in acidic (sulfuric acid, –2.8 and 1.92), neutral ($H_2PO_4^-$, 7.21), or alkaline (boric acid, 9.24) regions, the hydrogen atom is detected mainly in exactly these pH ranges, which is in line with electron capture by the acidic proton of the AOH group: $e + AOH \rightarrow [AOH]^{\bullet} \rightarrow H^{\bullet} + AO^{-}$. In the absence of distant acceptor R, the intensity of atomic hydrogen signals obtained for different oxyacids correlated with their acidity constants p*K*α, which confirms the presence of a linear relationship between the p*K*α of acids and the logarithm of the rate constants for their reactions with electron [24] (Figure 3).

Therefore, it was assumed that the acidic (electrophilic) proton is the attachment site for the low energy electron, and this allows detecting the product as the characteristic EPR spectrum of atomic hydrogen (the left hand part of Scheme 1).

According to Brønsted and Lowry, a stronger acid has a smaller p*K*a value whereas a stronger Lewis acid has a stronger capability to accept a pair of electrons. The findings reveal that a stronger Brønsted acid in most cases behaves as a stronger Lewis acid as well. An acid with a lower p*K*a value is expected to have a higher electron affinity, electronegativity and electrophilicity compared to other acids of similar type if the two theories need to correlate. Proton acidity should greatly influence direct electron attachment to the OH group, as well as acting as an efficient intermediary in transferring electrons from one portion of the molecular to another [27].

It was suggested that an electron acceptor relay process in the presence of the remote acceptor R in a low concentration occurs from site to site in the acids framework. Apparently, this relay process does not generate hydrogen atoms, but remote acceptors radical $R^{\cdot}$ (Scheme1).

In a typical experiment, a small concentration of photoexcited fluorophore (<10^{-3}M) as a source of donor electron, is embedded in a frozen solvent amid a higher concentration (>0,1M) of randomly distributed primary acceptors AOH molecules as a relay (phosphate, sulfate,.borate) and a low concentration (<0,01M) of distant terminal acceptors R (nitrate, thymine, vanadate). If the weaker primary electron captor AOH is present in large excess, acidic bound proton can be reduced (chemical trapping) and, the excess electron later transfers to the remote acceptor of greater electron affinity.

The oxyacids system is the ET reaction we are aware of for which the spectra of all species involved in the reaction – the reactants, intermediates relays, and products $R^{\cdot-}$ can be detected and well characterized (Scheme1).

In our experiments we obtained evidence for the fact that in the presence of remote acceptors R in a low concentration, the products of electron capture $[AO–H]^{\cdot}$ can transmit the electron to remote electron acceptors R (nitrate, vanadate, thymine), which is detected in the experiment as a decrease in the intensity of the EPR spectrum of hydrogen and appearance of the EPR spectra of R• (the right_hand part of Scheme 1). This effect is observed at about the highest concentration of the bound acidic proton having a high electron affinity (Figures 4, 5), which suggests that intramolecular transport occurs via successive attachment of the electron to the acidic proton as a relay removing the distance limitations of one electron transport.

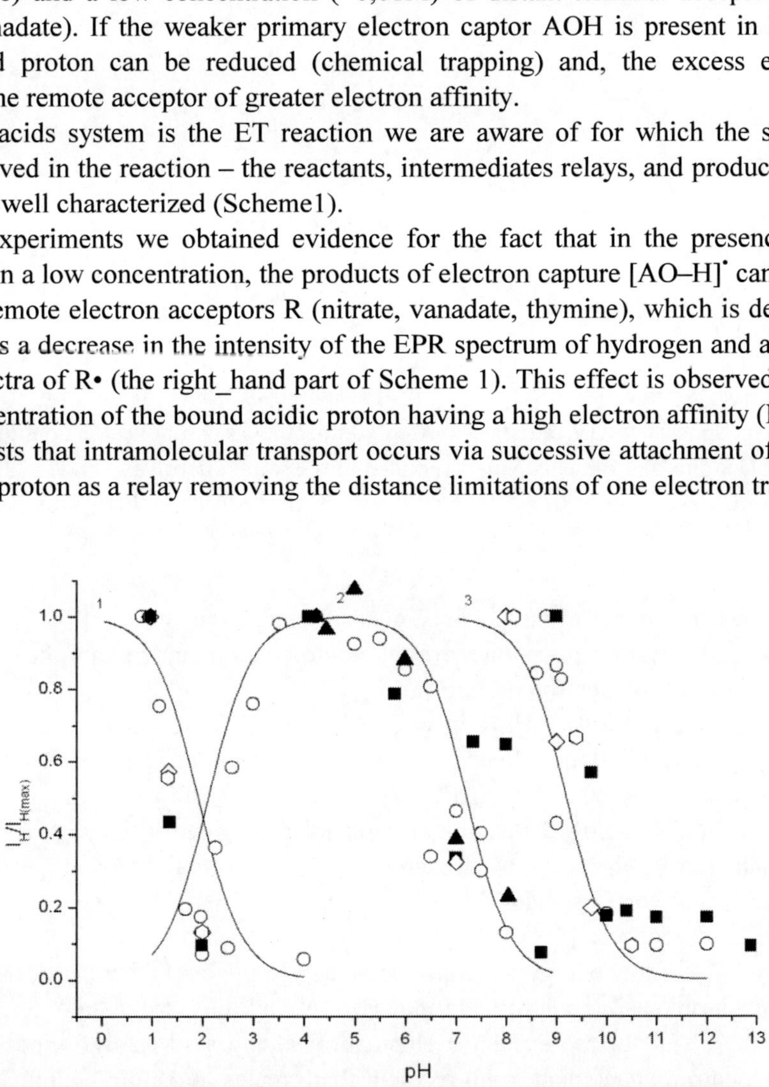

Figure 4. pH dependence of the yield of H atoms (circles) in the absence of acceptors R and corresponding yield of acceptors $R^{\cdot}$ for solutions of oxyacids with different p*K*α at 77 K(squares – NO_3^{2-}, triangles VO^{2+}). The curves are calculated pH dependences of mole fractions of AO-H (*1*) HSO_4^- , (*2*) $H_2PO_4^-$ and (*3*) H_3BO_3 ions.

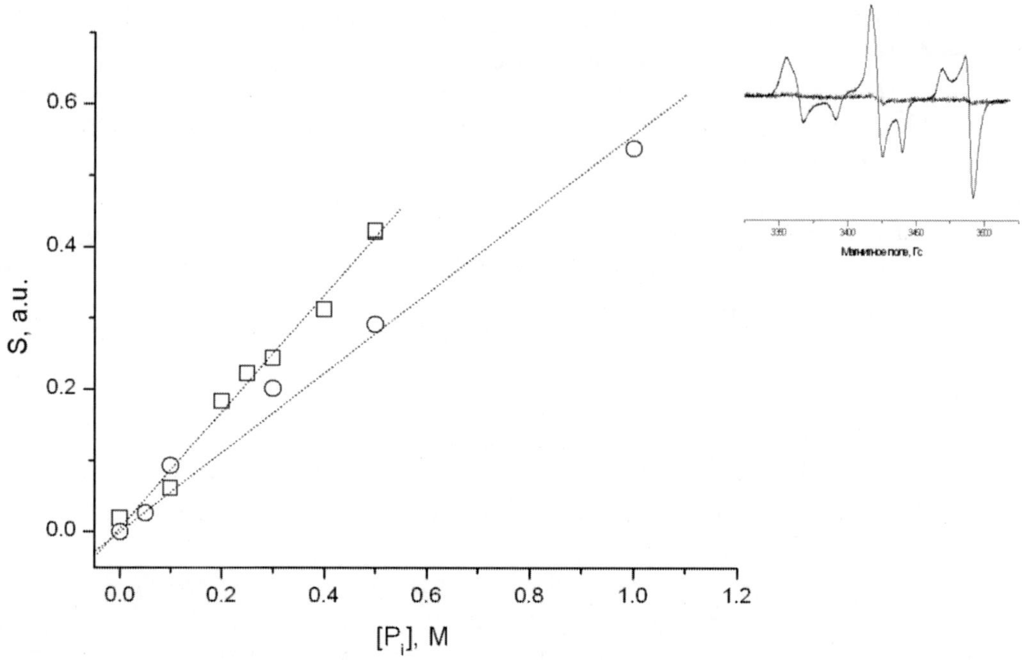

Figure 5. Yields of atoms (*1 circles*) H and (*2 squares*) NO_3^{2-} in Trp + $H_2PO_4^-$ systems without electron acceptor KNO_3 and Trp + $H_2PO_4^-$ + KNO_3 systems with electron acceptor vs. $H_2PO_4^-$ concentration on exposure to UV radiation at $\lambda \geq 240$ nm (the *y* axis shows the double integrals of EPR spectra of H atoms and radicals NO_3^{2-} reduced to the same microwave power). Reagent concentrations: Trp, 5 10^{-4}; $H_2PO_4^-$ 0.5 M; KNO_3 0.01 M (*3, 4*) Inset: EPR spectra of NO_3^{2-} at pH 5.0 recorded on UV irradiation of an aqueous solution of Trp and KNO_3 only in the presence of phosphate.

The effect was attributed to the enhancement of charge separation of the photoproduced primary ion pair due electron injection from the photoreaction centers into the electron relay system and the succeeding electron migration.

In this study, detection of the EPR spectra of atomic hydrogen $H^{\bullet}$ (g = 2.0023, A = 50.6 mT) in frozen aqueous solutions initiated by low energy photoinjected electrons was crucial for identification of the electron capture by oxyacids AOH. The detection of the hydrogen atom upon energy impact below the excitation thresholds of water or oxyacids (photon energy of not more than 4.5 eV, Figure 1) in a frozen aqueous medium at different pH values serves as unambiguous proof for dissociative capture of the low_energy photoinjected electron by AOH acids with different p*Kα* [6-8].

Thus our studies of low energy electron attachment to the bound acidic proton of oxyacid $AO^{\delta-}$–$H^{\delta+}$ containing easily controllable polarized bond OH following distant acceptor reduction can be useful for more precise elucidation of the acid catalysis mechanism. The data on direct electron attachment to an oxyacid that creates an acidic bound [H+] network under certain conditions [28] allows one to consider direct participation of the acidic bound proton in the intermolecular ET as one electron relay.

CONCLUSION

In our opinion, the mechanism of acid catalysis of redox reactions is based on activation of resonance capture of low energy (≤4 eV) electrons by the acidic proton of XH molecules (or their functional groups OH, NH, CH); the acidity (electrophilicity) of the bound proton controlled by specific catalysis determines the rate of ET reactions.

One of the most remarkable features of our results (anticipated from theoretical and experimental studies on the atomic hydrogen dynamics in pH-dependent photoelectron capture) is the observation that a bound acidic proton $AO^{\delta-}$ - $H^{\delta+}$ can be highly effective in mediating intramolecular electron transport.

ACKNOWLEDGMENTS

We are grateful to Academician A.E. Shilov for permanent attention to the work and fruitful discussion of the results.

This work was supported by the Program of the Division of Chemistry and Material Science of the RAS "Theoretical and Experimental Studies of the Nature of the Chemical Bond and Mechanisms of Chemical Reactions and Processes."

REFERENCES

[1] Bolton J.R. (1996) Solar photoproduction of hydrogen: a review. *Solar energy* 57, 37-50.

[2] Barbara P. F., Meyer T. J., Ratner M. A. (1996) Contemporary Issues in Electron Transfer Research. *J. Phys. Chem.* 100, 13148–13168.

[3] Esswein A. J. and Nocera D.G. (2007) Hydrogen Production by Molecular Photocatalysis. *Chem. Rev.* 107, 4022–4047.

[4] Closs G.L., Miller J.R. (1988) Intramolecular Long-Distance Electron Transfer in Organic Molecules. *Science* 240, 440 – 447.

[5] Paddon-Row M.N. (1994) Investigating Long-Range Electron-Transfer Processes with Rigid, Covalently Linked Donor-(Norbornylogous Bridge)-Acceptor Systems. *Acc. Chem. Res.* 27, 18-25.

[6] Gray H. B. and Winkler J. R. (2005) Long-range electron transfer. *Proc. Natl. Acad. Sci. U S A.* 102, 3534–3539.

[7] Rice S. A. (1980) Do Excess Electrons Migrate from Trap to Trap in Low Temperature Molecular Glasses? Some Energetic and Kinetic Considerations. *J. Phys. Chem.* 84, 1280-1286.

[8] Brzhevskaya O. N., Degtyarev E. N., Levin P. P., Lozinova T. A., and Nedelina O. S. (2005). Hydrogen Atoms in Frozen Phosphate-Containing Aqueous Media Photosensitized by Biological Cofactors. *Doklady Biochemistry and Biophysics* 405, 395–399. *Translated from Doklady Akademii Nauk,* 405, 259-263.

[9] Degtyarev E. N., Brzhevskaya O. N. and O. S. Nedelina (2006) Detection of the Products of Acceptor Interaction between Orthophosphate and Photoejected Electron by EPR Spectroscopy. *Doklady Biochemistry and Biophysics* 410, 286–291. *Original*

Russian Text © E.N. Degtyarev, O.N. Brzhevskaya, O.S. Nedelin, published in Doklady Akademii Nauk, 2006, 410, 542–547.

[10] Degtyarev E. N., Brzhevskaya O. N. and O. S. Nedelina (2009) Inorganic Oxyacids $AO^{\delta-}$-$H^{\delta+}$ as One_Electron Relay in Acid Catalysis of Redox Reactions *Doklady Chemistry,* 428, 233–237. *Original Russian Text © O.S. Nedelina, O.N. Brzhevskaya, E.N. Degtyarev, 2009, published in Doklady Akademii Nauk*, 2009, Vol. 428, No. 4, pp. 474–479.

[11] Lesclaux R.and Joussot-Dubien J. (1973), Electron photoinjection from aromatic molecules in condensed media. In *Organic Molecular Photophysics* edited by J.B.Birks (Wiley, London) Vol.1, p.457-487.

[12] *Massey* H. S. W., *Negative Ions* (Cambridge Univ. Press, Cambridge, England, 1976), 3rd ed.

[13] Ingólfsson O., Weik F. and Illenberger E. (1996) The reactivity of slow electrons with molecules at different degrees of aggregation: gas phase, clusters and condensed phase *Int.J.Mass Spectrom.* 155, 1-68.

[14] Simons J. (2005) Response of a molecule to adding or removing an electron. *Adv.Quantum Chem.* 50, 213-233.

[15] Liu B, Hvelplund P., Nielsen S.B., (2004) Hydrogen loss from nucleobase nitrogens upon electron attachment to isolated DNA and RNA nucleotide anions. *J. Chem. Phys.* 121, 4175-4179.

[16] Burrow P.D., Gallup G.A. , Modelli A. (2008) Are there π*Shape resonance in electron scattering from phosphate group? *J. Phys.Chem.A* 112, 4106-4113.

[17] Pelc A., Sailer W., Scheier P., Probst M., .Mason N.J., Illenberger E., Mark T.D. (2002).

[18] Dissociative electron attachment to formic acid (HCOOH) *Chemical Physics Letters.* 361, 277-284.

[19] Vasil'ev Y.V, Muftakhov M.V., Tuimedov G.M., Khatymov R.V., AbzalimovR.R., Mazunov V.A., Drewello T. (2001) Specific formation of $(M\text{-}H)^-$ ions from OH-group containing molecules. *Int.J.Mass Spectrom.* 205, 119-135.

[20] Prabhudesai V.S., Kelkar A. H., Nandi D, and Krishnakumar E. (2005) Functional Group Dependent Site Specific Fragmentation of Molecules by Low Energy Electrons. PRL *Phys.Rev.Lett.* 95, 143202 (1-4).

[21] Khrushch A.P., Shilov A.E., Vorontzova T.A. (1974) Role of acids in reduction of acetylene catalyzed by molybdenum-thiol complexes. *J.Amer.Chem.Soc.* 96. 4987-4989.

[22] Fukuzumi S., Ishikawa K., Hironaka K., and Tanaka T. (1987) Acid Catalysis in Thermal and Photoinduced Electron-Transfer Reactions. *J. Chem. Soc., Perkin Trans. 2*, 751-760.

[23] Irebo T., Reece S.Y., Sjodin M., Nocera D.G., Yammarstrom L. (2007) Proton-coupled Electron Transfer of Tyrosine Oxidation: Buffer Dependence and Parallel Mechanisms. *J.Amer.Chem.Soc.* 129, 15462-15464.

[24] Kevan L., Fine C. (1966) Trapped hydrogen yields and precursors in frozen solution radiolysis. *J. Amer. Chem. Soc.* 88, 896–872.

[25] Hart, E.J. and Anbar, M., *Hydrated Electron*, New York: Wiley, 1970. Translated under the title *Gidratirovannyi elektron*, Moscow: Atomizdat, 1973.

[26] Rustad J.R., Dixon D. A., Kubicki J. D. and Felmy A. R. (2000) Gas-Phase Acidities of Tetrahedral Oxyacids from ab Initio Electronic Structure Theory. *J. Phys. Chem. A*, 104, 4051–4057.

[27] Smiechowski M., Gojło E. and Stangret J. (2009) Systematic Study of Hydration Patterns of Phosphoric(V) Acid and Its Mono-, Di-, and Tripotassium Salts in Aqueous Solution *J. Phys. Chem. B*, 113, 7650–7661.

[28] Gupta K, D.R. Roy D.R., Subramanian V. and Chattaraj P.K. (2007) Are strong Brønsted acids necessarily strong Lewis acids? *Journal of Molecular Structure: THEOCHEM*, 13-24.

[29] Reece S.Y., Hodgkiss J.M., Stubbe J.A., Nocera D.A. (2006) Proton-coupled electron transfer: the mechanistic underpinning for radical transport and catalysis in biology, *Phil.Trans.R.Soc.B.* 361, 1351-1364.

In: Biology and Communication in the Information Age
Editor: Rafiqul Islam
ISBN: 978-1-62808-252-4

Chapter 21

Formation of Supramolecular Nanostructures Based on Catalytically Active Heteroligand Nickel Complexes with The Assistance of Intermolecular Hydrogen Bonds

L. I. Matienko**, L. A. Mosolova, V. I. Binyukov, E. M. Mil and G. E. Zaikov
Federal State Budget Institution of Science
N. M. Emanuel Institute of Biochemical Physics,
RAS, Moscow, Russian Federation

Abstract

The AFM method has been used for research of possibility of the stable supramolecular nanostructures formation on the basis of nickel $Ni_xL^1_y(L^1_{ox})_z(L^2)_n(H_2O)_m$ and $Ni^{II}(acac)_2$·NaSt·PhOH complexes with the assistance of intermolecular H-bonds.

Keywords: Nanostructures, heteroligand nickel complexes $Ni_xL^1_y(L^1_{ox})_z(L^2)_n(H_2O)_m$ and $Ni^{II}(acac)_2$·NaSt·PhOH, H-bonds, catalysis, ethyl benzene, oxidation, dioxygen, model of Ni^{II}-ARD dioxygenase

I. Aim and Background

Research of structure and catalytic activity of complexes of nickel in different reactions of oxidation with molecular O2 causes heightened interest of researchers last years in

* Fax (7-495) 137 41 01, e-mail: matienko@sky.chph.ras.ru (L.I. Matienko), e-mail: chembio@sky.chph.ras.ru (G.E. Zaikov).

connection with that nickel as an essential catalytic cofactor of enzymes is found in eubacteria, archaebacteria, fungi, and plants. These enzymes catalyze a diverse array of reactions that include both redox and nonredox chemistries [1].

> Last years development of researches in sphere of a homogeneous catalysis occurs in two directions – free radicals chain catalytic oxidation and a catalysis with complexes of the metals modeling action of enzymes. The majority of the reactions which are carried out commercially are based on autoxidation reactions. Low exits of products of oxidation counting on the spent hydrocarbon, caused by fast deactivation of the catalyst, are the main obstacle for commercially application of majority of biomimetic systems. At the same time conclusions about the mechanism of dioxygenases action and their models appear rather useful at treatment of the mechanism of catalysis with metals complexes in processes of oxidation of hydrocarbons by molecular O_2 [2,3]. On the other hand results of research of catalysis of processes of oxidation can be used at studying of the mechanism of enzymes operation.

We developed the method for enhancing the catalytic activity of transition metal complexes in the processes of alkylarens (ethyl benzene, cumene) oxidation with dioxygen to afford the corresponding hydro peroxides [2,3], intermediates in the large-scale production of important monomers. [4].

This method consists of introducing additional mono- or multidentate modifying ligands into catalytic metal complexes. The activity of systems $\{ML^1_n + L^2\}$ (M=Ni, Fe, L^1 = $acac^-$, L^2 = N-methylpirrolidon-2 (MP), HMPA, MSt (M=Na, Li, K), crown ethers or quaternary ammonium salts) is associated with the fact that during the ethyl benzene oxidation, the primary $(M^{II}L^1_2)_x(L^2)_y$ complexes and the real active catalytic heteroligand $M^{II}_xL^1_y(L^1_{ox})_z(L^2)_n(H_2O)_m$ ("A") complexes are formed to be involved in the oxidation process [2,3]. As it is established by us that the mechanism of formation of active complexes ("A") is similar to action of ARD enzymes, complexes "A" can be considered as models of Ni^{II}-ARD [5].

Thus, the conclusions, received in the present work, can be useful at discussion of mechanism of Ni^{II}-ARD dioxygenase operation. On the other hand we discovered unusual effect of significant increase in efficiency of binary systems $\{Ni^{II}(acac)_2+L^2\}$ (electron-donating ligand-modifier L^2 = MSt (M=Na, Li), MP (MP=N-methylpirrolidon-2), HMPA) as the selective catalyst of ethyl benzene oxidation into α-phenyl ethyl hydro peroxide (PEH), with the introduction in catalytic system of the third component – phenol. The advantage of these ternary systems is the long-term activity of the in situ formed complexes $Ni^{II}(acac)_2{\cdot}L^2{\cdot}PhOH$ [2,3].

The most effective binary and triple catalytic systems as catalysts of ethyl benzene oxidation into α-phenyl ethyl hydro peroxide (PEH) with dioxygen on parameters C ($C = \Delta[RH]/[RH]_0$ % (at $S_{PEH} \geq 90$ % ($S_{PEH,max}$=99-90% (L^2=18C6), $S_{PEH,max}$=97-90% (L^2=Me_4NBr), $S_{PEH,max}$=90% (L^2=MP, NaSt)) and $[PEH]_{max}$ (mass%) are presented on Figure 1. As one can see, the best results were obtained in the case of catalysis with system $\{Ni^{II}(acac)_2$+ NaSt + PhOH$\}$ (Figure 1).

The preservation of high activity of catalysts–heteroligand complexes $Ni^{II}_xL^1_y(L^1_{ox})_z(L^2)_n(H_2O)_m$ ("A") and heterobinuclear heteroligand triple complexes $Ni^{II}(acac)_2{\cdot}NaSt{\cdot}PhOH$, during oxidation seems to be due to formation of the stable supramolecular structures on the basis of ("A"), or triple complexes concluding PhOH, with assistance of intermolecular H-bonds.

This hypothesis is evidenced by AFM (Atomic Force Microscopy) data. Thus we have offered the new approach to research of mechanism of catalysis with use of AFM technique [6,7].

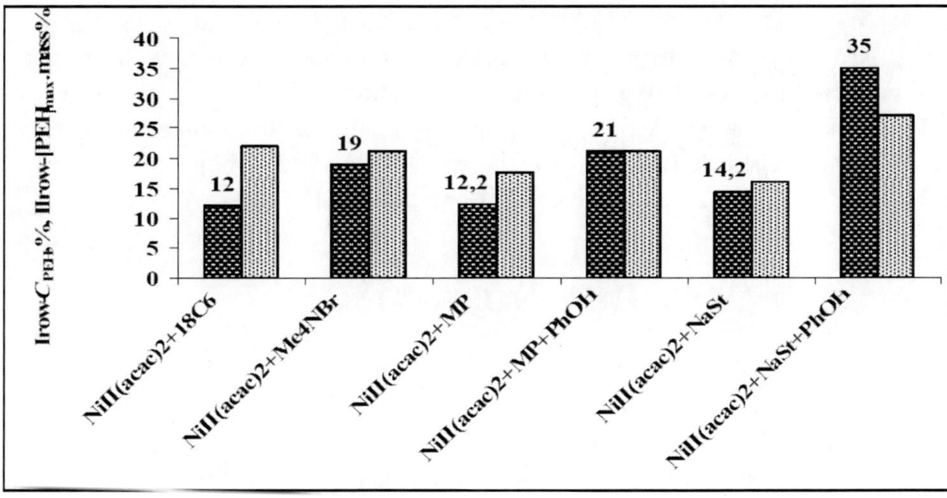

Figure 1. Values of conversion C_{PEH} (%), maximum values of hydro peroxide concentrations $[PEH]_{max}$ (mass.%) in reactions of ethylbenzene oxidation in the presence of the best binary and triple catalytic systems {Ni(II)(acac)$_2$+L^2} and {Ni(II)(acac)$_2$+L^2+PhOH} (L^2 = 18C6, Me_4NBr, MP, NaSt), 120°C.

II. INTRODUCTION

Nanostructure science and supramolecular chemistry are fast evolving fields that are concerned with manipulation of materials that have important structural features of nanometer size (1 nm to 1 μm) [8,9]. Nature has been exploiting noncovalent interactions for the construction of various cell components. For instance, microtubules, ribosomes, mitochondria, and chromosomes use mostly hydrogen bonding in conjunction with covalently formed peptide bonds to form specific structures. The self-assembled systems and self-organized structures mediated by transition metals are considered in connection with increasing research interest in chemical transformations with use of these systems [10].

Hydrogen bonds vary enormously in bond energy from ~15–40 kcal/mol for the strongest interactions to less than 4 kcal/mol for the weakest. It is proposed, largely based on calculations, that strong hydrogen bonds have more covalent character, whereas electrostatics are more important for weak hydrogen bonds, but the precise contribution of electrostatics to hydrogen bonding is widely debated [11]. Hydrogen bonds are important in non-covalent aromatic interactions, where π- electrons play the role of the proton acceptor, which are a very common phenomenon in chemistry and biology. They play an important role in the structures of proteins and DNA, as well as in drug receptor binding and catalysis [3,12].

H-bonding can be a remarkably diverse driving force for the self-assembly and self-organization of materials. H-bonds are commonly used for the fabrication of supramolecular assemblies because they are directional and have a wide range of interactions energies that are tunable by adjusting the number of H-bonds, their relative orientation, and their position in the overall structure. H-bonds in the center of protein helices can be ca. 20 kcal/mol due to cooperative dipolar interactions [13, 14]. One of the reasons of stability of heteroligand complexes $M^{II}{}_xL^1{}_y(L^1{}_{ox})_z(L^2)_n(H_2O)_m(L^1{}_{ox} = CH_3COO^-)$ ("A") and heteroligand triple complexes $Ni^{II}(acac)_2{\cdot}L^2{\cdot}PhOH$ in the conditions of oxidation may be a formation of intermolecular H-bonds. [2,3]. We used the AFM method in analytics aims, for research of possibility of the supramolecular nanostructures formation on basis of nickel heteroligand complexes $Ni_2(AcO)_3(acac)L^2{\cdot}2H_2O$ (L^2 = MP) and binuclear heteroligand $Ni^{II}(acac)_2{\cdot}NaSt{\cdot}PhOH$ with the assistance of H-bonding [6,7].

III. Experimental

AFM SOLVER P47/SMENA/ with Silicon Cantilevers NSG11S (NT MDT) with curvature radius 10 nm, tip height: 10 – 15 μm and cone angle ≤ 22° in taping mode on resonant frequency 150 KHz was used.

As substrate the polished Silicone surface special chemically modified was used.

Waterproof modified Silicone surface was exploit for the self-assembly-driven growth due to H-bonding of heteroligand complexes $Ni_2(AcO)_3(acac)MP{\cdot}2H_2O$ ("A") and $\{Ni^{II}(acac)_2{\cdot}NaSt{\cdot}PhOH\}_n$ with Silicone surface.

The saturated solution of complex ("A") in water was put on a surface, maintained some time, and then water deleted from a surface by means of special method – spin-coating process.

In the course of scanning of investigated samples it has been found, that the structures, based on ("A"), are fixed on a surface not strongly enough to spend measurements in contact mode. So samples are well measured in taping mode. At that in additional experiments it has been shown that at scanning there is a turning movement and orientation of investigated structures with cantilever.

The saturated chloroform ($CHCl_3$) solution of complex $Ni^{II}(acac)_2{\cdot}NaSt{\cdot}PhOH$ (1:1:1) was put on a surface, maintained some time, and then solvent was deleted from a surface by means of special method – spin-coating process.

In the course of scanning of investigated samples it has been found, that the structures are fixed on a surface strongly enough due to H-bonding. The self-assembly-driven growth of the supramolecular structures formation on the basis of complexes Ni(II)(acac)2·NaSt·PhOH due to H-bonds and perhaps the other non covalent interactions was observed on Silicone surface. One can watch these structures with big height and volume (Figures 4-6). In check experiments it has been shown that for binary systems {Ni(II)(acac)2+NaSt}, and {Ni(II)(acac)2+PhOH} the formation of the similar structures (exceeding on height of 2-10 nanometers) is not observed.

IV. Results and Discussion

Role of H-Bonding in Stabilization of Catalytic Complexes $Ni_x(acac)_y(AcO)_z(L^2)_n(H_2O)_m$ ("A")

It is well known that transition metal β-diketonates are involved in various substitution reactions. Methine protons of chelate rings in β-diketonate complexes can be substituted by different electrophiles. Formally, these reactions are analogous to the Michael addition reactions [3]. This is a metal-controlled process of the C–C bond formation. The complex $Ni^{II}(acac)_2$ is the most efficient catalyst of such reactions [3].

We have established that the electron-donating ligand L^2, axially coordinated to $M^{II}L^1_2$ (M=Ni, Fe, L^1 =acac$^-$), controls the formation of primary active complexes $M^{II}L^1_2{\cdot}L^2$ and the subsequent reactions in the outer coordination sphere of these complexes. The coordination of an electron-donating extra-ligand L^2 with an $M^{II}L^1_2$ complex favorable for stabilization of the transient zwitter-ion $L^2[L^1M(L^1)\cdot O_2^-]$ enhances the probability of regioselective O_2 addition to the methine C–H bond of an acetylacetonate ligand, activated by its coordination with metal ions. The outer-sphere reaction of O_2 incorporation into the chelate ring depends on the nature of the metal.

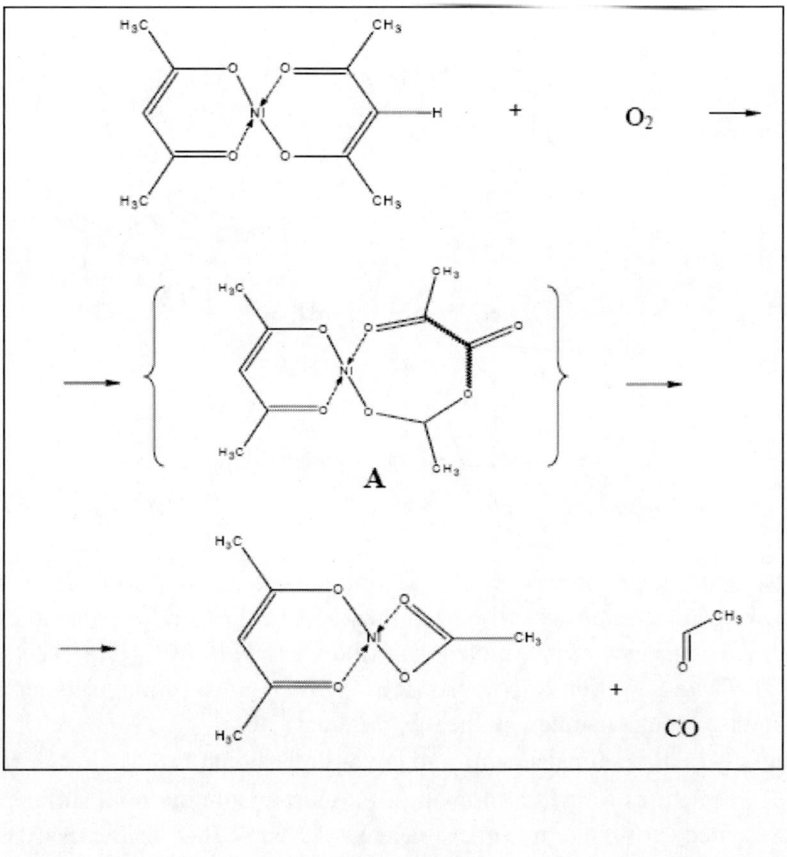

Scheme 1.

Transformation routes of a ligand $(acac)^-$ for Ni and Fe are various, but lead to formation of similar heteroligand complexes $M^{II}{}_xL^1{}_y(L^1{}_{ox})_z(L^2)_n(H_2O)_m(L^1{}_{ox} = CH_3COO^-)$ ("A") [2,3]. For nickel complexes, the reaction of acac-ligand oxygenation follows a mechanism analogous to those of Ni^{II}-containing dioxygenase ARD [5] or Cu- and Fe-containing quercetin 2,3-dioxygenases [15,16]. Namely, incorporation of O_2 into the chelate acac-ring was accompanied by the proton transfer and the redistribution of bonds in the transition complex leading to the scission of the cyclic system (A) to form a chelate ligand OAc^-, acetaldehyde and CO (in the Criegee rearrangement) (Scheme 1).

It is known that heteroligand complexes are more active in relation to reactions with electrophiles in comparison with homoligand complexes [3]. Thus the stability of heteroligand complexes $Ni_x(acac)_y(OAc)_z(L^2)_n(H_2O)_m$ with respect to conversion into inactive form $Ni(OAc)_2$ seems to be due to the formation of intermolecular H-bonds. The possibility of supramolecular nanostructures formation was investigated on the example of active complex of type ("A"), formed in the course of ethyl benzene oxidation, catalyzed with system $\{Ni^{II}(acac)_2 + MP\}$. Complex has been synthesized by us and its structure has been defined with mass spectrometry, electron and IR spectroscopy and element analysis [2,3]. The certain structure of a complex $Ni_2(OAc)_3(acac)MP\cdot 2H_2O$ ("A") (L^2=MP, x=2, y=1, z=3, n=1, m=2) corresponds predicted on the basis of the kinetic data [2,3]. Prospective structure of the complex $Ni_2(OAc)_3(acac)MP\cdot 2H_2O$ ("A") is presented with Scheme 2.

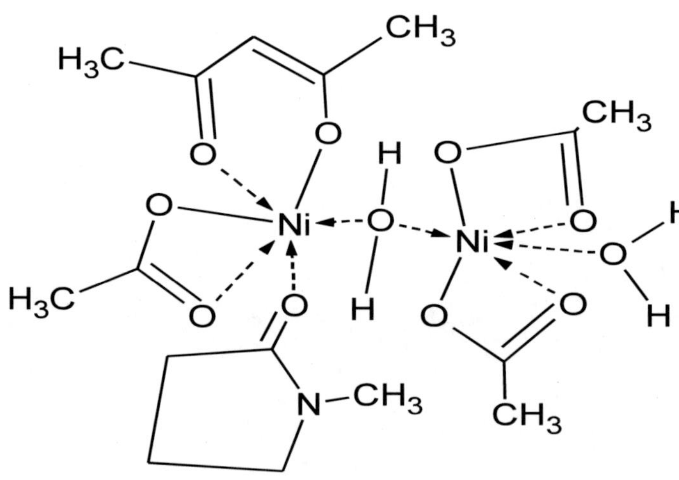

Scheme 2.

On the basis of the known from the literature facts it was possible to assume that binuclear heteroligand complexes $Ni_2(OAc)_3(acac)\cdot MP\cdot 2H_2O$ are capable to form macro structures with the assistance of intermolecular H-bonds (H_2O – MP, H_2O – acetate (or $acac^-$) group) [17, 18]. The association of $Ni_2(OAc)_3(acac)MP\cdot 2H_2O$ to supramolecular structures as result of H-bonding is demonstrated on the next Figures (1-3).

On Figure 1-3 three-dimensional and two-dimensional AFM image (10×10(μm), 2×2(μm)) of the structures formed at drawing of a uterine solution on a surface of modified silicone are presented (μm = micron (micrometer) = 10^3 nm). It is visible that the majority of the generated structures have rather similar form of three almost merged spheres.

As it is possible to see in Figure 1,a and 3, except particles with the form reminding three almost merged spheres (Figure 1,b), there are also structures of more simple form (with the height approximately equal 3-4 nm). On Figure 3,c a profile of one of the particles of the size 3-4 nm on height is presented. The distribution histogram (Figure 2,b) shows that the greatest number of particles - is particles of the size 3-4 nm on height.

From the received data the following is visible. It is important to notice that for all structures the sizes in plane XY do not depend on height on Z. They make about 200 nm along a shaft which are passing through two big spheres, and about 150 nm along a shaft crossing the big and smaller spheres. But all structures are various on heights from the minimal 3-4 nm to ~ 20-25 nm for maximal values. In distribution on height there is a small quantity of particles with maximum height 20-25 nm and considerably smaller quantity with height to 35 nm (Figure 1,b).

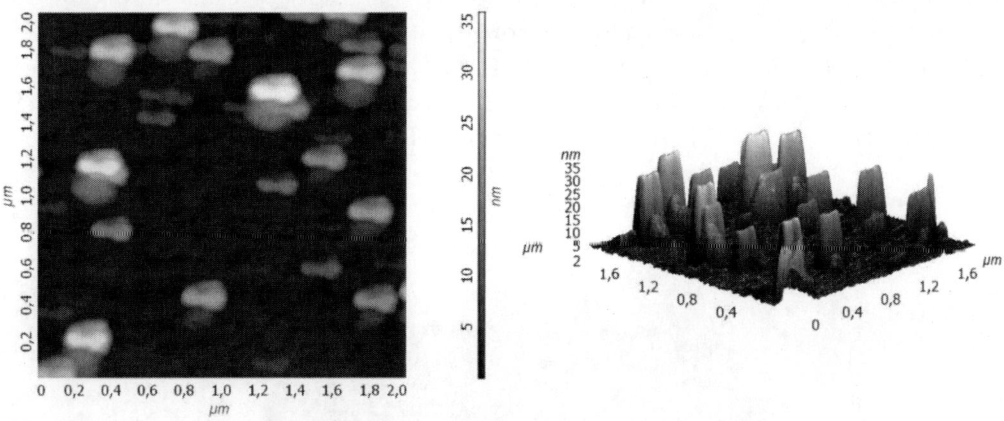

Figure 1. The AFM two- (a) and three-dimensional (b) images of nanoparticles on the basis $Ni_2(OAc)_3(acac)MP\cdot 2H_2O$, formed on the surface of modified silicone.

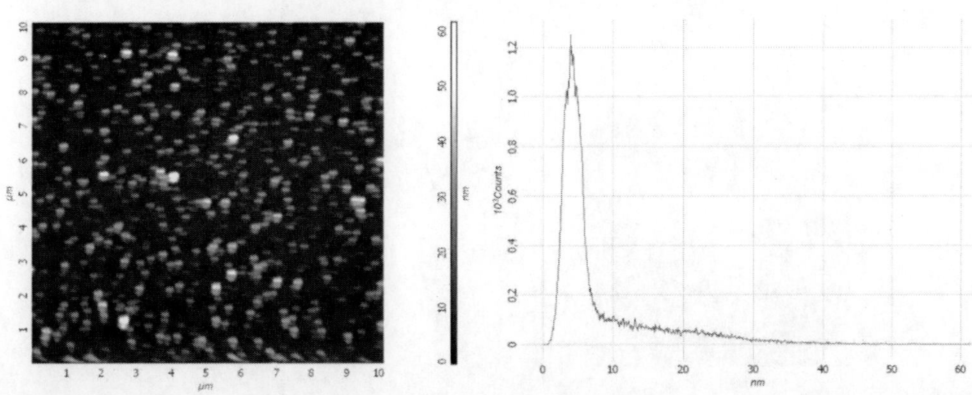

Figure 2. AFM image of structures (on the basis of $Ni_2(OAc)_3(acac)MP\cdot 2H_2O$) on the modified silicone surface 10×10 μm (at the left). The distribution histogram on height of nanoparticles (to the right).

Thus, we have shown what the self-assembly-driven growth seems to be due to H-bonding of binuclear heteroligand complex $Ni_2(OAc)_3(acac)\cdot MP\cdot 2H_2O$ ("A") with a surface of modified silicone, and further due to directional intermolecular H-bonds, apparently at participation of H_2O molecules, $acac^-$, OAc^- groups, MP [17, 18] (see Scheme 2).

Role of H-bonding in stabilization of triple catalytic complexes $Ni^{II}(acac)_2\cdot L^2\cdot PhOH$. Nanoparticles $\{Ni^{II}(acac)_2\cdot L^2\cdot PhOH\}_n$ *formation* as evidence of possibility of association of $Ni^{II}(acac)_2\cdot L^2\cdot PhOH$ complexes in supra molecular nanostructures in the course of the ethyl benzene oxidation.

Earlier we established H-bonding interactions in mechanism of formation of catalytic complexes on $Ni^{II}(acac)_2\cdot L^2\cdot PhOH$ (L^2 =NaSt(MP)) in the course of ethyl benzene oxidation in the presence of triple systems $\{Ni^{II}(acac)_2+L^2+PhOH\}$[3]. Besides the high activity (and stability) of triple complexes on $Ni^{II}(acac)_2\cdot L^2\cdot PhOH$ (L^2 =NaSt) as catalysts of the ethyl benzene oxidation into α-phenyl ethyl hydro peroxide (it is expressed in substantial growth of degree of conversion of RH and the yields of PEH) seems to be due to the formation of supra molecular structures $\{Ni^{II}(acac)_2\cdot NaSt\cdot PhOH\}_n$ in the course of ethyl benzene oxidation as a result of intra- and intermolecular (phenol–carboxylate) H-bonds [18].

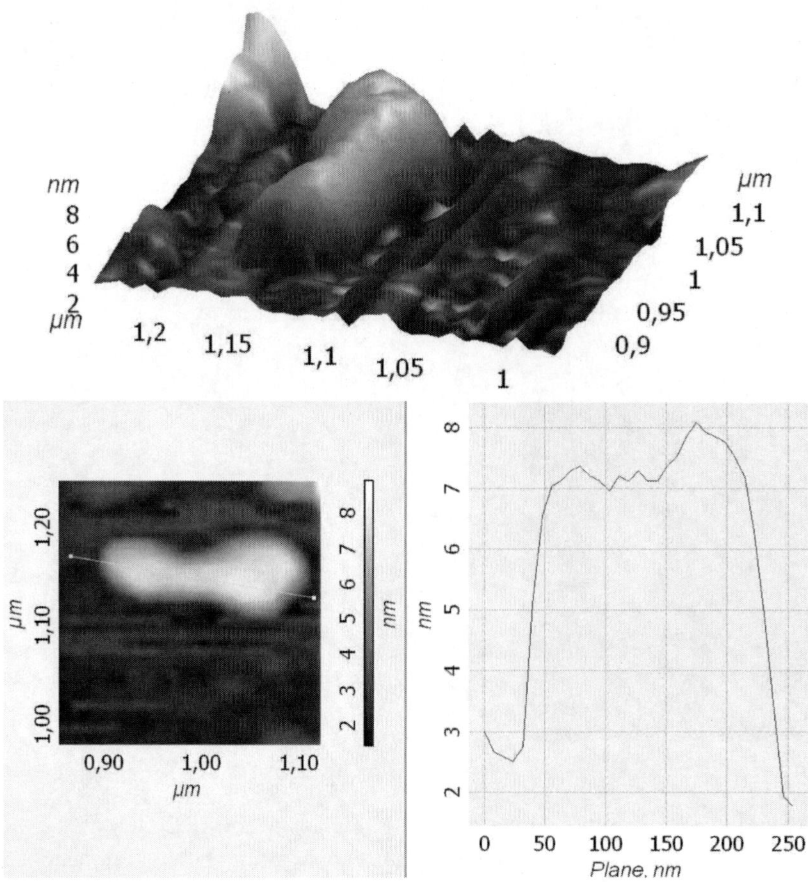

Figure 3. The AFM three- (a) and two-dimensional (b) image and profile of the structure (c) (on the basis of $Ni_2(OAc)_3(acac)MP\cdot 2H_2O$) with minimum height along the greatest size in plane XY.

The stability of triple complexes in relation to the processes leading to their deactivation during reaction of ethyl benzene oxidation, seems to be due to electronic and spatial factors operating against transformation of the catalyst in the inactive form as it takes place in the case of catalysis with binary systems {$Ni^{II}(acac)_2$+NaSt(MP)} at the absence of additives of PhOH [2,3].

The possibility of association of triple complexes on $Ni^{II}(acac)_2$·NaSt·PhOH to supramolecular structures as result of H-bonding interactions is followed from data of AFM analysis, received by us. Results are presented on the next Figures (4-6). Figures 4-6 demonstrated three-dimensional and two-dimensional AFM image (30×30(μm), and 10×10(μm)) of the structures on the basis of triple complexes $Ni^{II}(acac)_2$·NaSt·PhOH, formed at drawing of a uterine solution on a surface of modified silicone.

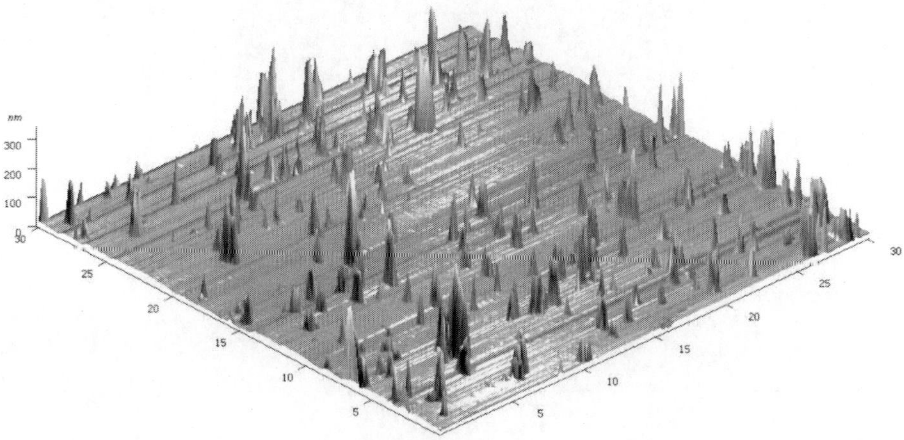

Figures 4, 5. The AFM three-dimensional image (30×30 and 10×10 (μm)) of the structures formed on a surface of modified silicone on the basis of triple complexes $Ni^{II}(acac)_2$·NaSt·PhOH.

In the Figure 7 the histogram of height mean values of the AFM images of nanoparticles on the basis of complexes $Ni^{II}(acac)_2 \cdot NaSt \cdot PhOH$ is presented. As one can see, structures are various at heights from 25 to ~ 250–300 nm for maximal values. The distribution histogram shows that the greatest numbers are particles of the mean size of 50-100 nm at height.

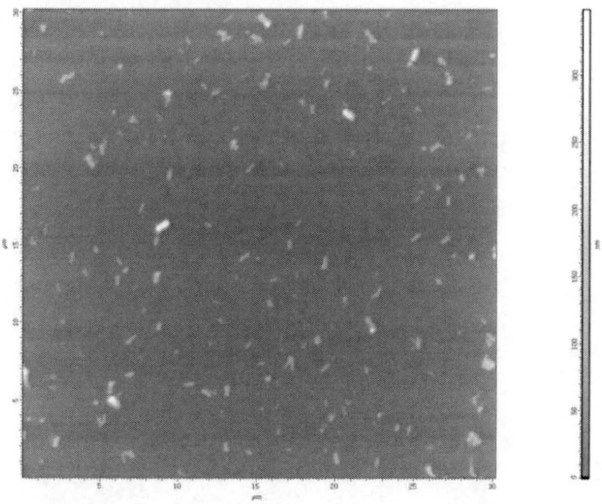

Figure 6. The AFM two-dimensional image of nanoparticles on the basis $Ni(II)(acac)_2 \cdot NaSt \cdot PhOH$, formed on the surface of modified silicone.

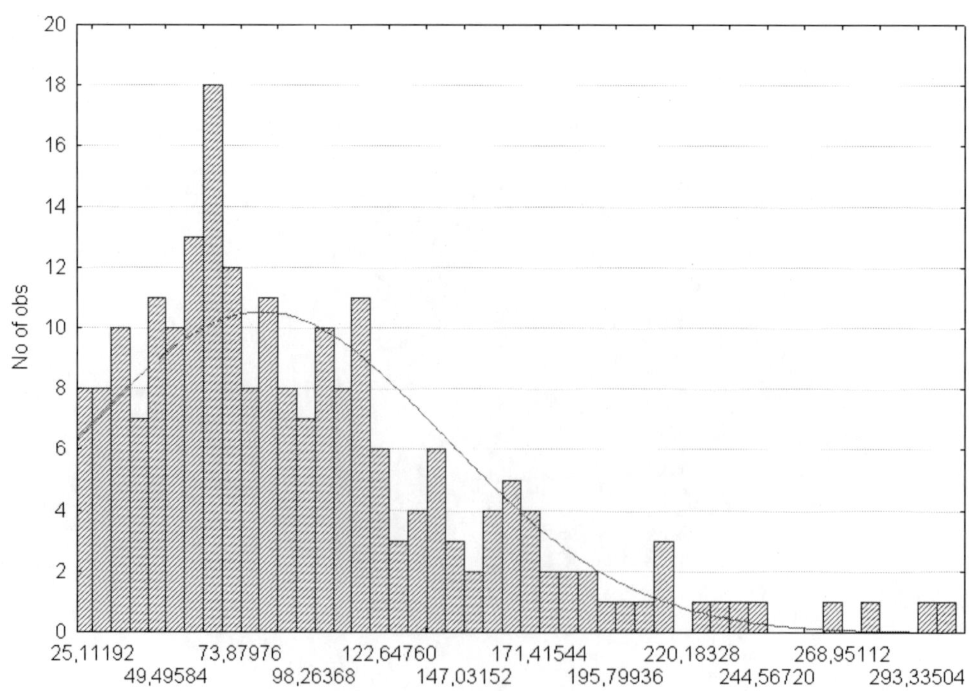

Figure 7. Histogram of height mean values (nm) of the AFM images of nanostructures based on $Ni^{II}(acac)_2 \cdot NaSt \cdot PhOH$, formed on the surface of modified silicone.

Table 1 shows the mean values of area, volume, height, width, length of nanoscale structures on the basis of triple complexes on $Ni^{II}(acac)_2{\cdot}NaSt{\cdot}PhOH$ formed on the surface of modified silicone.

We revealed an interesting fact that the length of the formed nanoparticles in the XY plane exceeds the width of nanoparticles about three times (Table 1).

Table 1. The mean values of area, volume, height, length, width of the AFM nanoparticles images on the basis of on $Ni^{II}(acac)_2{\cdot}NaSt{\cdot}PhOH$ formed on the surface of modified silicone

Variable	Mean values	Confidence -95.000 %	Confidence +95.000 %
Area, μm^2	0.13211	0.11489	0.14933
Volume, μm^3	14.11354	11.60499	16.62210
Height (*Z*), nm	80.56714	73.23940	87.89489
Length, μm	0.58154	0.53758	0.62549
Width, μm	0.19047	0.17987	0.20107

CONCLUSION

1. Thus, we have shown with use of AFM method the self-assembly-driven growth due to H-bonding of heteroligand complex $Ni_2(AcO)_3(acac)MP{\cdot}2H_2O$ ("A") with a surface of modified silicone, and further due to directional inter-molecular H-bonds, apparently at participation of H_2O molecules, acac, acetate groups, MP [17,18]. The role of van Der Waals-attractions and π-π bonding seems to take into account.

H-bonding seems to be one of the factors, responsible for the stability of real catalysts – heteroligand complexes $Ni_x(acac)_y(AcO)_z(L^2)_n(H_2O)_m$, in the course of alkylarens (ethyl benzene, cumene) oxidation by dioxygen into hydro peroxide (intermediates in the large-scale production of important monomers) at the presence of catalytic systems $\{Ni^{II}(acac)_2 + L^2\}$.

As we established, that mechanism of complexes "A" formation is analogous to Acireductone dioxygenase Ni^{II}-ARD action, the received results can be use in interpretation of biology effects in Ni^{II}-ARD operation. The transformation of catalytic complexes into supramolecular nanostructures in assistance of intermolecular H-bonding may be one of the factors of regulation of Ni^{II}-ARD activity in plants and living organisms.

2. With use of AFM method we have shown the self-assembly-driven growth on the basis of binuclear heteroligand triple complexes $Ni^{II}(acac)_2{\cdot}NaSt{\cdot}PhOH$ due to H-bonding with the modified silicone surface, and further the supramolecular nanostructures formation $\{Ni^{II}(acac)_2{\cdot}NaSt{\cdot}PhOH\}_n$ due to directional intermolecular (phenol-carboxylate) H-bonds [18], and, possibly, other non-covalent interactions (van Der Waals-attractions, π-π bonding).

These data support the very probable supramolecular structures appearance on the basis of heterobinuclear heteroligand triple complexes $Ni^{II}(acac)_2{\cdot}NaSt{\cdot}PhOH$ in the course of the ethyl benzene oxidation with dioxygen, catalyzed by three-component catalytic system $\{Ni^{II}(acac)_2{+}NaSt{+}PhOH\}$ and therefore the high values of the conversion of the ethyl benzene into α-phenyl ethyl hydro peroxide at the selectivity S_{PEH} preservation at the level not below S_{PEH} = 90 % in this process.

REFERENCES

[1] Y. Li, D.B. Zamble, (2009). *Chem. Rev.,* 109, 4617.

[2] Matienko, L.I. (2007). In: *Reactions and Properties of Monomers and Polymers*, D'Amore, A. and Zaikov, G. (Eds.), Chapter 2, Nova Sience Publ., Inc.: New York, 21.

[3] Matienko, L.I., Mosolova, L.A., Zaikov, G.E. (2010). Selective Catalytic Hydrocarbons Oxidation. New Perspectives, Nova Science Publ., Inc.: New York, USA. 158P.

[4] Weissermel, K., Arpe, H.-J. (1997). *Industrial Organic Chemistry*, 3nd ed., transl. by Lindley C.R. VCH: New York, USA. 427P.

[5] Dai, Y., Pochapsky, Th. C., Abeles, R.H. (2001). *Biochemistry*, 40, 6379.

[6] Matienko, L.I., Binyukov, V.I., Mosolova, L.A., and Mil, E.M. (2011). *Polymer Yearbook 2011.* Nova Science Publ., Inc.: New York, USA, 221.

[7] Matienko, L.I., Binyukov, V.I., Mosolova, L.A., Zaikov, G.E. (2011). *Polymer Research Journal,* 5, 151.

[8] Leninger, St., Olenyuk, B., Stang, P.J. (2000). *Chem. Rev*., 100, 853.

[9] Stang, P.J., Olenyuk, B. (1997). *Acc. Chem. Res*., 30, 502.

[10] Beletskaya, I., Tyurin, V.S., Tsivadze, A.Yu., Guilard, R., Stem, Ch. (2009). *Chem. Rev.,* 109, 1659.

[11] Saggu, M., Levinson, N.M., Boxer, S.G. (2011). *J.Am.Chem.Soc*., 133, 17414.

[12] Ma, J. C.; Dougherty, D. A. (1997). *Chem. Rev*. 97, 1303.

[13] Drain, C.M., Varotto, A., Radivojevic, I. (2009). *Chem. Rev*., 109, 1630.

[14] Chu, Cheng-Che, Raffy, G., Ray, D., Del Guerzo, A., Kauffmann, B., Wantz, G., Hirsch, L. and Bassani, D.M. (2010). *J. Am. Chem. Soc*., *132*, 12717.

[15] Gopal, B., Madan, L.L., Betz, S.F., Kossiakoff, A.A. (2005). *Biochemistry*, 44, 193.

[16] Balogh-Hergovich, É., Kaizer, J., Speier, G. (2000). *J. Mol. Catal. A: Chem*., 159, 215.

[17] Basiuk, E.V., Basiuk, V.V., Gomez-Lara, J., Toscano, R.A. (2000). *J. Incl. Phenom.* Macrocycl. Chem., 38, 45.

[18] Mukherjee, P., Drew, M.G.B., Gómez-Garcia, C.J., Ghosh, A. (2009). *Inorg. Chem.*, 48, 4817.

[19] Dubey M., Koner R.R., Ray M. (2009). *Inorg. Chem*., 48, 9294.

In: Biology and Communication in the Information Age
Editor: Rafiqul Islam
ISBN: 978-1-62808-252-4

Chapter 22

DAMAGES OF CHROMOSOMES AND DISRUPTIONS OF MITOTIC ACTIVITY IN SEEDLINGS OF *CREPIS CAPILLARIS* BY ALKYLATING ANTINEOPLASTIC AGENT

Larissa I. Weisfeld*
N.M. Emanuel Institute of Biochemical Physics RAS, Moscow, Russia

ABSTRACT

In this article we attempt to consider mechanism of origin of rearrangement of chromosomes in different phases of mitotic cycle. We are showing the appearance of rearrangements of chromosome under the influence of antineoplastic medicament «phosphemid» — di- (ethylene imid)-pyrimidyl-2-amidophosphoric acid, named mutagen. Seeds *Crepis capillaris* and fibroblasts of human and mouse were treated by this mutagen. Rearrangements of chromosomes were analyzing in seedlings on metaphase plates. The rearrangements were analyzed at the scheduled periods of time after appearance of seedlings named «arisings». Also: chromosomal rearrangements were examined at anaphases and telophases of fibroblasts.

Keywords: «arising», chemical mutagenesis, chromosome rearrangements, colchicine, ethylene imine, fibroblasts, karyotype, 2*n*-meristem cells, mitotic activity, mutagen, mutant, phosphemid (phosphasin), rearrangements of chromatid type, rearrangements of chromosomal type, seedling, super-mutagen

INTRODUCTION

Scientists draw attention to the chemical compounds that cause heritable changes. The mechanism of their influense on the chromosome remains unsolved. The action of many

* E-mail: liv11@yandex.ru.

compounds is similar to ionizing radiation; it causes mutations of genes, disruptions of cell division and rearrangement of chromosomes.

It is known that ethylene imine and its derivatives alkylate DNA and proteins (see review [1]). They induce mutations as it was shown in various model objects (*Drosophila*, higher plants, fungi, bacteria, viruses and others) and breakage of the chromosome apparatus (see, for example, the review [2]). Chemical compounds that cause mutations and breakage of chromosomes usually are called "chemical mutagens".

In the history of the study of chemical mutagens a large role belongs to works of J.A. Rapoport [3]. He is the discoverer of the phenomenon of the chemical mutagenesis. He revealed out super-mutagens and discovered possibilities their application in the breeding of crops and in other areas of agricultural production. J.A. Rapoport organized synthesis of super-mutagens. Every year since 1959, he organized the All-Union conferences for scientists and breeders on chemical mutagenesis and its application in agriculture. These meetings served in those years a good genetic school, especially for young breeders, who were trained and the unscientific method of Lysenko. On the basis of the methodology developed by Rapoport and with the help of mutagens, which he distributed free of charge, breeders created the source material of crops and introduced new varieties [see the review 3]. A number of investigations on the application of the chemical mutagen ethylene imine and on mechanism of its action in winter wheat was conducted and is still conducting by N.S. Eiges [see for example 6, 7], the follower of J.A. Rapoport.

The most available and affordable way to analyze the mutagenesis is a cytogenetic method, that is the study of rearrangements (aberrations) of chromosomes observed in dividing cells (mitosis, karyokinesis), and disorders in passing of the mitotic cycle. The mechanism of damage of the chromosomes and their relationship to the mitotic cycle and DNA synthesis at various biological objects was studied (in the cells of the meristem of plants onions, *Vicia faba*, *Crepis*, etc., embryonic cells of animals and human in vitro, at microbes, viruses and other objects). Ionizing radiation damages the chromosomes immediately after irradiation at all stages of the mitotic cycle — "undelayed" effect. In the cells that came into mitosis from G_2 and S phases aberrations of chromatid type arise, in cells that come in mitosis from phase G_1 (pre-synthesis) aberrations of chromosome type (double bridges, paired fragments at anaphase) occur.

Under treatment by chemical mutagens of asynchronous cell cultures no rearrangements are detected during the first 2—4 hours (depending on the duration of G_2 phase in different objects). This phenomenon is named "delayed" effect. Chromosomal rearrangements appear later — after the entry to mitosis of cells treated at the beginning or during DNA synthesis (phase S) or before phase S (named pre-synthetic phase G_1). Chromosome rearrangements are usually analyzed in the ana-telophases or meta-phases. A large number of chromosomes in many objects makes the identification of chromosome aberrations in metaphase plates difficult. In this case fragments of chromosomes or broken bridges cannot be identified. In this case, scientists analyze the anaphase and early telophases (ana-telophase method).

For the estimation of environmental contamination the method of analyzes of ana-telophases is sufficient. The estimation of pollution on the meristem of birch in the industrial area of the city of Staryj Oskol is in example [8].

In Moscow in the 60—70-th years N.P. Dubinin, his colleagues and followers, carried extensive cytogenetic studies of chemical and radiation mutagenesis. The work was begun in the laboratory of radiation genetics of Institute of biophysics of the Academy of Science of the USSR and was continued in the Institute of general genetics. A large number of articles and monographs on cytogenetic effects of ionizing radiation or chemical compounds were published (see review [9]). N.P. Dubinin formulated the idea of the mechanism of action of chemical mutagens: mutagens cause potential changes in the chromosomes at all stages of the mitotic cycle, which are realized in a number of cell generations. He called it "chain processes" in mutagenesis [10—12].

At the same time period the staff of the laboratory under the direction of B.N. Sidorov and N.N. Sokolov (Andreev V.S., Generalova M.V., Grinih L.I., Durymanova S.E., Kagramanyan R.G., Protopopova E.M., Shevchenko V.V., and others) were conducted extensive work on the induction of chromosomal damage, their localization in chromosomes, their association with the phases of the mitotic cycle and DNA synthesis after treatment of seedlings by ethylene imine, tio-TEPA, radiation and other mutagens on the model object *Crepis capillaris*. They studied chromosome aberrations in metaphase plates, which make it possible to take into account the polyploidy of metaphases.

In the early 60's, working with N.P. Dubinin, I investigated the cytogenetic effect of alkylating agent phosphemidum (lat. synonym phosphemid, phosphasin) — di-(ethylene imid)-pyrimidyl-2-amidophosphoric acid (Figure 1).

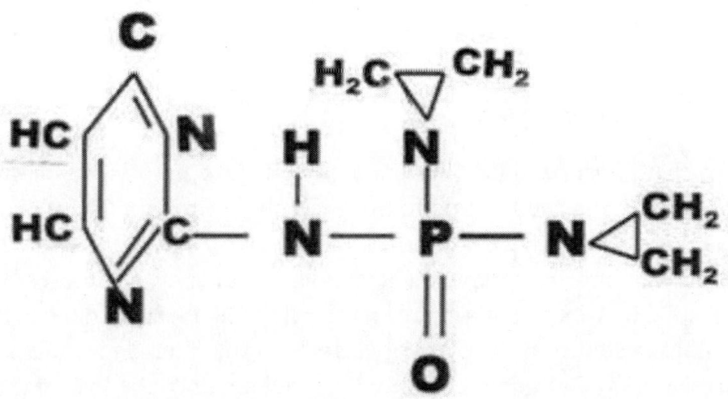

Figure 1. Phosphemidum (syn. phosphasin, phosphemid).

This compound is interesting because it consists of pyrimidine base and two molecules of ethylene imine. It was assumed that phosphemid will cause a lot of damages in DNA synthesis and thereby would inhibit tumor growth. The drug was synthesized in the laboratory of V.A. Chernov et al. [13], the All-Union Scientific Research Chemical-Pharmaceutical Institute (now the Center for Chemistry of drugs), and was given to N.P. Dubinin for cytogenetic studies. The drug is a white crystalline powder, dissoluble in water and alcohol. Chernov et al. [13-14] has shown that phosphemid (then it was named phosphasin) inhibits the growth of tumors in rats, mice, rabbits, but at the same time it causes leukopenia, leukocytosis, and suppressed erythropoiesis. But currently the drug is used in medicine for the treatment of certain tumors such as leukemia, lymphoma. In the early work with phosphemid (1963—1964gg), it was important to approach the mechanism of its antineoplastic action. It was assumed that due to pyrimidine bases and ethylene imine groups of the drug it will directly affect DNA during a synthesis and thereby destroys actively dividable tumor cells.

The work was carried out in the primary culture of embryonic tissues — the mouse and human fibroblasts [15, 16]. We analyzed chromosomal aberrations in cells at the stages of anaphase or telophases (ana-telophases), after treatment culture by phosphemid in a concentration of $1 \cdot 10^{-4} M$.

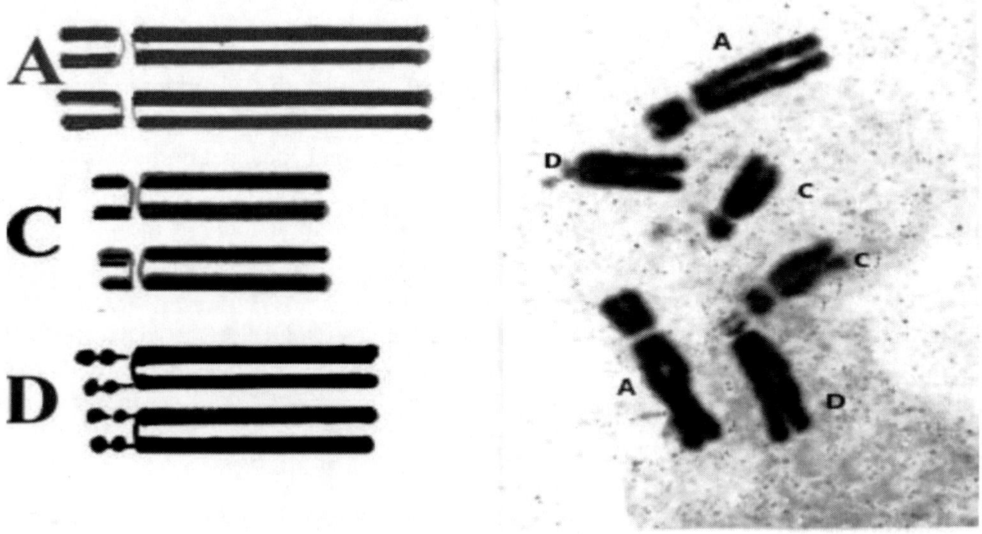

Figure 2. Karyotype *Crepis capillaries.*

It was shown that phosphemid delays entry of cells into mitosis and inhibits the mitotic activity of fibroblasts during the total period of the culture growth. The waves of fall and of rise of mitotic activity generally are repeating the waves of mitotic activity in control group (without treatment), but at a lower level. The average frequency of mitoses made 54% from control. Lesions spindle was not observed, as a rule, but in individual cells were visible clumping of the chromosomes in the form of "stars". At the later stages of fixing the number of nuclei was decreased by 3-4 times, that indicates cell death or the loss of their contact with surface of glass. At consideration of the types of chromosomal rearrangements, it became clear that in the ana-telophases the chromatid type of rearrangements mainly appears. At the later stages of fixation after 26 hours of the growth of culture double bridges (chromosomal type) were observed. They could have arisen as a result of doubling of the chromatid bridges in the second cycle (as a result of their passing to one of poles during the first division) or as a result of the breaking-fusion of chromosomes before synthesis DNA (phase G_1) with the subsequent doubling during the synthesis of DNA, in accordance with the thesis of N.P. Dubinin.

It was necessary to set connection with the phases of mitotic cycle. To do this, it would be desirable to find an object that, firstly, would be synchronized (or at least partially synchronized) with the terms of the phases of the mitotic cycle, and secondly, it would be convenient for the analysis of chromosomes in metaphase. To do this, it is desirable to find an object which, firstly, would be synchronized (or at least partially be synchronized) with phases of the mitotic cycle, and secondly, was suitable for the analysis of chromosomes in metaphase.

In these aspects *Crepis capillaris* (L.) Wallr. serves as an ideal object. Mikhail S. Navashin studying chromosomes in the metaphase plates of seedlings published a number of works about taxonomy of genus *Crepis* [17]. He showed emergence of a large number of chromosome rearrangements when seeds became older.

At metaphase plate *Cr. capillaris* three pairs of homologous chromosomes clearly identified (Figure 2). Analysis of rearrangements here is certain.

According to the results of experiments with ionizing radiation it was assumed, that the cells of seeds (not seedlings!) *Cr. capillaris* are in phase G_1. The phase of G_1 in germinal cells of seed of *Crepis* is heterogeneous, as was shown in [18] the first marked cellular nuclei appear in the seedlings through 10 hours after the treatment of grain by thymidine H^3.

As far as the seeds germinate the cells from the phase of S come in mitosis at first and then from the phase of G_1. If the chemical mutagen interacts with the chromosome before the start of germination, i.e. before the start of DNA synthesis - in phase G_1, then in metaphases of seedling germinating rearrangements of chromosome type must appear. The data of our experiments, presented below, show that after the treatment of seeds in the 2*n*-methaphases by phosphemid appear rearrangements of chromatid type.

MATERIALS AND METHODOLOGY

Air-dried seeds of *Crepis capillaris* of crop of 1967 were analyzed in 1968: April (8 months of storage after harvest), June (10 months of storage), and July (11 months of storage). In the control and in the experiments we used distilled water. The chromosome aberrations were analyzed in seedlings (meristem of root tip).

A certain amount of seeds (100 pieces) were treated in an aqueous solution phosphemid of concentrations: $1 \cdot 10^{-2}M$ (22.4*mg* was dissolved in 10*ml* water), $2 \cdot 10^{-2}M$ (22.4*mg* was dissolved in 50*ml*) or $2 \cdot 10^{-3}M$ (2.24*mg* dissolved in 50*ml* of water). The treatment was carried out at room temperature (19—21°*C*) for three hours. Then the seeds were washed with running water for 45 minutes. The washed seeds were placed in Petri dishes on filter-paper moistened with a solution of colchicine (0.01%). Seeds were germinating in a thermostat at 25°*C*, but in July 1968 the temperature in the thermostat could reach 27°*C* because of hot weather. In parallel control experiments, the seeds were treated with aqua distillate.

In 24, 27 and 31 hours after soaking of seeds we choose "arisings". The term "*arising*" refers to those seedlings that emerge after the beginning of soaking of dry seeds and have a size of less than 1mm. These seedlings were selected in a Petri dish for further germination and subsequent fixation. In Russian the term "arising" is called "proklev".

For the analysis of chromosomes in metaphase plates, root tips (0.5—1*cm*) were cut off with a razor and placed in the solution: 96% ethanol 3 parts + 1 part of glacial acetic acid. Solution was poured out through 3—4 hours. Seedlings were washed 45 minutes in 70% alcohol. These seedlings were kept in 70% alcohol. We were preparing temporary pressure preparations: fixed root tips were stained acetous carmine and crushed in a solution of chloral hydrate between the slide and cover slip. We analyzed chromosome aberrations in metaphase plates of seedlings in the first division after treatment of seeds (2*n*-karyotype). In each seedling all metaphases were counted up. Intact seeds of harvest of 1966, 1067 and 1969 years served as control. The seedlings were fixed at different time intervals from 3 to 24hrs.

Mitotic activity in the seedlings was determined by two criteria:

- tthe number of metaphase plates, in relation to the number of nuclei in seedlings in the control and experiment (after seed treatment phosphemid). We used the seedlings after 2, 4, 6, 8, 10 and 20 hours after the "arising". In each seedling, we counted between 500 and 1,000 nuclei; in an each seedling counted from 500 to 1000 nuclei;
- tthe number of seedlings with metaphase to all watched seedlings at all stages of fixation.

In all the experiments was estimated standard deviation from the mean value.

RESULTS AND DISCUSSION

The natural level of mitotic activity - control was estimated by the criterion of the number of metaphases in seedlings: of the 18 000 counted nuclei metaphase plates were 2.83 ± 0.12%. Table 1 shows that level of natural mitotic activity for 1966, 1967, and 1969, determined by criterion of number of metaphases in all studied roots, is in a range of 90 — 99 %.

Table 1. The natural level of chromosome rearrangements in the 2*n*-meristem cells of *Crepis capillaris* after soaking the seeds in water and germinating in 0.01% solution of colchicine. Harvested in 1966, 1967, 1969 years

Year of harvest	Month, year studies	Number of investigating seedlings		Metaphases		Rearrangements of chromosomal type	
		Σ	with metaphases	Σ	with rearrangements, %±	Σ	%±
1966	XII, 1966	25	25	1125	0,44±0,198	0	-
	I, 1967	27	27	2120	0,24±0,105	2	-
	III, 1967	43	42	1692	0,24±0,118	0	-
Average:		*95*	94/99,0%	*4937*	*0,31*±0,008	2	0,04±0,029
1967	IV, 1968	*49*	48	1169	0,77±0,256	0	-
	VI, VII, 1968	20	14	350	1,14±0,569	3	0,86±0,492
	III, 1969	20	18	1316	2,96±0,468	24	1,73±0,369
Average:		89	80/89,9%	2835	1,62±0,214	27	0,81±0,155
1969	IV, 1970	*33*	32	1384	1,59±0,006	21	1,52±0,329
Total (1996-1969гг):		*217*	*206(94,9 ±1,49%)*	*9156*	*1,05*±0,134	*50*	*0,55*±0,080

Note. We don't take into account some metaphases with numerous damages of the spindle and chromosomes.

On the average, over the years and months, were studied 217 seedlings. Of these, 94,9% contained metaphases. The frequency of metaphases with rearrangements in 1966 was in the range 0.24—0.44%. In 1967 there was a clear tendency to increase the frequency of rearrangements, as they are stored. Mitotic activity in 1967 was about 90%. The number of rearrangements of chromosomal type was various and small. For all the years we have discovered 50 such changes (see Table 1). It should be noted that there is a tendency to increase their number during the aging of seeds (harvest 1967, 19 months of storage). The average level of metaphases with rearrangements in different years was less than 2%, rearrangements of chromosomal type in average of 0.55%; the largest number of them was in 1967.

As the count the nuclei in the seedlings after treatment by phosphemid at concentrations of $1 \cdot 10^{-2}$ to $5 \cdot 10^{-4}$ *M* of 68 500 counted nuclei were 743 metaphases, an average of 1.08 ± 0.06%, thus phosphemid suppressed mitotic activity more than twice (see above 2.82% in control).

After treatment of seeds by phosphemid we analyzed 17 513 seedlings with metaphases among all concentration and terms of fixings. Among them 2306 metaphases with rearrangements (13.17%) are revealed. Alterations of chromosomal type were less than 1% — 0.113%. This magnitude is similar to the natural frequency. At the concentration of phosphemid of $2 \cdot 10^{-3} M$ mitotic activities was on the average 51.5% (Table 2), i.e., two times lower than in the controls (94,9%) (cf Table 1).

Table 2. Aberrations of chromosomes in 2n-meristem cells of *Crepis capillaris* after 24 and 27 hours from the start of treatment of seeds in a solution of phosphemid $2 \cdot 10^{-3} M$ (2.24*mg*, 50*ml* of water). Fixation through 3—8 hours after «arising». Seeds were germinating in 0.01% solution of colchicine. Harvested in 1967 year. The analysis in April 1968

Time, in hours		Number of investigating seedlings			Metaphases		Rearrangements of chromosomal type
from soaking up to «arising»	from soaking up to fixation	Σ	with metaphases		Σ	with rearrangements, %±	
			Σ	%			
24	3	57	24	42.1	741	9.4±1,08	0
	6	56	32	57.1	1270	17.2±1.06	1
	8	54	36	66.7	1815	14.0±0.82	0
27	3	47	16	34.0	373	20.6±2.10	1
	5	19	12	63.2	313	15.3±2.04	0
Total		233	120	**51.5**±3.28	4512	**14.8**±0.53	**0.043**±0.031%

Data of table 2 show that at the weak concentration of preparation - $2 \cdot 10^{3}$ with the increase of term of fixation in every “arising” mitotic activity increases, and frequency of alterations diminishes.

At max concentration - $1 \cdot 10^{-2} M$ (table 3), we registered a large number of rearrangements – nearly 60% through 27 hour after «arising» in term 24hr, of which 16.4% were mitosis with multiple rearrangements (it is marked with asterisk). On the average, considerably significant is increasing up to 13.3±2.9%. Rearrangements of chromosomal type also were not found. Average frequency of mitotic activity after phosphemid in a concentration of $1 \cdot 10^{-2} M$ at the same quantity of seeds, as shown in Table 2, was lower almost twice - 25% (table 3). Average number of metaphases with rearrangements increased – nearly 22% versus 14.8%. Thus the average level of metaphases with chromatid aberrations was increasing with increasing of dose of preparation.

During storage of untreated seeds and after subsequent treatment of seeds - namely in June and July of 1968 patterns of mitotic activity and frequencies of rearrangements were different, despite the use of the same concentration of phosphemid: 2 • $10^{-2}M$ (Figures 3a, b and 4a, b). Mitotic activity was than in April. With increasing time from «arising» to fixation after 12 hours frequency of seedlings with mitoses close to 90% and was slightly below of control.

Table 3. Rearrangements of the chromosomes in the 2*n*-meristem cells of *Crepis capillaris* seedlings at 24 and 27 hours from the start of treatment of seeds in a solution of phosphemid 1 • $10^{-2}M$ (22.4*mg*, 10*ml* water). Fixation through 3-24hr after «arising». Seeds were germinating in 0.01% solution of colchicine. Harvested in 1967. The analysis: April 1968

Time, in hours		Number of investigated seedlings		Metaphases		
from soaking up to «arising»	from soaking up to fixation					
		Σ	with metaphases	Σ	with rearrangements	
					Σ	%±
24	3	21	6	148	25	16.9±3,09
	6	19	3	365	24(2*)	6.6±1,23
27	3	20	2	73	7(2*)	9.59±3,47
	24	12	7	185	110(18*)	59.46±3,62
Total		72	16/25.0±5.14%	771	166(22*)	21.53±2,85

Note. (*) Number of greatly damaged metaphases with multiple rearrangements.

Through 24 and 27 hours from the start of soaking of seeds the mitotic activity was steadily increasing to depending on time of fixation in each "arising». At a later «arising» — 36 hours (see Figure 3*a*) or 31 hours (see Figure 4a) mitotic activity was fluctuating at the higher level as compared to the earlier stages of «arising» and terms of fixation from «arising». These data on figures 3 and 4 are statistically significant. The frequency of metaphases with rearrangements increased steadily in the fraction of the 24hr from «arising» from 3 to 12 hours (see Figure 3*b*) and from 3 to 9 hours (see Figure 4*b*).

Through 24hr after the soakage of seeds the frequency of metaphases with rearrangements was growing steadily with time of fixing after "arising", average frequency of rearrangements was higher through 27, 31 and 36 hours (see Figure 3*b* and 4*b*). At the 36hr «arising» frequency of metaphases with rearrangements increases with time from «arising» before fixation (see Figure 3*b*). Through 12hr after «arising» the level was almost 23% in 31hr «arising» (see Figure 4*b*), at 31hr «arising» (see Figure 4b), at a fixations through 3 and 4hr frequency of metaphases with rearrangements was at the high level — 17-18%.

Rearrangements of chromatid type was observed at all terms of fixation , and only solitary rearrangements of the chromosomal type were found - an average of less than 1%, that is similar to the control level. This fact is important, because shows, that the chemical preparation doesn't breaks chromosomes before synthesis DNA.

From the data of figures 3 and 4 at the first «arising» (24hr) under any fixation and concentration the frequency of aberrations increased from 3hr and more. The level of rearrangements in the early fixations was lower than in the later ones, apparently due to the fact that the mutagen affected the chromosome during the shortest period of time.

It is possible to explain the facts reflected in pictures 3 and 4 by 1) the preparation does not influence on chromosomes during phase G1, but remains in seeds; 2) chromosomes are damaged by the preparation regardless of the stage of mitotic cycle, but this damage shows up only during synthesis of DNA (presence exclusively chromatid rearrangements). The number rearrangements of the broken areas of chromosomes increases with the increase of terms of fixation; 3) existence of both factors is possible. Below the works of B.N. Sidorov and N.N. Sokolov will be described, they convincingly explaining this effect exceptionally by chromatid alterations.

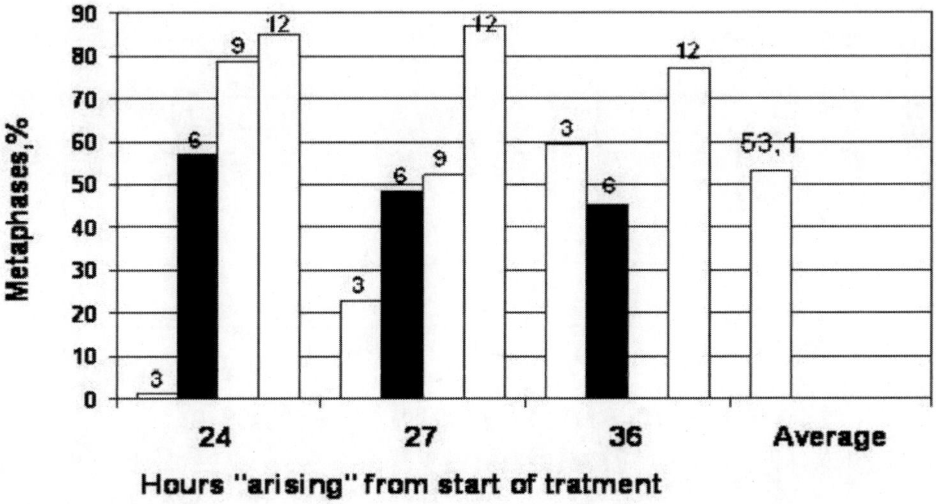

(a)

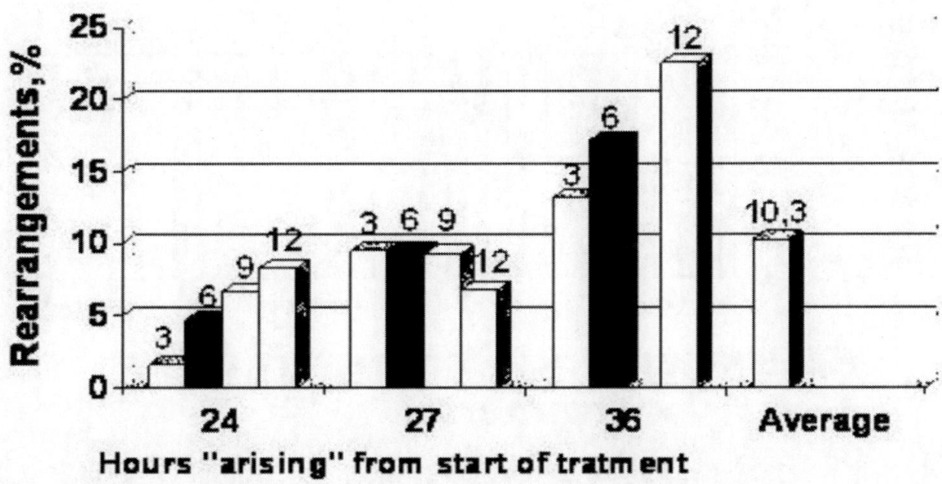

(b)

Figure 3. *Crepis capillaris*: the mitotic indexes in seedlings (*a*) and rearrangements of chromosomes in metaphase (*b*) after 24, 27 and 36 hours from the start of seed soaking in a solution of phosphemid $2 \cdot 10^{-2}M$ (22,4*mg*, 50*ml* of water). Fixation: through 3—12 hours after “arising”. Seeds germinate in 0.01% colchicine. Seeds are of harvested in 1967. Analysis: June 1968.

Increased mitotic activity of the seedlings with increasing terms of fixation indicates the possibility of washout of the preparation from the cells during the growth of seedlings. Increasing number of rearrangements in «arisings» while conformance with increasing time before fixation and keeping of high level of metaphases with rearrangements in the seedlings show that phosphemid penetrating in the seeds from the beginning of their treatment, is included in the metabolism of cells and stores there for a long time.

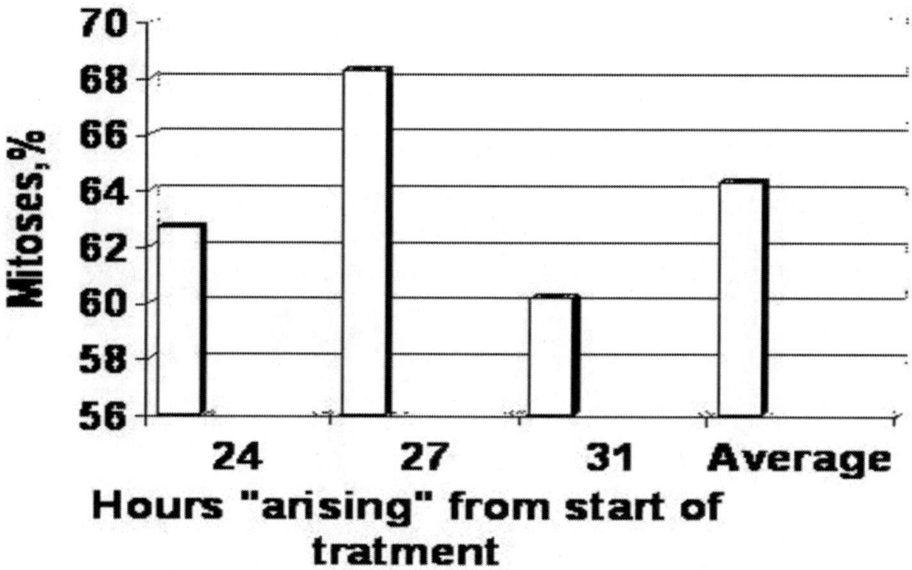

(a)

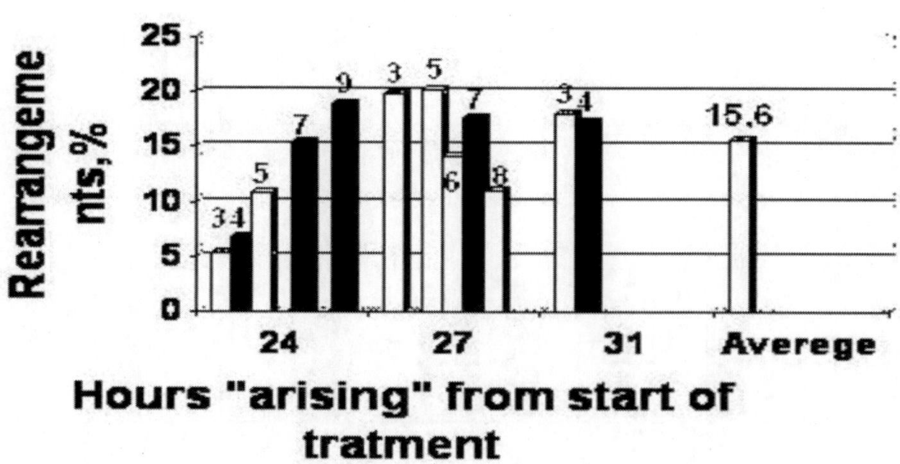

(b)

Figure 4. *Crepis capillaris*: the mitotic indexes in seedlings (*a*) and rearrangement of chromosomes in metaphase 2n-meristem cells (*b*) after 24, 27 and 36 hours from the start of seed soaking in a solution of phosphemid 2 • 10^{-2}*M* (22,4*mg*, 50*ml* of water). Fixation: through 3—9 hours after seedling. Seeds were germinating in 0.01% colchicine. Seeds are of harvest 1967. Analysis: July 1968.

Phosphemid suppressed mitotic activity the stronger the concentration is higher (see Table. 2, 3, Figs 3, 4). Phosphemid also interacts with proteins of spindle, as at higher concentrations of the preparation the number of heavily damaged mitoses significantly increased (see table 3). Such metaphases often form a sort of a star at the center of the cell. A similar pattern sometimes was observed in the culture of fibroblasts. In addition to these disruptions, in some cells we have seen that all the chromosomes were fragmented.

Mitotic activity estimated as the number of metaphases to the number of non-dividing nuclei through 2—10 hours after «arising» in the control amounted to 2.11% at 16 000 nuclei. After treatment of the seeds the number of metaphases in the seedlings over the same period (2—10 hours) was lower and depended on the concentration of phosphemid. At a high concentration of $1 \cdot 10^{-2} M$ at an average around 16 000 nuclei was 0.73% of metaphases, with preparation concentration $5 \cdot 10^{-3} M$ – 0.82% of metaphases around 26 500 nuclei, at a concentration of $5 \cdot 10^{-4} M$ observed 1.42% of metaphases around 21 000 nuclei. ***However, by 20hr after the start of treatment the frequency of mitoses increased both in control – 3.40%, and at concentration of phosphemid $5 \cdot 10^{-3} M$ - 2.38%. These data also reflect the decline of mitotic activity with the increase of concentration of phosphemid and it's increasing with reduction of concentration of preparation

Mutagen may be remains in the seeds in conjunction with other cellular proteins, and at that brakes transitions of phases of mitotic cycle, thus influencing on the chromosomes longer and therefore stronger is damaged the synthesis of larger number of loci of chromosomes.

In the 60th—70th year's outstanding scientists N.N. Sokolov, B.N. Sidorov [19—21] realized a series of studies on effects of ethylene imine on seedlings *Cr. capillaris*. They cultivated seedlings in a solution of colchicine for five cell generations. They found in tetraploid and higher polyploidy cells rearrangements of chromatid type ("not multiplied") in tetraploid and higher polyploidy cells under influence of colchicine.

The authors explained this phenomenon by the fact that the mutagen is saved in the cells and there are new rearrangements. In a next article [20] seedlings were treating of ethylene imine, these seedlings were washed in running water within 2hr. From these after 48hr "thin gruel" were preparing. Intact seedlings of *Cr. capillaris* were treated by this thin gruel. In these seedlings treated by thin gruel rearrangements of chromatid type appeared.

The authors suggested that ethylene imine formed active secondary mutagens, connected with the components of the cell, including nucleic acids. In the third article of this series [21], the authors added ethylene imine to amino acids *in vitro*: glycine and histidine, to the hexamine (hexamethylenetetramine), to vitamins: thiamin (vitamin B_1), nicotinic acid.

Consequent treatment of seedlings by these preparations did not cause rearrangements. The frequency of them was at the level of control. Ethylene imine (concentration 0.05%) caused about 15% of rearrangements, while in mixture with thiamine caused significantly more rearrangements (+22.27%). Treatment by ethylene imine in mixture with nicotinic acid, glycine, and histidine showed even some protective effect. Treatment by ethylene imine together with adenine, guanine (derivative of purine), cytosine (derivative of pyrimidine) showed absence of effect or a small excess of rearrangements above the level of rearrangements of pure ethylene imine.

Significant excess frequency of rearrangements was received under the influence of mixtures: uracil + ethylene imine gave an increase nearly 19%, thymine + ethylene imine caused nearly 61% of rearrangements (+37.25%). Guanine and cytosine in a mixture with ethylene imine showed an insignificant action. In the mixture with thio-TEPA (three ethylene imine groups), only thiamine gave significant excess of frequency of rearrangements (+25.97%) over control (ethylene imine). After treatment of seedlings by a mixture of ethylene imine with thymine, the excess of frequency of rearrangements was not found. All the rearrangements were chromatid type. The authors explain this phenomenon by the formation of secondary mutagens in cells. Thusly these experiments [19—21] showed that mutagens do not cause aberrations of chromosomes before or after phases of the mitotic cycle G_2, G_1 and causes only chromatid aberrations effecting on the chromosomes in the course of DNA synthesis. Modern microscopy suggests that mutagen penetrating into the cell is remaining in "space around a chromosome" (see review [22]) in bonding with proteins or DNA, but its effect is revealing when mutagen passes through phase of DNA synthesis and becomes visible in the form of chromosome rearrangements during mitosis. The same can be evidence of the phenomenon of fragmentation of chromosomes, strong destruction of mitosis and the spindle during treatment by high doses of the alkylating agent (in our case by phosphemid) in culture of fibroblasts of human and mouse or in germinal cells of seeds *Cr. capillaris*. Perhaps the same principle is working in the course of "chain process" of Dubinin. Someday, 4D-microscopy will reveal the mechanism of interaction of chromosomes with the chemical mutagens.

Understanding the mechanism of chemical mutagenesis in the present time is fundamentally important in view of the global contamination of the surrounding nature by different chemicals that damage the genetic structure of organisms.

CONCLUSION

1. We have previously shown above that in cultivated fibroblasts phosphemid was suppressing mitotic index, induces rearrangements of chromosomes.
2. In seedlings *Crepis capillaris* phosphemid also causes inhibition of the mitotic cycle. The average number of metaphases on the number of nuclei in seedlings after treatment phosphemid decreased twice.
3. Phosphemid after treatment of seeds *Cr. capillaris* causes rearrangements in the cells of seedlings, regardless of the age of the treated seeds, but depending on the concentrations of the drug. The greatest number of rearrangements occurs when with the highest concentration of the preparation.
4. After treatment of the seeds *Cr. capillaris* by phosphemid only chromatid type rearrangements were found in the seedlings. The number of rearrangements of chromosomal type was at the level of controls or smaller. This means that the preparation works as well as other chemical mutagens, i.e., chromosomes break during DNA synthesis.
5. Phosphemid in treatment seeds *Cr. capillaris* showed heterogeneity of germinal cells in the seeds during the G1 phase. The frequency of chromosomal rearrangements varies depends on the time between fixation and "arising".
6. At high concentrations ($1 \bullet 1^{-2}M$) phosphemid caused the destruction of mitotic spindle and multiple fractures of the chromosomes.

The author is grateful to F. Tatarinov, N. Brakker and A. Bukhonin for their assistance in the preparation of the text.

REFERENCES

[1] Ross W. Biological alkylating agents/ Fundamental chemistry and design of compounds for selective toxicity. London. 1962. Translation edited by A. Berlin's. Moscow. - *Medicine.* 1964. - 259 p.

[2] Loveless A. Genetic allied effects of alkylating agents / London. 1966. Translation edited by N.P. Dubinin's. Moscow. - *Nauka*. 1970. - 255 p.

[3] Stroeva O.G. Josef Abramowitz Rapoport, 1912-1990 // Moscow. - *Nauka.* - 215 p.

[4] Josef Abramowitz Rapoport – scientist, warrior, citizen. Essays, memoirs, materials. Compiled by O.G. Stroeva // Moscow. - *Nauka.* - 2003. - 335 p.

[5] Kihlman B.A. Aberrations induced by radiomimetic compounds and their relations to radiation induced aberrations // In: *Radiation-induced chromosome aberrations*. N.Y.-London. - 1963. – 260 p.

[6] Eiges N.S. *Characteristic features of chemical mutagenesis method* I.A. Rapoport and its use in breeding of winter wheat. Penza. - 2008. - P. 14-17.

[7] Eiges N.S., Vaysfel'd L.I., Volchenko G.A. Volchenko S.G. Some aspects of the securities chemomutant characters a collection of winter wheat and characterization of these characters// International teleconference number 1: Basic Science and Practice. January 2010. Website: http://tele-conf.ru/nasledstvennyie-morfologicheskie-kletochnyie fakto/aspektyi-ispolzovaniya-tsennyih-hemomutantnyih-priznakov-kollektsii-ozimoy-pshenitsyi-i-ih-har-ka-chast-1.html

[8] Kalaev V.N., Butorina A.K., Sheluhina O.Yu. Assessment of anthropogenic pollution areas Staryj Oskol on cytogenetic parameters of birch seed family// *Ecological genetics*. - 2006. – V. IV, issue 2. – P. 9-21.

[9] Dubinina L.G. Structural mutations in the experiments with *Crepis capillaris //* Moscow - *Nauka*. - 1978. – 187 p.

[10] Dubinin N.P. Some key questions of the modern theory of mutations // *Genetika.* - 1966. - № 7. – P. 3- 20.

[11] Dubinin N.P. Unresolved issues of modern molecular theory of mutations // *Proc. USSR*. - 1971. - №2. – P. 165-178.

[12] Dubinin N.P. Unresolved issues of modern molecular theory of mutations // *Proc. USSR*. - 1971. - №3. – P. 333-344.

[13] Chernov V.A. Cytotoxic substances in chemotherapy of malignant tumors // Moscow. - *Medicine*. – 1964. – 320 p.

[14] Chernov V.A., Grushina A.A., Lytkina L.G. Antineoplastic activity of phosphasin // *Pharmacology and Toxicology*. - 1963. – V. 26, № 1. - P. 102-108.

[15] Weisfeld L.I. Cytogenetic phosphasin effect on human cells and mice in tissue culture // *Genetika*. - 1965. – № 4. – P. 85-92.

[16] Weisfeld L.I. Influence phosphasin on the duration of phases of the mitotic cycle of human cells and mice in culture // Genetika. - 1968. – V. IV, № 7. – P. 119-125.

[17] Navashin M.S. Problems karyotype and cytogenetic studies in the genus *Crepis. //* Moscow. - *Nauka*. - 1985. - 349 p.

[18] Protopopova E.M., Shevchenko V.V., Generalova M.V. Beginning of DNA synthesis in seeds *Crepis capillaris* // *Genetika*. - 1967. - № 6. - P. 19-23.

[19] Sidorov B.N., Sokolov N.N., Andreev V.A. Mutagenic effect of ethylene imine in a number of cell generations // *Genetika*. - 1965. - № 1. - P. 121-122.

[20] Andreev V.S., Sidorov B.N., Sokolov N.N. The reasons for long-term mutagenic action of ethylenimine // *Genetika,* - 1966. - № 4. - P. 28-36.

[21] Sidorov B.N., Sokolov N.N., Andreev, V.A. Highly active secondary alkylating mutagens // *Genetika*. - 1966. - № 7. - P. 124-133

[22] Rubtsov N.B. Human chromosome in four dimensions // *Priroda*. - 2007. - № 8. - P. 1-8.

In: Biology and Communication in the Information Age
Editor: Rafiqul Islam
ISBN: 978-1-62808-252-4

Chapter 23

CHANGE OF SOME PHYSICO-CHEMICAL PROPERTIES OF ASCORBIC ACID AND PARACETAMOL HIGH-DILUTED SOLUTIONS

F. F. Niyazy*[*], *N. V. Kuvardin and E. A. Fatianova
Department "General and Inorganic chemistry",
South-West State University, Russian Federation

ABSTRACT

The article considers regularities of meniscus height changing in the capillary and ultraviolet spectra of ascorbic acid and paracetamol high-diluted solutions, and also solutions, containing both compounds. There have been detected concentration ranges of these compounds action in supersmall doses.

Keywords: Supersmall doses, concentration, solution, ascorbic acid, paracetamol, dilution, meniscus height in the capillary, electronic spectra

INTRODUCTION

During last decades there is a tendency of the growing interest to the study of high-diluted solutions of bioactive substances. Besides, concentration ranges under study are related to the category of supersmall or, in other words, 'illusory' concentrations. Such solutions, unlike more saturated ones, but with pretherapeutic content of active substance, may possess high biological activity.

Use of bioobjects to reveal supersmall doses effect (SSD) in the substances is the most exact method today allowing not only to define the effect existence and to find out how it shows itself, but to determine concentration ranges of its action. However, the use of this method will entail great difficulties. In this connection it is necessary to search for other

[*] E-mail: farukhniyazi@yandex.ru.

methods, including physico-chemical ones, allowing to define presence of SSD effect in the compounds vif only at the stage of preliminary tests. Study of physico-chemical bases of SSD effect display is one of the most interesting questions in the given sphere of research and attracts attention of many scientists [1-4].

Antineoplastic and antitumorous agents, radioprotectors, neutropic preparations, neupeptides, hormones, adaptogenes, immunomodulators, antioxidants, detoxicants, stimulants and inhibitors of plants growth and so on are included into the group of bioactive substances possessing SSD effect. Study of high-diluted solutions of bioactive compounds was carried out on one-component solutions that are ones containing only one solute. But now, mainly multicomponent medical preparations, possessing several therapeutic actions, are used in medicine. So, preparations of analgesic-antipyretic action are possibly used at sharp respiratory illnesses, accompanied by muscular pain and rise of temperature. It is possible that effects of multicomponent medical preparation in supersmall concentrations and its separate components will differ.

We studied some physico-chemical properties of high-diluted aqueous solutions of paracetamol and ascorbic acid with the purpose of finding out peculiarities of their change at solutions dilution and also of definition of possible concentration ranges of SSD effect action. Paracetamol and ascorbic acid are the components of combined analgesic-antipyretic and antiinflammatory preparations [5]. Ascorbic acid is used as fortifying and stimulating remedy for immune system. Paracetamol (acetominophene) has anaesthetic and febrifugal effect.

There have been prepared one-component solutions of ascorbic acid and paracetamol, two-component solutions of paracetamol with ascorbic acid (relation of dissoluted compounds in solutions is 1:1), in the following concentrations of dissoluted substances (mole/l): 10^{-1}, 10^{-3}, 10^{-5}, 10^{-7},10^{-9}, 10^{-11}, 10^{-13}, 10^{-15}, 10^{-17}, 10^{-19}, 10^{-21}, 10^{-23}. Water cleansed by reverse osmosic was used as solvent. Solutions were prepared by successive dilution by 100 times using classical methods. Initial solution was 0,1M one. Before choosing of solution portion for the following dilution the sample was subjected to taking antilogs.

Prepared solutions were studied by cathetometric method of substances screening, the ones acting in supersmall concentrations, and also by method of electronic spestroscopy.

RESULTS AND DISCUSSION

Cathetometric method of substances screening is based on the study of the change of solution meniscus height in capillary [6]. Results of measuring meniscus height of ascorbic acid, paracetamol, paracetamol with ascorbic acid solutions are given in Figures 1, 2, 3, accordingly.

Meniscus height of dilution with ascorbic acid concentration of 10^{-3} mole/l was 0,8 mm (Figure 1). During further dilution value of meniscus height in the capillary reduces, but changes are not uniquely defined. The most lowering of meniscus height is observed in samples in which content of ascorbic acid is 10^{-9}, 10^{-13}, 10^{-15}, 10^{-17} mole/l.

Meniscus height reduces on an average by 13,75%. Lowering of meniscus height by 23,7%, in comparison with more concentrated solution, is also observed for the sample with ascorbic acid content of 10^{-23} mole/l.

Equivalent lowering of meniscus in the capillary has been stated for solutions with ascorbic acid concentration of 10^{-9} mole/l and 10^{-13} – 10^{-17} mole/l. Between these concentration ranges there is concentration range in which there are no essential changes. Being based on literature data and also on cathetometric method for screening of substances activity in supersmall doses, we can assume that dilutions of ascorbic acid in concentrations of 10^{-9} mole/l and 10^{-13} – 10^{-17} mole/l show biological activity regarding bioobjects.

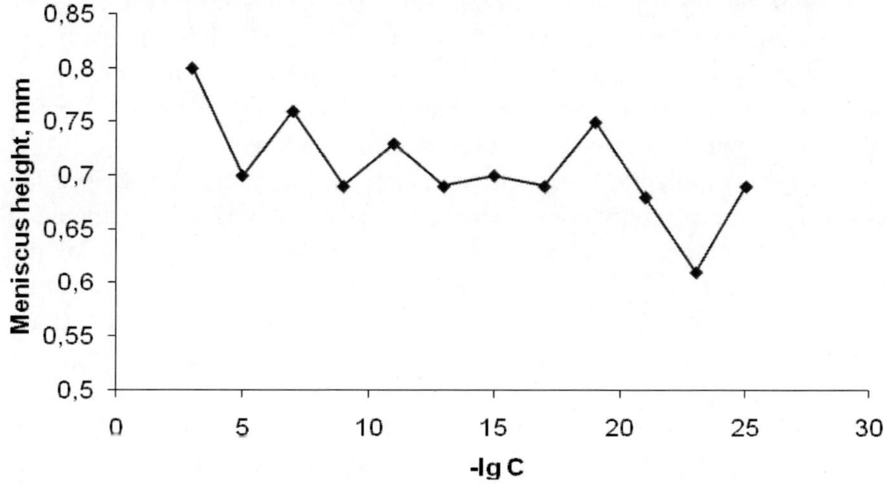

Figure 1. Values of meniscus height in the capillary of ascorbic acid solutions (concentration, mole/l).

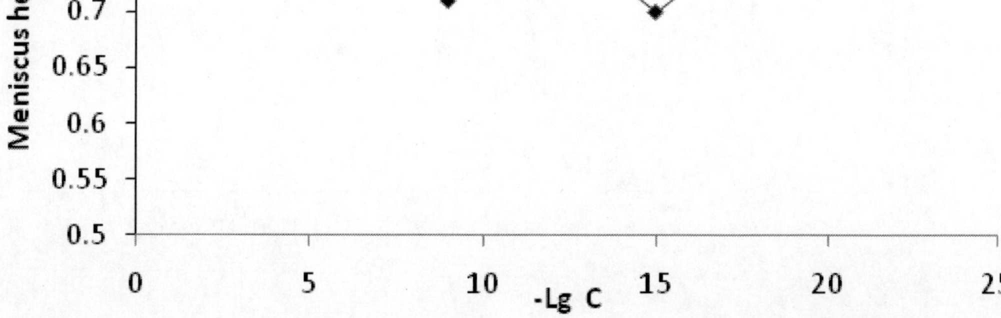

Figure 2. Values of meniscus height in the capillary of paracetamol solution (concentration, mole/l).

While studying solutions of paracetamol by cathetometric method it can be observed that in first dilutions by 100 times (paracetamol concentrations being 10^{-3} – 10^{-7} mole/l) height of meniscus reduces slightly, maximum 5,7%, regarding meniscus height of the first dilution. This slight lowering of meniscus height in the capillary is caused by rather large dose of active substance in these dilutions. However, sudden lowering of meniscus height in the capillary up to 0,71mm, which is by 19,3% lower than meniscus height of the first dilution, is observed at diluting paracetamol solution up to the concentration of 10^{-9} mole/l (Figure 2).

The same dependence is observed for dilution of paracetamol solution with concentration of 10^{-15} mole/l. So, at this concentration meniscus height is 0,7mm. Growth of meniscus height in the capillary is observed further for solutions with the following dilution. This process is motivated by very high dilution, that is the solution, according to its composition and properties, tries to attain the state of pure solvent.

These changes are polymodal dependence effect – concentration, described for different substances and different properties in domestic and world scientific literature. From the Figure and its description it is clearly seen that there are concentrations of paracetamol solution of 10^{-9} mole/l and 10^{-15} mole/l, for which there has been stated essential lowering of meniscus in the capillary regarding other concentrations. Between these concentration ranges there is a concentration range in which there are no essential changes, this range being the so-called "dead zone".

We have studied solutions of ascorbic acid and paracetamol mixture. It is necessary to note that not uniquely defined change of meniscus height is observed in solutions containing simultaneously two active substances. Meniscus of one-component solutions is more narrow than that of water, but in two-component solutions both reduction and increase of meniscus height values are possible in comparison with water (Figure 3).

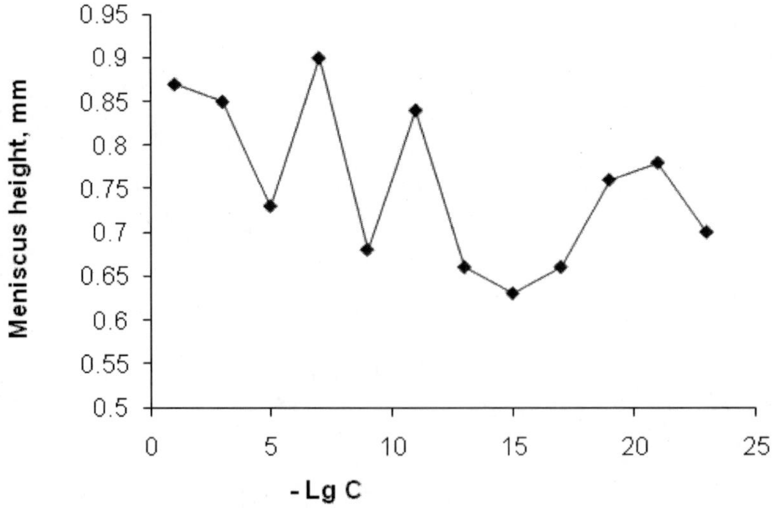

Figure 3. Values of meniscus height in the capillary of paracetamol solutions with ascorbic acid.

Table 1. Results of study of ascorbic acid, paracetamol, paracetamol with ascorbic acid solutions of wide concentration range by method of electronic spectroscopy and cathetometric method

Method of study	Concentration ranges, mole/l		
	Ascorbic acid solutions	Paracetamol solutions	Solutions of paracetamol with ascorbic acid
Cathetometric method of screening	10^{-9}, 10^{-13} – 10^{-17}, 10^{-23}	10^{-9}, 10^{-15}	10^{-9}, 10^{-13} – 10^{-17}
Electronic spectroscopy	10^{-9}, 10^{-15}, 10^{-19}	10^{-9}, 10^{-13}, 10^{-15}, 10^{-21}	10^{-9}, 10^{-13}, 10^{-15}, 10^{-21}

Table 2. Maximum values of absorption of ascorbic acid solutions in ultra-violet field

Concentration, mole/l	Absorption maximum	
	Wave length, nm	A
10^{-3}	230-270	0,79
10^{-5}	265	0,04
10^{-7}	220-230	0,012
10^{-9}	225	0,03
10^{-11}	235	0,01
10^{-13}	220	0,015
10^{-15}	220	0,04
10^{-17}	220-240	0,005
10^{-19}	220	0,03
10^{-21}	220	0,005
10^{-23}	220	0

Lowering of meniscus height is observed in solutions with concentration of paracetamol with ascorbic acid 10^{-5}, 10^{-9}, 10^{-13}, 10^{-15}, 10^{-17} mole/l on 11,5 %, 17,5%, 20%, 23,6% and 20% correspondingly. It is possible to distinguish two concentration ranges in the field of supersmall concentrations where reduction of meniscus height in the capillary takes place: this is – 10^{-9} mole/l and 10^{-13} – 10^{-17} mole/l. Received concentration ranges coincide with data of cathetometric studies of ascorbic acid and paracetamol solutions (Table 1).

Insrease of meniscus height in comparison with the solvent is observed in solutions with concentrations of diluted compounds of 10^{-1}, 10^{-3}, 10^{-7}, 10^{-11} mole/l. Growth of meniscus height above such value of the solvent can be explained by the change in SSD effect display.

We have analyzed ultra-violet spectra of solutions of ascorbic acid, paracetamol, paracetamol with ascorbic acid. All the spectra were read from spectrophotometer Cary 100,

UV-Visible Specrtophotometer in the range of 200-350 nm. The most absorption in all solutions takes place in short waves part of the effective wave band. More concentrated solutions (10^{-1}, 10^{-3} mole/l) of ascorbic acid have maximum absorption in the length interval of 220-280 nm, of paracetamol – 220-310 nm, paracetamol with ascorbic acid – 200-320 nm. Tendency to narrowing of absorption field, reduction of optical density value and quantity of tops up to one or two is registered in all solutions, under study, as content of diluted substance reduces. Solutions spectra with ascorbic acid concentration of 10^{-15} and 10^{-19} mole/l are equal in form and differ from spectra of other dilutions. Lowering of ascorbic acid concentration is not accompanied by uninterrupted reduction of optical density value.

Irregular growth of absorption in comparison with more concentrated solutions is observed in solutions with ascorbic acid concentrations of 10^{-9}, 10^{-15}, 10^{-19} mole/l (Table 2).

Community in given structures may be assumed taking into account coincidence of waves length and values of optical density for solutions with concentrations 10^{-9}, 10^{-15}, 10^{-19} mole/l. Gradual reduction of paracetamol concentration in solutions is not accompanied by the same reduction of optical density value. Growth of absorption in comparison with more concentrated solutions is observed for solutions with concentrations 10^{-9}, 10^{-13}, 10^{-15}, 10^{-21} mole/l (Table 3). All this allows to assume the rise of structural changes in these solutions.

Spectrum of ascorbic acid and paracetamol solution in concentrations of 10^{-1} mole/l is characterized by wide absorption band in the field of 200-320nm. Dilution of 0,1M solution by 100 times is accompanied by reduction of absorption field width up to 200-280nm without changing of spectrum form and intensity of absorption on peaks.

Table 3. Maximum values of absorption of paracetamol solutions in ultra-violet field.

Concentration, mole/l	Absorption maximum	
	Wave length, nm	A
10^{-1}	230	0,58
10^{-3}	270	0,53
10^{-5}	245	0,15
10^{-7}	240	0,02
10^{-9}	235	0,015
10^{-11}	240	0,007
10^{-13}	245	0,007
10^{-15}	220-245	0,01
10^{-17}	235	0,003
10^{-19}	290	0,001
10^{-21}	225	0,02
10^{-23}	230	0,005

Table 4. Maximum values of absorption of ascorbic acid with paracetamol solutions (1:1) in ultra-violet field

Concentration, mole/l	Absorption maximum	
	Wave length, nm	A
10^{-1}	237	4,651
10^{-3}	235	4,208
10^{-5}	243	0,095
10^{-7}	205	0,242
10^{-9}	205	0,293
10^{-11}	207	0,084
10^{-13}	206	0,367
10^{-15}	205	0,186
10^{-17}	207	0,086
10^{-19}	207	0,089
10^{-21}	206	0,261
10^{-23}	207	0,105

Changing of spectrum form accompanied by essential reduction of absorption from 4 to 0,095 with the maximum on the length 243nm takes place while diluting solution of ascorbic acid with paracetamol up to the concentration 10^{-5} mole/l (Table 4).

Two peaks on lengths 205-206 nm and 270-273nm are shown on spectra of solutions with paracetamol and ascorbic acid concentrations 10^{-7}, 10^{-9},10^{-13}, 10^{-15}, 10^{-21} mole/l. One peak on the length 207 nm is shown on spectra of solutions with paracetamol and ascorbic acid concentrations 10^{-11}, 10^{-17}, 10^{-19}, 10^{-23} mole/l. Display of maximum on the length 270nm occurred under conditions that absorption on maximum 205nm was not less than 0,1.

Increase of optical density is observed in spectra of solutions with paracetamol and ascorbic acid concentrations 10^{-7}, 10^{-9}, 10^{-13}, 10^{-19}, 10^{-21} mole/l.

CONCLUSION

As a result of this work we, by cathetometric method, have defined concentration ranges of possible display of medical compounds biological activity that forms 10^{-9} mole/l and 10^{-13} – 10^{-17} mole/l for ascorbic acid and 10^{-9}, 10^{-15} of paracetamole, and also 10^{-9}, 10^{-13} – 10^{-17} for the mixture of ascorbic acid - paracetamol.

This allows to assume compatibility of components data in given concentration ranges. While studying solutions in super-small doses of ascorbic acid, paracetamol and also at their joint presence by method of electronic spectroscopy it has been found out that irregular growth of absorption is observed for solutions with ascorbic acid concentrations 10^{-9}, 10^{-15},

10^{-19} mole/l in comparison with more concentrated solutions and for paracetamol solutions with concentrations 10^{-9}, 10^{-13}, 10^{-15}, 10^{-21} mole/l.

But for the mixture ascorbic acid - paracetamol such interval is 10^{-9}, 10^{-13}, 10^{-15}, 10^{-21} mole/l. This fact allows to speak about appearance of medical compounds structures with water in this solutions which differ by relatively high absorption.

Information received by method of ultra-violet spectroscopy, agrees with data of cathetometric screening both for solutions of ascorbic acid and of paracetamol.

ACKNOWLEDGMENTS

The article is based on the materials of scientific –research work within the framework of realization of Federal special-purpose programme "Scientific and scientific-pedagogical personnel of innovation Russia" for 2009-2013 years.

REFERENCES

[1] Konovalov A.I. Physico-chemical mystery of super-small doses. // *Chemistry and life*. 2009. V2, p.5-9.

[2] Kuznetsov P.E., Zlobin V.A., Nazarov G.V. On the question about physical nature of superlow concentrations action.// *Heads of reports at III International Symposium "Mechanisms of super-small doses action*", Moscow, December, 3-6, 2002, p.229.

[3] Chernikov F.R. Method for evaluation of homoeopathic preparations and its physico-chemical foundations.// *Matherials of Congress of homoeopathists of Russia*. Novosibirsk, 1999, p.73.

[4] Pal'mina N.P. Mechanisms of super-small doses action// *Chemistry and life*. 2009. N2, p.10.

[5] Maslikovsky M.D. *Combined analgesic-febrifugal and anti-inflammatory preparations*. M., 1995, p.208.

[6] Niyazi F.F., Kuvardin N.V. Method for determining abilities of bioactive substances to display "super-small doses" effect. Patent N 2346260 of 10.02.2009.

In: Biology and Communication in the Information Age
Editor: Rafiqul Islam
ISBN: 978-1-62808-252-4

Chapter 24

Activation of Cereal Seeds by Para-Aminobenzoic Acid

***S. A. Bekuzarova*[1], *N. A. Bome*[2] *and L. I. Weisfeld*[3]**
[1]Gorsky State Agrarian University, Vladikavkaz, Republic of North Ossetia Alania, Russia
[2]Tyumen State University, Tyumen, Russia
[3]N.M. Emanuel Institute of Biochemical Physics of Russian Academy of Sciences, Moscow, Russia

Abstract

Para-aminobenzoic acid (PABA) by way of the well-known activator of growth and development was applied for treatment of ears of winter wheat and winter barley. The varieties of these cultures are destined for North-Caucasian region. Also panicles of selection samples of millet – the varieties are introduced in North Ossetia - were subject to treatment by PABA. PABA was dissolved in ac. a., but not in water, as usual. Advantages of this mode are described. This method allows accelerating a plant-breeding process in case of a long period of harvesting.

Keywords: Dissolve, ear, North-Caucasian region, North Ossetia, panicle, inflorescences, para-aminobenzoic acid, sowing, treatment, variety, winter barley, winter wheat

Introduction

In the process of selection for the speed-up study of material it is necessary to get the sufficient amount of seed in the earliest possible dates. For that inflorescence of plants of grain crops, which we chose for research, were planted in opened soil without trashing. The method of sowing by ears was previously tested for two years in Tyumen research station of N.I. Vavilov Research Institute of Plant Industry on more than 100 samples of winter wheat from world collections. The productivity of winter forms of cereals depends on a set of biotic (pathogenic microorganisms) and abiotic (temperature, amount of precipitation etc.) factors.

We obtained positive results for winter hardiness, resistance to snow mold, which main pathogen is *Microdochium nivale* (Fr.) Samuels and I.C. Hallett (= *Fusarium nivale* Ces. ex Berl. and Voglino).

Para-aminobenzoic acid (PABA) was discovered by J.A. Rapoport as modifier of metabolic processes as early as in the 1940s. He showed on the example of experimental object, *Drosophila*, that PABA evokes positive changes of non-hereditary character (i.e., it is not a mutagen) in the organism development [1]. Rapoport [2] proposed a scheme of relations between genotype, ferments and phenotype: «genes → their hetero catalysis (on the substrate of ribonucleic acid (RNA) molecule) → messenger RNA (mRNA) → mRNA catalysis (substrate of amino acids) → ferments → their catalysis (different substrates) → phenotype».

Studies of application of PABA in agriculture are continued and deepened. The experiments performed in the Republic of North Ossetia [4-9] showed, that the addition of PABA to the nutritive substances (potassium humate, irlit, leskenit, corn extract, juice of ambrosia, melted snow water and others) gives a positive effect. Treatment with PABA seeds of pea and honey plant sverbiga east (*Bunias orientalis*) before planting [4], extra nutrition of clover [5], seed of leguminoze grasses [6] stimulates germination of plants more, then application of nutritious solutions without PABA addition. A positive result was got at treatment of sprouts of potato by the mixture of the melted water with juice of ambrosia, leskenit and PABA [7]. This method resulted in the increase of harvest of potato and decline of disease of Fusarium. Addition PABA in nutrient medium for treatment of seed of triticale [8] results in the increase of maintenance of protein in green mass. Protracted treatment by PABA of handles of dogwood [9] increased engraftment of grafts. PABA was used to get potato without viruses [10]. In the joint research of scientists of the Institute of Biochemical Physics RAS and the North Caucasus Research Institute of mountain and foothill agriculture it was shown that the treatment by PABA of potatoes tubers with the subsequent enveloping in ash [11] increase yield of potato. Treatment of seed and seedlings of vegetable cultures by solution of PABA in mixture with boric acid and permanganate potassium [12] improves resistance to diseases of young plantlets and increases harvest of carrot, beet, cucumbers, and tomatoes, seeds and seedlings of vegetables. After treatment by PABA of binary mixture of seed of winter wheat and winter vetch [13] the productivity and quality of green feed increases at mowing of mixture in a period from the beginning of exit in a tube of wheat and beginning of budding of vetch to forming of grain of milky ripeness of wheat.

In the conditions of the northern forest-steppe of Tyumen region, where saline occurs beside of fertile soils, jointed impact of salt solutions and PABA (0,01% solution) on the seeds of three barley varieties with low salinity resistance considerably increased the salinity resistance of germs independently on NaCl concentration [14]. The positive results were obtained under spraying of inflorescences of mother plants of barley by the solutions of PABA before crossing. In a series of hybrid combinations the exceeding over control of length and width of flag leaf, number of leaves by plant and plant height was observed [15]. Spraying of inflorescences of four amaranth samples by PABA solutions increased seed productivity, the concentration of 0.02% showed to be the most effective [15].

In the present research we studied the effect of PABA on germination, germination energy and winter-hardiness of cereals, namely ears of winter cereals and panicles of millet species. In contrast with the most of above-mentioned studies, where PABA was dissolved in hot water following the method developed by Rapoport with his research associates [3], in the present study PABA was dissolved in an acetic acid. The combination of PABA and acetic acid creates an acid media and allows preventing of a set of fungi illnesses and at the same time preserves genotype of samples.

Below we present the results of experiments performed on the experimental base of the North Caucasus Research Institute of mountain and foothill agriculture where perennial crops of legumes herbs were precursor.

MATERIAL AND METHODS

We have used as material cereals the most widely applied in the agricultural production and selection. We tested the following cultures of yield of 2009: winter wheat varieties Ivina, Vassa, Bat'ko, Don 107, Kollega, Kalym and winter barley variety Bastion – the varieties designed for North-Caucasian region. We also tested the following varieties introduced in North Ossetia: selection samples of millet: Japanese millet (*Echinochloa frumentacea*), panic (*Setaria italica Panicumitalicum*) Italian millet (*Setaria italica*).

PABA represents a fine-grained powder; it dissolves in 3% solution of an acetic acid easily and completely without heating. We dissolved one teaspoon of dry PABA powder (10 g) in a small volume (20-25 ml) of 3% solution of an acetic acid. Then we dissolved the obtained mixture in 1 liter of tap water under room temperature. Thus we obtain the PABA concentration of 0.1%. In order to obtain the concentration of 0.2% we dissolve 20 g of dry powder. We soaked 10 samples for the inflorescences without trashing of cereals with mature seeds, namely ears of cereals and panicles of millet species, in the solutions of PABA (0.1 or 0.2%) during 2–2.5; 3–4 and 5–6 hours. Then the inflorescences were planted in the in the soil in the open field. We kept the distance of 20–25 cm between samples. Untreated inflorescences, soaked in water during 3-4 hours, were applied as control (control I). When preparing control II we dissolved PABA in hot water and soaked inflorescences during 2—2.5 hours. In tables the variants of experiments were marked: Control I is presented in variant "0" where inflorescences were soaked in water (without treatment by PABA); Control of II is presented in variant "1", where inflorescences were soaked in PABA dissolved in hot water; variants 2-5 are a soakage of inflorescences in PABA dissolved in an acetic acid of concentrations of 0.1-0.2% through 2-6 hours. Simultaneously we placed thrashed seeds for germination in a Petri dish with PABA dissolved in hot water (control I) or in an acetic acid (control II). For the control II the data averaged for PABA concentrations of 0.1% and 0.2% are presented because their difference was insignificant. Experiments were conducted with participation of F.T. Tsomartova (PG student of Gorsky State Agrarian University), and Luschenko G.V (PG student of North Caucasian Research Institute of mountain and foothill agriculture).

RESULTS

The PABA concentration of 0.1% or 0.2% in an acetic acid was sufficient for the penetration of substance in the embryo when treating cereal inflorescences. High concentrations (9–11%) inhibit height and development of plants [10]. Germination of threshed seeds soaked in PABA dissolved in an acetic acid, appeared 3-4 days earlier than in the seeds soaked in water. On seedlings from seed germinated in an aqueous solution of

PABA developed fungal microflora and they died. When dissolved in an acetic acid PABA seedlings were preserved for a long time, they continued to grow.

Germinating capacity of seeds in inflorescences under soaking in PABA, dissolved in an acetic acid, considerably exceeded germinating capacity in control variants (Table 1). The effectiveness of treatment by PABA solutions depended on its duration and concentration of solutions. The same trend concerning germination energy was observed for all five cereals under study.

Table 1. Germinating capacity of seeds of winter grain crops and millet species under different ways of treatments of cereal inflorescences by para-aminobenzoic acid

Variant of experiment	Winter wheat		Winter barley		Japanese millet		Panic		Italian millet	
	%	+/%	%	+/%	%	+/%	%	+/%	%	+/%
0. Control I without treatment: soaking in water for 3-4 h	62	0	75	0	67	0	72	0	68	0
1. Control II: soaking in PABA (0.1-0.2%) dissolved in hot water for 2-2.5 h	74	12/16	84	9/11	76	9/12	82	10/1.2	72	4/6
2. Soaking in PABA (0.1%) dissolved in an ac. a. for 2-2.5 h	82	20/24	93	18/19	87	20/23	92	20/22	84	16/19
3. Soaking in PABA (0.2%) dissolved in an ac. a. for 2-2.5 h	86	24/28	95	20/21	92	25/27	92	20/22	87	19/24
4. Soaking in PABA (0.2%) dissolved in an ac. a. for 3-4 h	96	34/35	98	23/23	95	28/29	95	23/23	92	24/25
5. Soaking in PABA (0,2%) dissolved in an ac. a. for 5-6 h	82	20/24	86	11/13	90	23/26	90	18/20	86	18/21
LSD_{05}	4.5		2.4		2.8		2.1		2.2	4.5

Note. LSD - the least substantial difference; ac. a. – an acetic acid.

The middle percent of wintering as compared to control of I (water without PABA) increases with the increase of concentration of PABA and durations of treatment. However at more long-continued soakage of plants - during 5-6 hours (variant 5) increase in relation to both Controls was less, than at a soakage during 3-4 hours (variant 4). At soakage in water solution of PABA of 0.1% increase was also observed but on lower level, then at dissolution in an acetic acid.

The energy of germination, presented to the table 2, exceeded Control I without treatment of PABA (variant “0”) in all variants of dissolution of PABA dissolved in an acetic acid and in a variant with treatment of PABA, dissolved in hot water (control of II, variant 1), and was higher, than in control without addition of PABA (variant “0”) at all five investigated cultures. Exceeding of indexes above control of I in control of II was lower, than in variants 1-5.

Winter-hardness of winter cereals (table 3) - wheat and barley in all variants with treatment of PABA dissolved in an acetic acid (variants 2-5) and in control of II (dissolution PABA is in hot water, variant "2") was higher then in control of 1 (without PABA, variant "0").

Table 2. Germination energy of seeds of winter grain crops and millet species under different ways of treatments of cereal inflorescences by para-aminobenzoic acid

Variant of experiment	Winter wheat		Winter barley		Japanese millet		Panic		Italian millet	
	%	+/%	%	+/%	%	+/%	%	+/%	%	+/%
0. Control I without treatment: soaking in water for 3-4 h	54	0	68	0	53	0	60	0	54	0
1. Control II: soaking in PABA (0.1-0.2%), dissolved in hot water 2-2.5 h	65	11/17	73	5/7	62	9/15	73	12/16	63	8/13
2. Soaking in PABA (0.1%), dissolved in an ac. a. 2-2.5 h	72	55/76	81	13/16	75	22/29	81	22/31	73	19/26
3. Soaking in PABA (0.2%), dissolved in an ac. a. 2-2.5 h	80	27/34	84	16/19	79	26/33	87	26/30	76	22/29
4. Soaking in PABA (0.2%), dissolved in an ac. a. 3-4 h	85	32/38	89	21/24	83	30/36	89	29/33	79	25/32
5. Soaking in PABA (0.2%), dissolved in an ac. a. 5-6 h	78	25/32	80	11/14	78	26/33	84	24/29	75	21/28
LSD_{05}		3.4		2.1		2.6		2.9		2.5

Table 3. Winter-hardiness (% of plants overwintered) of winter cereals under different ways of treatments of ears by para-aminobenzoic acid

Variant of experiment	Winter wheat		Winter barley	
	%	+/%	%	+/%
0. Control I without treatment: soaking in water for 3-4 h	78	0	84	0
1. Control II: soaking in PABA (0.1-0.2%), dissolved in hot water 2-2.5 h	82	4/4.8	88	4/4.8
2. Soaking in PABA (0.1%), dissolved in an ac. a. 2-2.5 h	86	8/9.3	95	11/22.6
3. Soaking in PABA (0.2%), dissolved in an ac. a. 2-2.5 h	90	12/13.3	97	13/13.4
4. Soaking in PABA (0.2%), dissolved in an ac. a. 3-4 h	98	20/20.4	98	14/14.5
5. Soaking in PABA (0.2%), dissolved in an ac. a. 5-6 h	88	10/11.4	92	8/8.7
LSD_{05}		3.1		0.8

As well as other experiments indexes at dissolution of PABA in hot water in control II were lower, than in other variants (3-5). The dynamics of activating of phenotype under act of PABA is evidently shown on a histogram (see below), reflecting the middle indexes of levels of increases of energy of germination, germination and winter-hardness. The tendency of growth of middle indexes is evidently.

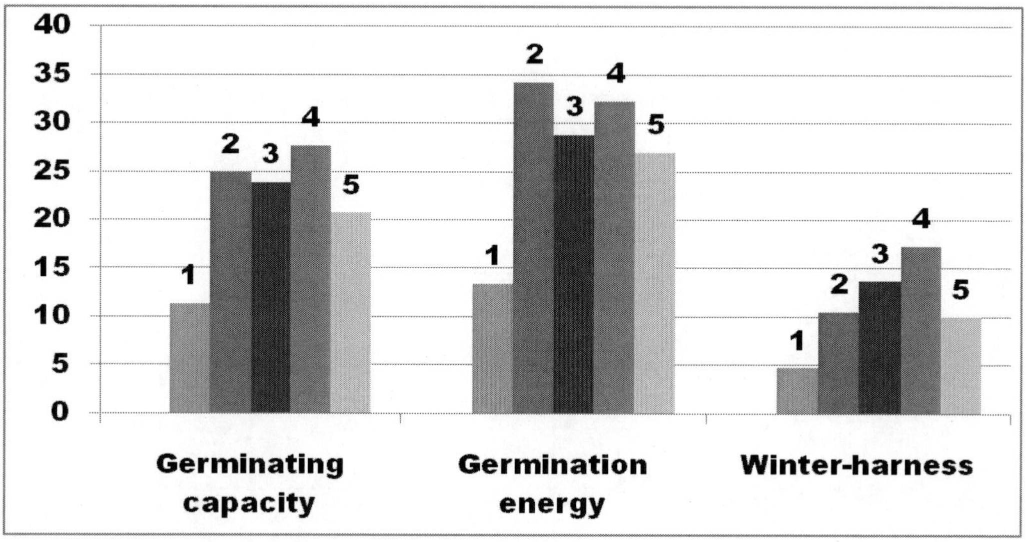

Figure. Comparison of averages (totally on all cultures, %) of germinated capacity, energies of germination of seed and winter-hardness of plants at the different methods of treatment of ears of winter crops and panicles of kinds millet by para-aminobenzoic acid in the variants of experiments from 1 to 5 (see tables 1-3) by comparison to control of I (without treatment, in tables - a zero variant). Ordinate axis, %.

Discussion

The higher indexes of germinating capacity, energy of germination at all cultures and winter-hardness cold of winter crops are marked - winter wheat and barley in variants (2-5 in diagram) after treatment of PABA, dissolved in an acetic acid, as compared to control variants - without treatment of PABA (variant “0” - Control I without PABA) and at dissolution of PABA in hot water (variant 1 - control II) in analogous variants (see tables 1-3, figure) were got. All indexes at a soakage in PABA in an acetic acid for longer time - 5-6 hours (variant 5) as compared to analogical treatment during 3-4 hours (variant 4) has a tendency to go down (see a table 1-3, picture). Maybe in this case braking of development of plantlets takes place [10].

This work executed in connection with the necessity of speed-up creation of new varieties steady to the ecologically diverse conditions and at the same time responsive on the additional fertilizing. This work is important from the economic point of view. Sowing of inflorescences without threshing considerably simplifies work of breeder both in a scientific aspect and in organizational aspect: the methods of selection are simplified, reproduction of the best plants is accelerated for selection and forming of new varieties, diminish expenses of labor. Usually at the unfavorable conditions of growing, at introduction of plants, at the high doses of mutagens or during distant hybridization of cultural plants with wild sorts fall down number in inflorescences, germination of seeds, viability fall down. At sowing of inflorescences (ears or panicles) possibility to distinguish the most perspective families on the number of germinating seed in them appears with mature grain. From every separate inflorescence it is possible to collect more seeds for the receiving of posterity, the terms of estimation of material and terms of reproduction of valuable selection samples grow short. The method especially important for freshly material, when at preparation to sowing of winter crops in the year of harvest time not enough for threshing of ears and additional treatment of grain. This quite often happens in the districts with short vegetation period, for example, in the Tyumen area.

CONCLUSION

1. The method of acceleration of plant-breeding process offers at sprouting of mature seed in inflorescences of cereal cultures (ears of grain growing and panicles of millet). The method plugs the soakage of inflorescences in PABA (concentrations of 0.1% and 0.2%) dissolved in an acetic acid.
2. Dissolution of PABA in an acetic acid is more effective, than dissolution in hot water (control II). Activating of phenotype at soakage in PABA, dissolved in an acetic acid, in variants (3-5) was higher then in control II – at soakage in PABA dissolved in hot water.
3. Soakage of inflorescences during 3-4 hours in PABA dissolved in acetic acid, gives the best result. Longer soakage (5-6 hours) reduces the results a little bit.
4. Concentrations of PABA applied to the treating of inflorescences were higher than in the cited works in which trashed seeds were treated.

The authors are grateful to F. Tatarinov, N. Brakker, and A. Bukhonin for their assistance in the preparation of the text.

REFERENCES

[1] Rapoport I.A. Phenogenetic analysis of dependent and independent differentiation // *Proceedings of the Institute of Cytology, Histology, Embryology of Academy of Sciences.* 1948. V. 2, № 1. P. 3-128.

[2] Rapoport I.A. The action of PABA in connection with the genetic structure // Chemical mutagens and para-aminobenzoic acid to increase the yield of crops. Moscow. - *Science.* - 1989. - P. 3-37.

[3] Rapoport I.A., editor of Chemical mutagens and para-aminobenzoic acid in enhancing the productivity of crops // Moscow. - *Science*. - 1989. 253 p.

[4] Bekuzarova S.A., Abiyeva T.S., Tedeeva A.A. *The method of pre-treatment of seeds* // Patent № 2270548. Published 27. 02. 2006.

[5] Bekuzarova S.A., Farniyev A.T., Basiyeva T.B., Gaziyev V.I., Kaliceva D.N. *The method of stimulation and development of clover plants* // Patent № 2416186. Published on 20.04.2011.

[6] Bekuzarova S.A., Shtchedrina D.I., Farnoyev A.T., Pliyev M.A. *Method of additional fertilizing of leguminous grasses* // Patent № 2282342. Published on 27.08.2006.

[7] Ikaev B.V., Marzoev A.I., Bekuzarova S.A., Basayev I.B., Bolieva Z.A, Kizinov F.I. *The method of treatment of pre-plant shoots of potato tubers* // Patent № 2385558. Published on 10.04.2010.

[8] Bekuzarova S.A., Antonov O.V., Fedorov A.K. *Method of increase of content of protein in green mass of winter triticale* // Patent № 2212777. Published on 27.09.2003.

[9] Cabolov P.H., Bekuzarova S.A., Tigiyeva I.F., Tadtayeva E.A., Eiges N.S. *The method of reproduction of dogwood drafts* // Patent № 2294619. Published on 10.03.2007.

[10] Shcherbinin A.N., Soldatova T.B. *The nutrient medium for micropropagation of potato* // Patent № 2228354. Published 10.05.2004.

[11] Eiges N.S., Weisfeld L.I., Volchenko G.A., Bekuzarova S.A. *Method of pre-treatment of tubers of potatoes* // Patent № 2202701. Published on 02.10.2007.

[12] Eiges N.S., Weissfeld L.I., Volchenko G.A., Bekuzarova S.A. *The method of pre-treatment of seeds and seedlings of vegetable crops* // Patent № 2200392. - Published 20. 03. 2003. Bull. №

[13] Eiges N.S., Weissfeld L.I., Volchenko G.A., Bekuzarova S.A., Pliyev M.A., *Hadarceva M.V. The method of receiving of feeds in green conveyer* // Patent № 2330410. – Published 10. 08. 2008. Bull. № 22.

[14] Bome N.A., Govorukhin A.A. The effectiveness of the influence of para-aminobenzoic acid on the ontogeny of plants under stress / *Bulletin of Tyumen State University. Tyumen: Tyumen State University Publishing House*, 1998, № 2. S. 176-182.

[15] Bome N.A, Bome A.Ja, Belozerova A.A. Stability of crop plants to adverse environmental factors. *Monograph* // Tyumen: Tyumen State University Publishing House, 2007. 192 p.

INDEX

A

B

C

D

E

I

J

K

L

M

N

O

P

Q

R

S

T

U

V

W

Y